Mastering PLC Programming
Second Edition

The software engineering survival guide
to automation programming

M. T. White

‹packt›

Mastering PLC Programming
Second Edition

Portfolio Director: Rohit Rajkumar
Relationship Lead: Kaustubh Manglurkar
Project Manager: Sandip Tadge
Content Engineer: Anuradha Joglekar
Technical Editor: Tejas Vijay Mhasvekar
Copy Editor: Safis Editing
Indexer: Tejal Soni
Proofreader: Anuradha Joglekar
Production Designer: Aparna Bhagat
Growth Lead: Namita Velgekar
Marketing Owner: Nivedita Pandey

First published: March 2023
Second edition: January 2026

Production reference: 1241225

Published by Packt Publishing Ltd.
Grosvenor House
11 St Paul's Square
Birmingham
B3 1RB, UK.

ISBN 978-1-83664-255-8
www.packtpub.com

To the Big Man upstairs who's been guiding my every step

– M.T.

Contributors

About the author

M. T. White has been programming since the age of 12. His fascination with robotics flourished when he was a child programming microcontrollers, such as Arduinos. He currently holds an undergraduate degree in mathematics, a master's degree in software engineering, a master's in cybersecurity and information assurance, as well as an MBA in IT Management. M.T. is currently working as a senior DevOps engineer and is an adjunct CIS instructor at ECPI University, where he teaches Python, C, and an array of other courses. His background mostly stems from the automation industry where he programmed PLCs and HMIs for many different types of applications. He has programmed many different brands of PLCs over the years and has developed HMIs using many different tools. Other technologies M.T is fluent in are Linux, C#, Java, and Python. Be sure to check out his channel **AlchemicalComputing** on YouTube.

I want to thank my mom for putting up with the 4-am writing sessions.

About the reviewer

Oleg Osovitskiy is a senior firmware engineer with over 25 years of expertise in industrial automation. He holds IEC 61508 FS Eng certification (#11605/15) and IEC 62443 Cybersecurity Specialist certification (#658/22). His career includes extensive work as a control engineer, where he designed and implemented technological and emergency algorithms for diverse industrial facilities.

He possesses deep technical proficiency in PLCs and I/O drivers, with substantial experience across industrial communication protocols, such as Modbus, HART, CANopen, EtherNet/IP, and others. Based in Quebec, Canada, he currently leads firmware development for multiple mission critical safety PLCs, ensuring compliance with rigorous operational and safety standards.

I want to thank my wife and two wonderful daughters for their unwavering support and understanding of the dedication required to pursue new skills and knowledge in our ever-changing, demanding world. They remain my greatest source of purpose and joy.

Table of Contents

Chapter 4: Object-Oriented Programming: Reducing, Reusing, and Recycling Code   91

Chapter 5: OOP: The Power of Objects 123

Chapter 6: Best Practices for Writing Incredible Code   153

Part II: Software Engineering for Automation   197

Chapter 8: Getting Started with Git   199

Chapter 15: Alarms: Avoiding Catastrophic Issues with Alarms   389

Preface

Industry 4.0 is shaking up the automation industry. The days of only needing to know Ladder Logic are coming to an end. As new technologies such as AI take the world by storm, automation programmers are going to need to adapt their code to this ever-changing world. To survive in the new landscape, programmers are going to have to master object-oriented programming.

For years automation programming has skirted around adopting modern software engineering practices. Many industries and companies have been using the same technologies and practices for decades. However, Industry 4.0 is shaking that philosophy to the core.

Quality software engineering is the backbone of modern program development. OOP and its relatively new introduction to PLC programming have rocked the automation world. This book will bridge the gap between the modern programming landscape and the controls world by teaching applied object-oriented and software engineering practices. Along the way this book will explore other concepts like

- Version Control
- UML design
- Emerging technologies
- Best practices
- HMI development

This book is designed to apply concepts that are usually reserved for traditional programming to PLC programming in a vendor-neutral and language agnostic manner. The goal of this book is to demystify techniques that are often ignored by the automation industry to build the future of manufacturing.

Who this book is for

This book is for automation programmers with a background in software engineering topics such as object-oriented programming and general software engineering knowledge. Automation engineers, software engineers, electrical engineers, PLC technicians, hobbyists, and upper-level university students with an interest in automation will also find this book useful and interesting. Anyone with a basic knowledge of PLCs can benefit from reading this book.

What this book covers

Chapter 1, Advanced Structured Text: Programming a PLC in Easy-to-Read English, details the basics of advanced Structured Text such as error handling, state machines, and expert systems. The basics of IEC 61131-3 are also introduced.

Chapter 2, Complex Variable Declaration: Using Variables to Their Fullest, revolves around complex variables such as arrays, constants, and more. This chapter also introduces structs and their uses.

Chapter 3, Functions: Making Code Modular and Maintainable, shows you how to build modular code. This chapter will introduce you to functions, arguments, return types, and more.

Chapter 4, Object-Oriented Programming: Reducing, Reusing, and Recycling Code, introduces the foundations of object-oriented programming. Methods, objects, and function blocks are covered.

Chapter 5, OOP: The Power of Objects, takes a deep dive into the more advanced topics of object-oriented programming. This chapter will also cover the four pillars of OOP, composition, the inheritance chain, and more.

Chapter 6, Best Practices for Writing Incredible Code, introduces you to the best practices of software engineering. Topics covered are technical debt and how to avoid it, naming conventions, and much more.

Chapter 7, Libraries: Write Once, Use Anywhere, covers the basics of creating software libraries. This chapter covers best practices, documentations, and more.

Chapter 8, Getting Started with Git, illustrates the art of using Git. This chapter includes basic Git commands, how to interface with GitLab, branching, and other practices.

Chapter 9, SDLC: Navigating the SDLC to Create Great Code, provides an overview of the software development lifecycle including popular implementation methodologies.

Chapter 10, Architecting Code with UML, explains how to use UML to design object-oriented programs.

Chapter 11, Testing and Troubleshooting, details the art of debugging and testing. This chapter will cover testing and debugging techniques including how to use modern AI to troubleshoot issues.

Chapter 12, Advanced Coding: Using SOLID to Make Solid Code, explores the principles of SOLID programming and how they can be leveraged in automation.

Chapter 13, Industrial Controls: User Inputs and Outputs, explore how to create HMIs. This introductory chapter explores basic controls, design techniques, and more.

Chapter 14, Layouts: Making HMIs User-Friendly, explores how to get the most of an HMI by making it as user friendly as possible. You will explore multi-screen layouts, navigation, and more.

Chapter 15, Alarms: Avoiding Catastrophic Issues with Alarms, provides a deep dive into alarms and their usage. This chapter covers colors, alarm acknowledgement, banner setup, and much, much more.

Chapter 16, DCSs, PLCs, and the Future, contains an exploratory look at basic networking technologies, distributed control systems, and the future of Industry 4.0.

Chapter 17, Putting It All Together: The Final Project, is a real world simulation for a broken project. You will apply the skills you learned to build and repair the codebase.

To get the most out of this book

To get the most of this book, you will need to have the knowledge of Structured Text or text-based programming.

Answers to all questions at the end of each chapter can be found in the Answer Sheet provided at the end of the book.

Download the example code files

The code bundle for the book is hosted on GitHub at `https://github.com/PacktPublishing/Mastering-PLC-Programming-Second-Edition`. We also have other code bundles from our rich catalog of books and videos available at `https://github.com/PacktPublishing`. Check them out!

Download the color images

We also provide a PDF file that has color images of the screenshots/diagrams used in this book. You can download it here: `https://packt.link/gbp/9781836642558`.

Conventions used

There are a number of text conventions used throughout this book.

CodeInText: Indicates code words in text, database table names, folder names, filenames, file extensions, pathnames, dummy URLs, user input, and Twitter handles. For example: "The VAR_ INPUT section is used for variables that will be used for what are called arguments or parameters."

A block of code is set as follows:

```
x = input
If x > 100 Then
        Fan = on
Elseif x < 90 then
        Fan = off
```

Any command-line input or output is written as follows:

```
git clone <url>
```

Bold: Indicates a new term, an important word, or words that you see on the screen. For instance, words in menus or dialog boxes appear in the text like this. For example: "Now, create a blank project, fill out the project name, and click **Create**."

Warnings or important notes appear like this.

Tips and tricks appear like this.

Get in touch

Feedback from our readers is always welcome.

General feedback: If you have questions about any aspect of this book or have any general feedback, please email us at customercare@packt.com and mention the book's title in the subject of your message.

Errata: Although we have taken every care to ensure the accuracy of our content, mistakes do happen. If you have found a mistake in this book, we would be grateful if you reported this to us. Please visit http://www.packt.com/submit-errata, click **Submit Errata**, and fill in the form.

Piracy: If you come across any illegal copies of our works in any form on the internet, we would be grateful if you would provide us with the location address or website name. Please contact us at copyright@packt.com with a link to the material.

If you are interested in becoming an author: If there is a topic that you have expertise in and you are interested in either writing or contributing to a book, please visit http://authors.packt.com/.

Share your thoughts

Once you've read *Mastering PLC Programming, Second Edition*, we'd love to hear your thoughts! Scan the QR code below to go straight to the Amazon review page for this book and share your feedback.

https://packt.link/r/1836642555

Your review is important to us and the tech community and will help us make sure we're delivering excellent quality content.

Free Benefits with Your Book

This book comes with free benefits to support your learning. Activate them now for instant access (see the "*How to Unlock*" section for instructions).

Here's a quick overview of what you can instantly unlock with your purchase:

PDF and ePub Copies

Next-Gen Web-Based Reader

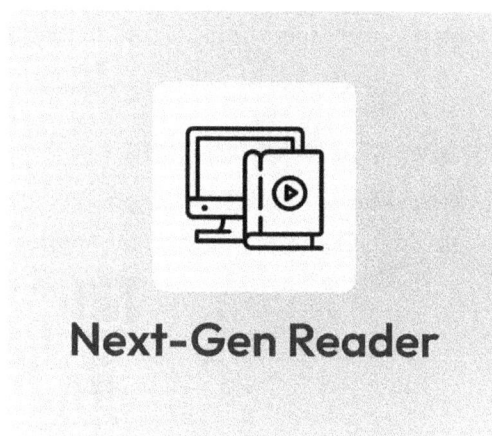

Free PDF and ePub versions

Next-Gen Reader

Access a DRM-free PDF copy of this book to read anywhere, on any device.

Use a DRM-free ePub version with your favorite e-reader.

Multi-device progress sync: Pick up where you left off, on any device.

Highlighting and notetaking: Capture ideas and turn reading into lasting knowledge.

Bookmarking: Save and revisit key sections whenever you need them.

Dark mode: Reduce eye strain by switching to dark or sepia themes.

How to Unlock

UNLOCK NOW

Scan the QR code (or go to packtpub.com/unlock). Search for this book by name, confirm the edition, and then follow the steps on the page.

Note: *Keep your invoice handy. Purchases made directly from Packt don't require one.*

Part 1

Advanced Structured Text

In this section, you'll build a strong foundation in advanced Structured Text (ST) programming and modern PLC software design. You'll move beyond the basics to explore the tools and techniques that allow you to write clean, modular, and maintainable industrial automation code. You'll learn how to structure complex logic, design robust architectures using functions and object-oriented programming, and apply professional best practices used by experienced PLC engineers. By the end of this part, you'll be equipped with the knowledge needed to craft scalable solutions, reduce technical debt, and develop reusable software components that can be applied across projects.

This part of the book includes the following chapters:

- *Chapter 1, Advanced Structured Text: Programming a PLC in Easy-to-Read English*
- *Chapter 2, Complex Variable Declaration: Using Variables to Their Fullest*
- *Chapter 3, Functions: Making Code Modular and Maintainable*
- *Chapter 4, Object-Oriented Programming: Reducing, Reusing, and Recycling Code*
- *Chapter 5, OOP: The Power of Objects*
- *Chapter 6, Best Practices for Writing Incredible Code*
- *Chapter 7, Libraries: Write Once, Use Anywhere*

1

Advanced Structured Text: Programming a PLC in Easy-to-Read English

Software engineering is a pivotal, yet often overlooked, aspect of **programmable logic controller (PLC)** programming. PLC software development often takes a backseat to hardware development. Unfortunately, many in the modern automation landscape see PLC software as a disposable component. Contrary to this belief, the software that controls the PLC is the true heart and soul of the system. The cold reality is that, without properly written software, fancy hardware is little more than very expensive paperweights.

Object-oriented programming (OOP) has dominated the IT landscape for decades. Most general-purpose programming languages, such as Java, C++, C#, Python, and so on, support the paradigm. Even some functional-first programming languages, such as Microsoft's F#, support OOP to some extent. Though often overlooked in the automation world, certain PLCs that follow the IEC 61131-3 standard can utilize the paradigm to some extent. This book will be unique as it will explore PLC programming from an OOP perspective.

This book is not a beginner's book. It assumes a certain level of proficiency with programming logic and PLC programming in general. If you are not comfortable with program flow, logic and design, and the basics of **Structured Text (ST)**, this book could be hard to follow. However, if you have programmed PLCs in the past using the ST language or have a background with a text-based programming language such as C++, Java, C#, Python, or some other text-based, OOP language, you should be able to use this book.

Almost all modern programming languages are text-based. Most PLC programming systems are no different. Many programming systems will allow you to choose between multiple programming languages. The word *language* is used loosely, as some of these so-called languages are actually graphic programming interfaces that allow users to use symbols and other graphical depictions to write programs. An example of this is **Ladder Logic** (**LL**). Though each of these interfaces has its time and applications, this book is going to predominantly focus on ST.

Though LL rules the PLC world due to its visual representation of circuits, which makes it more intuitive for programmers with a background in electronics, ST can greatly reduce the overall complexity of a program. Unfortunately, many PLC programmers only have a sparse understanding of ST. Though ST is not necessarily a prerequisite to understanding OOP, it will greatly help. Therefore, before we do a deep dive into OOP, we need to first explore some of the more advanced programming capabilities that ST has to offer. For this, we're going to:

- Explore the IEC 61131-3 standard
- Explore the needed software and learning approach
- Learn about error handling
- Explore state machines
- Explore expert systems

Free Benefits with Your Book

Your purchase includes a free PDF copy of this book along with other exclusive benefits. Check the *Free Benefits with Your Book* section in the Preface to unlock them instantly and maximize your learning experience.

Technical requirements

To get the most out of this chapter, a Windows computer and a working copy of an **IEC 61131-3**-compliant programming environment that supports OOP will be needed. For this book, the recommended programming system is **CODESYS**: https://us.store.codesys.com/.

This chapter will have multiple code examples. The code examples can be downloaded from GitHub by following the link: https://github.com/PacktPublishing/Mastering-PLC-Programming-Second-Edition.

The projects will utilize CODESYS and the book will assume you're using that system. If you opt to use a different programming system, you will need to copy and paste the examples into the system of your choice. For the most part, an IEC 61131-3-compliant system should require minimal to no modifications to the source code; however, this will depend on which feature the system has adopted and how the vendor chose to implement the features. Regardless, the principles and techniques explored in this book can be applied to any programming system that you use, including general-purpose programming languages such as C# or Java.

Note

As with all software, CODESYS will be updated from time to time. So, if you download the code and it gives a compatibility issue, the easiest workaround will be to simply create a new project in the updated version and copy the code there.

Exploring the IEC 61131-3 standard

Most advanced PLCs are IEC 61131-3-compliant to some degree, especially if they are PC-based, such as Beckhoff PLCs. IEC 61131-3 is a standard that essentially governs the programming environments that the PLC supports and the general functionality for the programming system. This means aspects such as syntax, semantics, typing, memory management, error handling, modularity, code organization, and so on should be mostly consistent between PLCs. The standard also governs the following programming languages:

- ST
- LL
- **Sequential Function Chart (SFC)**, often called *Sequential Flow Charts* in slang
- **Function Block Diagram (FBD)**

The IEC 61131-3 standard is just that, a standard. IEC 61131-3 is not a programming language, PLC brand, or anything of the sort. The standard promotes uniformity across programming systems and ultimately promotes vendor neutrality.

Though the standard is vendor-neutral, you typically cannot take a project written for one brand, and sometimes one model, and use the code as is with another brand or model. Different PLC brands will often use different system architectures, such as processors, and have their own locks that prevent a project written for *Brand A* from being compiled and run on *Brand B*. Though the syntax will generally be the same, it is not unusual for different brands to put their own touches on the standard.

If you attempt to port a codebase, you will have to recompile the source code for the new brand. In other words, if you attempt to run a program meant for one PLC brand on another, you will, at a minimum, have to copy the code to the proper programming environment and tweak the code as necessary. This is because it's not unusual for manufacturers to implement their own custom function blocks, functions, features, and even tweak the programming syntax. It is also the norm for manufacturers to choose not to implement certain features that the standard governs.

> Note
>
> A new technology called **PLCOpen XML** has recently been introduced to help interoperability between compliant systems.

This may seem all doom and gloom as the standard is supposed to support vendor agnosticism, and from what was just explored, we can't necessarily port code, at least not easily. This couldn't be farther from the truth. The standard provides us with a general set of rules to follow. Meaning if there are differences between the different environments, they are typically minor and will mostly consist of **custom function blocks** and **functions**, which we're going to explore in later chapters.

This is where many programmers often fall off the wagon. Most programmers look at programming through a very language-specific lens, especially automation programmers. However, this is a very poor philosophy as the principles and techniques that govern programming are language-agnostic, meaning that they apply to most languages. This means that regardless of what you're doing, if you focus on the principles presented in this book, you can apply them to any programming system and create hyper-advanced codebases.

IEC 61131-3 and OOP

One capability that is governed by the standard, at least in terms of automation, is a concept called OOP. OOP is not a programming language, nor is it something akin to a function or function block; it is known as a **programming paradigm**. This means that it's a way to conceptualize and ultimately architect code. OOP is not unique to automation programming, nor is it by any means exotic to the programming industry. It is a widely adopted programming paradigm that has been used in almost every traditional programming language and in a vast majority of software applications since its inception in the 1980s.

In 2013, the automation industry followed suit when the third edition of IEC 61131-3 introduced object-oriented programming to PLCs, allowing programmers to apply modern software engineering techniques to their control code. IEC 61131-3 standardizes OOP and the components that are generally associated with the paradigm for compliant PLCs. By understanding what these components do and how they behave, you can apply time-tested OOP practices to greatly enhance your codebase.

Not every programming system will support OOP. OOP is still considered novel to the automation world, and many automation programmers still view it with a level of skepticism, misunderstanding, and fear. It must be understood that even though OOP is still novel to automation, it is, as stated before, a 40+ year-old, time-tested paradigm. Though many automation programmers typically do not like to venture outside of what they're familiar with, it is highly recommended to try using OOP with a compliant device at least once. To follow along with this book, we need software that supports the paradigm. In the next section, we're going to explore what software we need to start writing advanced code!

Needed software and learning approach

The biggest hurdle to being a programming instructor is getting students to understand that the key to being a good engineer is not memorizing patterns or programming commands. To be successful as a developer, whether it be a developer for traditional apps or PLCs, lies in the ability to take established programming principles and apply them to any system. The key to mastering the material presented in this book is to approach it from a software engineering perspective. As you go through this book you will be tempted to think the material presented is only for CODESYS or even PLCs in general. However, this is a misnomer as the techniques explored in this book can be used with any OOP based language or system.

Programming software

To follow along with this book, an IEC 61131-3 programming environment that supports OOP will be required. The CODESYS environment will support most, if not all, of the IEC 61131-3 features and is free to download and use. Under no circumstances should this book be considered a CODESYS book! This means that you can essentially use whatever you want to follow along with, as long as it supports OOP. Nonetheless, this book will assume you are using the CODESYS software, which can be downloaded for free on the CODESYS website. Installing the software is a straightforward process. All you need to do is download the software and follow the wizard. Many other systems, such as TwinCAT, are just as easy. Once you get a software system installed, you should be able to follow along.

Once you get a system stood up it is important to understand that no matter how good a programmer you are, your program will always face potential issues that could interrupt its execution. This could be something as simple as the denominator of a number being rounded to zero before division or as complex as the mismanagement of memory. Either way, your program must gracefully be able to handle the error without crashing. In the following section, we're going to explore how to gracefully handle otherwise fatal errors!

Error handling

Errors can kill the execution of a program, which, in turn, can lead to injury or death in extreme cases. **Exceptions** occur when the PLC encounters a problem that it cannot handle at runtime. When one such error occurs, the PLC program will lock up or crash, and the PLC will typically need to be rebooted. On top of all that, if the condition that caused the error occurs again, the program will again crash and cause another lock up. In essence, the only safe way to handle the condition is to modify the code to ensure that the erroneous condition is handled gracefully.

Exception errors will not show up during the compilation process. Instead, exceptions occur when the program is running. Due to their nature, it is often difficult or impossible to fully predict when an exception will occur. To make matters worse, some exceptions can take very specific conditions to trigger, and, as a result, there might be long intervals between occurrences. To compound the issue, certain exceptions may not show up during development. Therefore, due diligence must be given to possible errors when developing the software. In other words, as a developer, you need to expect the unexpected!

Many different things can cause an exception and crash a program. A common exception that can often occur is a *division by zero* error. This error occurs when a divisor is accidentally set to 0 or gets extremely close to a decimal point, which will cause the PLC to round it to 0. A common reason for this type of error is a malfunctioning sensor. Other common errors that can throw an exception are **null pointers** or an *array out of index* error. Depending on what you're working on, there could be others as well. Generally, it is good to use some form of error handling when working with any of the following:

- Division
- Pointers
- Arrays

To explore what an error looks like, let's create a simple program that will attempt to divide a number by 0.

For this program, the required variables are as follows:

```
PROGRAM PLC_PRG
VAR
    dividend : INT;
    divisor  : INT;
    division : INT;
END_VAR
```

For this example, we are going to have a dividend and divisor variable, and the quotient of the two is going to be assigned to division.

The logic for the program that will go in the main section of the PLC_PRG POU file is as follows:

```
dividend := 5;
divisor   := 0;
division := dividend / divisor;
```

The code in the file will attempt a division by zero calculation. Since division by zero is an illegal operation in any form of computer programming, the PLC program will crash, and an error such as the one in *Figure 1.1* will be produced.

Expression	Type	Value
dividend	INT	5
divisor	INT	0
division	INT	0

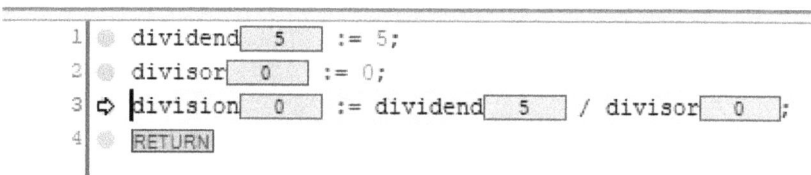

```
1    dividend[   5   ] := 5;
2    divisor[   0   ] := 0;
3 ⇨  division[   0   ] := dividend[   5   ] / divisor[   0   ];
4    RETURN
```

Figure 1.1: Division by 0 error

After you run the program, you should notice two things. The first is that the line that does the computation is now highlighted. The highlight means there is an error present on that line of code. The second thing you should notice is that the program automatically stops. If you watch the **Play** button, it will automatically reenable.

If you try to change the number to a value that is not zero and attempt to log back in with the default login selection, **Login with online change**, you will either be met with the pop-up error in *Figure 1.2* or the program will fail to run.

Figure 1.2: Download failed popup

When an error such as a division by zero occurs, restarting the program can be problematic. The easiest way to fix the issue is to simply fix the error. In this case, change the zero to any other non-zero value, then restart. Any code change will trigger the options in *Figure 1.3*. To restart the virtual hardware, you must press the **Login** button again. This time, instead of selecting **Login with online change**, you must select the **Login with download** option and ensure that **Update boot application** is selected as well.

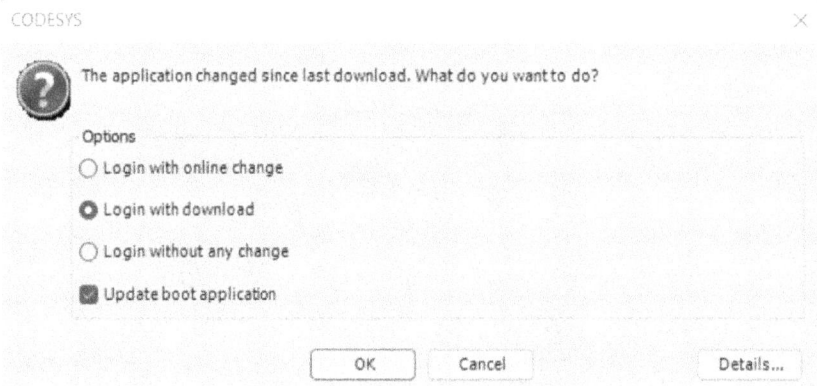

Figure 1.3: Necessary selections to reset the PLC

Once **OK** is clicked, the application should be reset, and you will be able to rerun your PLC program. As can be deduced, in a fast-paced production environment, having to perform these steps every time a value is set to 0 can easily become a major issue.

There are a couple of ways to handle division by zero errors. One way is to use a TRY-CATCH block, which we will explore later, or a simple IF statement. To get our feet wet with error handling, we're going to start exploring how simple conditional statements can be used to handle issues such as division by zero. To explore this concept, implement the following code:

```
dividend := 5;
divisor := 0;
IF divisor <> 0 THEN
    division := dividend / divisor;
END_IF
```

This code performs a simple check on the value stored in the divisor variable. If the value of the divisor variable is not 0, the program will perform the computation; however, if the value of the divisor is 0, it will not perform the operation. This code is an applicable solution when there are only a few values that need to be checked, or you're working with a system that does not have more advanced error-handling capabilities.

Problems like these are very easy to gloss over during development. To make matters worse, when your program must perform a lot of calculations, it can be easy to miss a check. A more applicable solution is to use what's called a TRY-CATCH block.

Understanding the TRY-CATCH block

A better and more formal solution is to use a TRY-CATCH block. TRY-CATCH blocks are like safety nets. When code is in a TRY block, it is essentially tested for errors. If an error is found, the code in the CATCH block will be executed. A TRY-CATCH block is a much more eloquent solution as it can be used to detect more faults without a dedicated IF statement. The pattern for a TRY-CATCH block is as follows:

```
__TRY
    <code to test>
__CATCH
    <code to run when there is an error>
__ENDTRY
```

Three statements are required to implement a TRY-CATCH block. As can be seen, the three blocks are a TRY, CATCH, and ENDTRY statement. The TRY section will test the code, and the CATCH block will run if there is an error. To end the TRY-CATCH block, the ENDTRY keyword is used. To demonstrate the TRY-CATCH block, let's look at an example:

```
PROGRAM PLC_PRG
VAR
    dividend : INT := 5;
    divisor  : INT := 0;
    division : INT;
    error    : WSTRING;
END_VAR
```

For this demonstration, we're going to preset dividend and divisor in the variable section. We're also going to include an error variable that will hold a message to help us track our location in the program. The following is the TRY-CATCH code that we're going to use in the PLC_PRG POU file:

```
error := ""; //Ensure error is cleared
__TRY
    division := dividend / divisor;
__CATCH
    division := -999;
    error := "Error Caught";
__ENDTRY
```

The computation in the TRY block will throw a division by zero error. When the error is thrown, the code in the CATCH block will run. When the program is run, the division variable will be set to -999, and our error variable will be set to Error Caught, as in *Figure 1.4*:

Device.Application.PLC_PRG		
Expression	Type	Value
🔷 dividend	INT	5
🔷 divisor	INT	0
🔷 division	INT	-999
🔷 error	WSTRING	"Error Caught"

Figure 1.4: The TRY-CATCH program output

If the divisor number is set to a value that will not cause a division by zero error, the code will not need to be reset, and the computation will execute without issues, as in *Figure 1.5*.

Device.Application.PLC_PRG		
Expression	Type	Value
◈ dividend	INT	5
◈ divisor	INT	1
◈ division	INT	5
◈ error	WSTRING	""

Figure 1.5: TRY-CATCH with no exception

The computation was executed without any issues. The overall takeaway is that even if an exception occurs, the program will not crash. Therefore, when valid values are passed back in, the program will execute normally without needing to restart the PLC.

The true power behind a TRY-CATCH block is that it can handle multiple errors. In other words, if you were trying to compute 20 different equations with a divisor that could possibly be set to 0, you wouldn't have to use 20 IF statements to check whether any of the divisors are set to that value. In all, you can test as much code as you need in a single TRY-CATCH block. The only real drawback to this technique is that not every PLC programming system will support TRY-CATCH; however, each PLC will usually have something that is similar in nature.

> Note
>
> A TRY block can and usually will contain multiple lines of code. Once one of the lines throws an error, the following lines of code will not be executed. Depending on what the code is for, this could cause errors in the code further downstream.

FINALLY statements

There is one additional block that can be used with TRY-CATCH statements. This command is known as a FINALLY statement. A FINALLY block is an optional block that is used in conjunction with TRY-CATCH. The code in a FINALLY block will execute regardless of whether an exception occurs or not. Essentially, the code that goes into a FINALLY block is used to do things that must be executed regardless of whether there is an error or not.

The following code is the syntax for a TRY-CATCH-FINALLY block:

```
__TRY
    <Code to test>
__CATCH
    <Code to run when there is an error>
__FINALLY
    <Code that will run whether there is an exception or not>
__ENDTRY
```

As can be seen, adding a FINALLY block is as simple as adding the extra keyword.

Identifying and handling errors

The TRY-CATCH blocks that we have explored so far did not specify what the error was. In practice, this usually isn't a preferred behavior. The type of TRY-CATCH block that we have explored so far can be called a **generic except block**. It is important to remember that many different things can throw an error. Therefore, if you want to address the issue, you will most likely need unique logic to handle it. For real-world applications, you generally do not want to use generic except blocks. In a real-world application, you want TRY-CATCH to have logic that can handle specific errors. For example, if you find yourself with a division by zero error, you may want to switch the dividend to 1 or conduct some other logic that will alleviate the situation so that the error does not occur again.

The first step in creating specific logic is setting up an Exception variable.

Exception variables

This is how an Exception variable is declared:

```
PROGRAM PLC_PRG
VAR
    exc : __SYSTEM.ExceptionCode;
END_VAR
```

To demonstrate this, we're going to explore a simple example. These are the necessary variables for TRY-CATCH with Exception:

```
VAR
    dividend : INT;
    divisor  : INT;
    division : INT;
```

```
    exc        : __SYSTEM.ExceptionCode;
END_VAR
```

These are mostly the same variables that were used for the division by zero programs, except the exc variable that will hold the exception.

This code will store the error in the exc variable and will be implemented in the PLC_PRG POU file:

```
__TRY
    division := dividend / divisor;
__CATCH(exc)
    division := -999;
__ENDTRY
```

This code is nearly the same as the code we used for the original TRY-CATCH program. The only difference between the programs is the (exc) code next to the CATCH statement. This variable will store the exception in the exc variable, as can be seen in *Figure 1.6*:

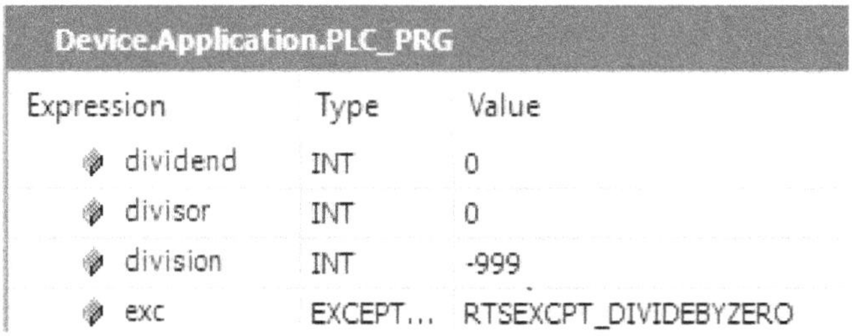

Device.Application.PLC_PRG		
Expression	Type	Value
dividend	INT	0
divisor	INT	0
division	INT	-999
exc	EXCEPT...	RTSEXCPT_DIVIDEBYZERO

Figure 1.6: Error output

Figure 1.6 shows that the error the code picked up is RTSEXCPT_DIVIDEBYZERO. This means that the code picked up a division by zero error.

As was stated before, it is usually considered a best practice and a good idea to implement custom logic to handle the specific error. For our purposes, we're going to set the division variable to -999 only when a division by zero exception is thrown.

Handling custom exceptions

The following code is one way of implementing logic to respond to unique exceptions:

```
__TRY
    division := dividend / divisor;
__CATCH(exc)
    IF (exc = __SYSTEM.ExceptionCode.RTSEXCPT_DIVIDEBYZERO) THEN
```

```
        division := -999;
    END_IF
__ENDTRY
```

This code has an IF statement that checks for a division by zero exception. This code will only change the division variable to -999 when a division by zero exception occurs.

> **Note**
>
> This code serves two purposes. The first is that it protects the program's execution from all errors. The second benefit this code provides stems from specifically handling division by zero.

In all, many types of exceptions can be thrown. There is no magic bullet to determine when and where you should use a TRY-CATCH block. However, a good rule of thumb is to wrap things such as arrays, math equations, and so on in TRY-CATCH blocks.

It is important to understand that if an error code is triggered, there is a problem. Just because the program doesn't crash does not mean that everything is okay. Error handling is just a means of allowing your program to gracefully handle an error without crashing. If your program is consistently triggering a TRY-CATCH block, the root cause of the error needs to be addressed. It is also important to understand that if an error is caught, it doesn't mean that everything is going to work as intended downstream of the error. All we did here was set a default value, which may or may not produce the correct results downstream.

As was mentioned before, pointers are often the cause of fatal PLC program errors. However, what is a pointer? The following section will explore pointers and references so you can understand how they work and how they can cause issues in a program.

Understanding pointers

To understand a pointer, it is first necessary to understand the basics of how variables are stored in memory. For many PLC programmers, creating a variable or a tag is simply inputting a name and assigning it a data type; however, some mechanics take place under the hood. For starters, a variable is much more than just a name and a data type that holds a value. A variable is a dedicated memory block that the computer (in this case, the PLC) uses to hold a value of a specific type. The memory block is generally not human-readable; as such, the variable name is just a human-readable façade that makes accessing and manipulating the data in the memory block easy while adding context to the value.

Representing PLC memory

Figure 1.7 is a graphical representation of a PLC's memory. It is a simplified way of conceptualizing how the PLC sees its memory addresses and the values that reside in those blocks:

0x01	0x02	0x03	0x04	Address
Hello World	12.3	0	12	Value

Figure 1.7: A graphical representation of computer memory

As you have probably deduced, working with the raw addresses would be very confusing and probably lead to bugs in the program. This is the reason why variable names are so important. In short, a variable name adds a layer of abstraction over the memory address.

Variables are not the only type of data that has a memory address. As we will see later, function blocks, methods, and more all have memory addresses when the program is running. A general rule of thumb is that if it has a name that you provide, it has a memory address.

More times than not, you'll want to work with a human-readable name over the memory block address. However, you can still directly access the memory address of a variable or anything else with what is known as a pointer. Pointers are declared in a similar way to regular types; however, the value they hold is the memory address of a variable, function block, or whatever else it might be.

General syntax for pointers

This is the syntax that is used to declare a pointer:

```
PROGRAM PLC_PRG
VAR
    pt : POINTER TO <TYPE>;
END_VAR
```

This code is just declaring a variable with the POINTER and TO keywords. This variable declaration will be able to hold the address of a function, variable, or so on of any type. In short, this is all that is needed to declare a pointer.

Although the working version of the code above will produce a pointer, it won't do anything meaningful. Essentially, we created a pointer that points to nothing. For a pointer to be of use, we need to explore the ADR operator.

The ADR operator

The ADR operator will provide the address of whatever is passed into it. Many times, the ADR operator is used with pointers. It is the main way to retrieve address information. So, it is usually assumed that if you're going to use a pointer, you're going to use the ADR operator as well.

To explore the ADR operator, we're going to create a small program that will display the memory address of a variable. The following are the variables we will need:

```
PROGRAM PLC_PRG
VAR
    pt : POINTER TO INT;
    testVal : INT := 10;
END_VAR
```

The pt variable is the variable that holds the memory address. The testVal variable is the important variable for this program. This is the variable whose memory address we're going to read with the following code:

```
pt := ADR(testVal);
```

The testVal variable is passed into the ADR operator, and that output is assigned to the pt variable. When the code is run, you should see an output similar to what is shown in *Figure 1.8*.

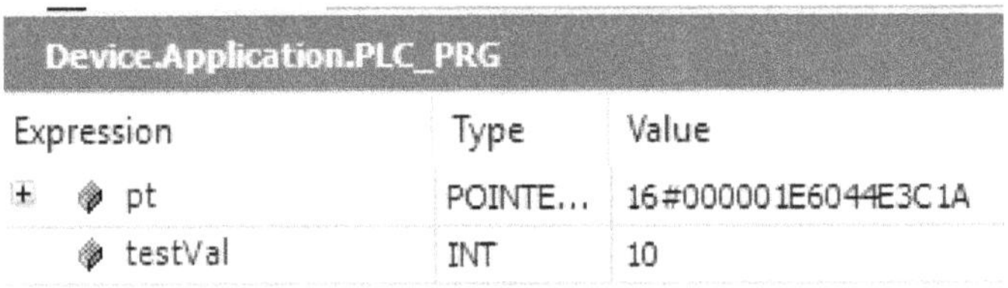

Device.Application.PLC_PRG		
Expression	Type	Value
+ ◈ pt	POINTE...	16#000001E6044E3C1A
◈ testVal	INT	10

Figure 1.8: Memory address output

The memory address that you get when you run this program will probably be different from the one in the screenshot. Nonetheless, *Figure 1.8* shows the memory address of testVal for this execution cycle.

Getting the memory address alone won't accomplish much. To do something meaningful, we have to **dereference** the pointer.

Dereferencing pointers

Obtaining a value out of a pointer is called dereferencing. This is accomplished by appending the ^ symbol to the pointer variable. The ^ symbol gives you the ability to access or manipulate the data in a pointer. To demonstrate dereferencing, create a new program and implement the following variables:

```
PROGRAM PLC_PRG
VAR
    testVal_pt : POINTER TO INT;
    testVal    : INT := 10;
    testVal2   : INT;
END_VAR
```

This code creates a pointer variable to hold the memory address of testVal. The testVal variable is initialized with a value of 10. The testVal2 variable is not initialized, and after the logic is run, it will be assigned the value that lives inside of testVal.

To power the project, implement the following logic in the PLC_PRG POU file:

```
testVal_pt := ADR(testVal);
testVal2 := testVal_pt^;
```

When the program executes, the first line assigns the address of testVal to the testVal_pt variable. The second line accesses the data in testVal_pt and assigns it to the testVal2 variable. After the program runs, the value of testVal2 should be 10, similar to what is shown in *Figure 1.9*:

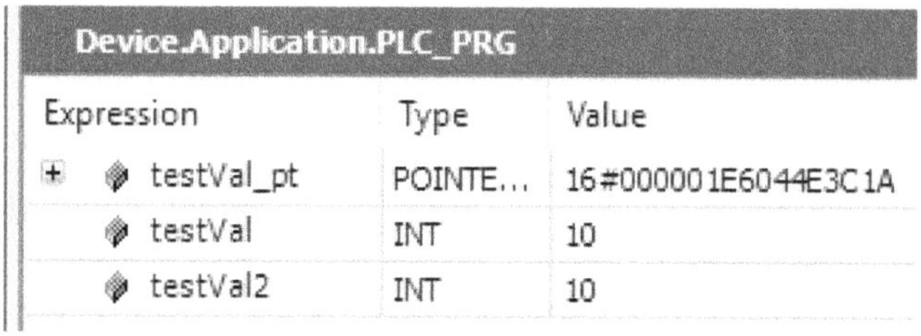

Device.Application.PLC_PRG		
Expression	Type	Value
+ ◈ testVal_pt	POINTE...	16#00000 1E6044E3C 1A
◈ testVal	INT	10
◈ testVal2	INT	10

Figure 1.9: Dereferencing output

When a pointer is not properly configured, it can become an **invalid pointer**. An invalid pointer is kind of like a null pointer in a traditional language. Though they occur less frequently in PLC programming, you need to know how to handle them.

Handling invalid pointers

If you've ever programmed in C/C++, Java, C#, or any traditional programming language, chances are you've run across a null pointer before. PLC programming is no different. If you're working with a pointer, you want to check that the pointer is pointing to something. For example, you may try to assign a value to a pointer variable, but if the variable isn't pointing to anything, you have the PLC equivalent of a null pointer. Consider the following code, which declares a pointer:

```
PROGRAM PLC_PRG
VAR
    testVal_pt : POINTER TO INT;
END_VAR
```

These are the variables we are going to use in the invalid pointer program. As can be seen, all we have is our standard pointer variable, testVal_pt.

The following is the logic that will be used to demonstrate the invalid pointer:

```
testVal_pt^ := 2;
```

As you can tell from the code, testVal doesn't point to anything. As it stands, the code is attempting to assign the number to an invalid pointer, as the pointer isn't pointing to a memory address. When the code is run, the output should match what is shown in the following figure:

Figure 1.10: Invalid pointer

The program will instantly fail when someone tries to run it. Essentially, this code is trying to assign the value of 2 to an empty pointer. Since the pointer does not point to a memory address, the PLC will not know how to handle the situation, and the program will crash.

Catching an invalid pointer

Depending on what you're trying to accomplish with your code, an easy way to check for an invalid pointer is to check the memory address. If the pointer is not pointing to anything, then the value will be 0. An easy way to check whether a pointer is valid is to perform a simple IF check on it.

IF statements for invalid pointers

This logic will only try to assign a value to a pointer if the pointer address is not 0:

```
IF testVal_pt <> 0 THEN
    testVal_pt^ := 2;
END_IF
```

If the memory address is 0, then the program will ignore the assignment and not crash.

Compared to the other logic, this code did not crash. The IF statement prevented the assignment that caused the previous code to crash by not allowing the erroneous line of code to be executed.

Using an IF statement to check for an invalid pointer is an excellent and very common way to detect invalid pointers. Many developers will always wrap their pointer code in a control statement as a best practice. However, as mentioned before, you can also use a TRY-CATCH block.

TRY-CATCH for invalid pointer variables

These are the necessary variables to demonstrate the TRY-CATCH invalid pointer program:

```
PROGRAM PLC_PRG
VAR
    pInvalid     : POINTER TO INT;
    exc          : __SYSTEM.ExceptionCode;
    catch_value  : WSTRING;
END_VAR
```

The following is the logic to catch the error:

```
__TRY
    pInvalid^ := 10;
__CATCH(exc)
    catch_value := "Caught error";
__ENDTRY
```

As can be seen, this is just a basic TRY-CATCH block. When this code is executed, you should see an output similar to the following figure:

Device.Application.PLC_PRG		
Expression	Type	Value
+ pInvalid	POINTE...	16#0000000000000000
exc	EXCEPT...	RTSEXCPT_ACCESS_VIOLATION
catch_value	WSTRING	"Caught error"

Figure 1.11: TRY-CATCH invalid pointer output

The output in *Figure 1.11* is a little more descriptive than just wrapping the pointer in an IF statement; however, the drawback to this method is that to remedy the underlying problem, you will still need custom IF statements to handle the logic. On the other hand, the TRY-CATCH blocks will provide a blanket of protection for multiple pointers. Essentially, both code blocks will prevent the program from crashing. It will ultimately be up to you as the developer to choose which method is more appropriate.

Pointers are fine to use, and there are many codebases that still use them. However, modern PLC programming has introduced a more user-friendly way of working with pointers that requires less syntax. For new codebases, it is usually a good idea to favor what is known as a **reference** over pointers.

Discovering references

A reference is a type of pointer that is more user-friendly and requires less syntax than a traditional pointer. A few major advantages of using a reference are that you do not have to use the ^ symbol, you do not have to use the ADR operator, and references are type safe.

References share many similarities with pointers, including similar syntax. Like pointers, a reference must be declared. Therefore, the first step in learning how to use references is to understand how to declare them.

Declaring a reference variable

Declaring a reference is almost the same syntax as declaring a pointer. Essentially, references can be thought of as a shorthand way of using a pointer. The major difference is that the REFERENCE keyword is used as opposed to the POINTER keyword.

The syntax to declare a REFERENCE variable is as follows:

```
<variable> : REFERENCE TO <data type>
```

Putting this syntax into practice, we can create a reference to an integer, as in the following code. These are the variables we will use for the reference demonstration:

```
PROGRAM PLC_PRG
VAR
    A : REFERENCE TO INT;
    B : INT;
END_VAR
```

This example will only use two variables. The A variable will be the REFERENCE variable that will be assigned the value that is in B when the program is run.

Example program

This is the code to demonstrate references:

```
A REF= B;
A := 33; //Comment this line after you run it
//B := 1111; //uncomment this line after you run A
```

With this code, the two variables are essentially linked. In a more fundamental sense, A is now an alias for B.

Run the program as is and observe your output. Then, comment out the second line and rerun it with the third uncommented. After following this procedure, you should see an output similar to what is in the following figure:

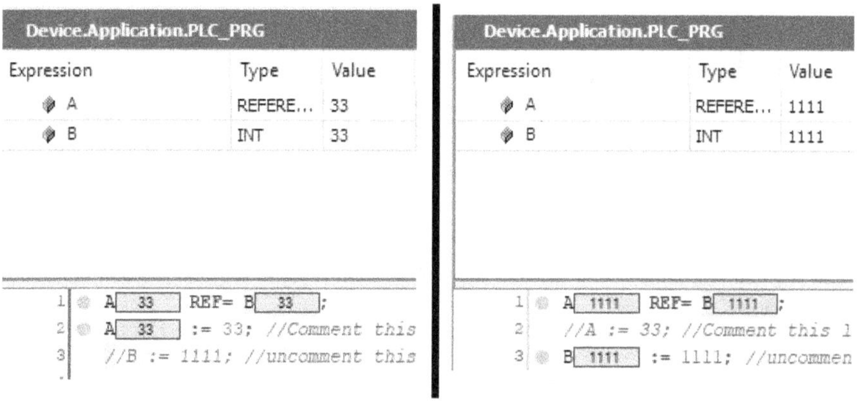

Figure 1.12: Reference program output

Essentially, the reference program does the same as the pointer program; however, the reference program uses a much simpler and more intuitive syntax. The first line is of vital importance. Since this line is equivalent to the ADR operator in the pointer program, if this line is neglected, you will get an invalid reference.

Checking for invalid references

It is important to check for invalid references in the same way we do for invalid pointers. There is an easy operator that can be used to test whether a reference is valid or not. The most effective way to do this is to use the __ISVALIDREF operator. This operator will return TRUE if the reference is valid, or FALSE if it is not.

These are the bare minimum variables to test for an invalid REFERENCE variable:

```
PROGRAM PLC_PRG
VAR
    A     : REFERENCE TO INT;
    B     : INT := 3;
    valid : BOOL;
END_VAR
```

The valid variable isn't always necessary, as the operator returns a TRUE or FALSE value, so it can often be embedded in a control statement. However, for this example, we are going to store the return in the valid variable. This is the logic we will use to check whether the reference is valid:

```
A REF= B;
valid := __ISVALIDREF(A);
```

All we are doing is passing the A reference variable into the operator and assigning the output of the operator to the valid variable. When the program is run, you should see the following output:

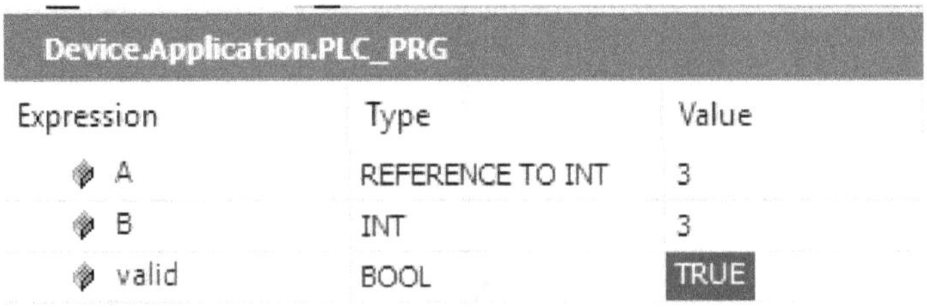

Device.Application.PLC_PRG		
Expression	Type	Value
⚑ A	REFERENCE TO INT	3
⚑ B	INT	3
⚑ valid	BOOL	TRUE

Figure 1.13: The __ISVALIDREF output

Since we have a valid reference, the operator's output is TRUE. If you were to comment out the first line of code and rerun the program, the valid variable will be set to FALSE.

Generally, whether to use a reference or a pointer will boil down to the developer's preference.

Another relatively simple yet confusing concept that many automation programmers do not have a good grasp of is the concept of state machines. So, in the next section, we're going to explore the concept and how to use one.

State machines

State machines sound like a complex and scary concept. Depending on the context of the state machine, they can be complex; however, fundamentally, they aren't. Fundamentally, a state machine can be defined as a structure where some type of input and a current state dictate some type of output state.

There are many types of state machines. In terms of automation, the most common is what's known as a **finite state machine (FSM)**. Essentially, an FSM is a state machine with a finite number of states. In a very basic sense, there are a few key components to an FSM:

- **Finite number of states**: A **state** can be thought of as a **mode**. In terms of automation, a mode can be something like a machine being in an on, off, or standby mode.
- **Transitions**: A transition can be thought of as the change from one state to another. For example, a machine going from on to off can be thought of as a transition.
- **Input**: An input can be some type of action or event that triggers a transition. This could be a reading from a sensor or some type of information that a user will input.
- **Output**: An output can be thought of as what happens as a result of the current state or a transition between states. This could be something like a machine turning on after you flip a switch.

The simplest way to conceptualize a state machine is with a lightbulb. Consider *Figure 1.14*:

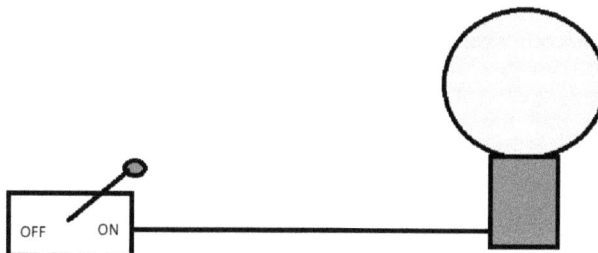

Figure 1.14: State machine (lightbulb on)

In this example, we have a lightbulb attached to a switch. When the switch is in the on position, the lightbulb will be turned on. When the switch is in the off position, the light will be off, as in *Figure 1.15*.

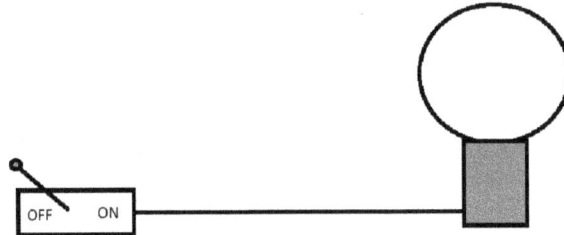

Figure 1.15: State machine (lightbulb off)

For this example, the switch can be thought of as the input. When the input is changed, whether it is switched on or off, the light will transition to mirror that state. When the light is finished transitioning, its output will be either on or off.

State machine code

Visually seeing a state machine can only go so far. So, in this section, we're going to explore the basics of writing a simple state machine. To start, we're going to look at the basic mechanics of creating a state machine.

State machine mechanics

In terms of coding, there are a couple of ways we can easily implement an FSM. The first is with an `IF-ELSIF` statement block, and the other is with a `CASE` statement. If we need to create a state machine for a piece of equipment that will put the machine into either an on, off, or standby mode, we could use a program like the following:

- `IF-ELSIF`: This code example uses a simple `IF-ELSIF-ELSE` block to implement a state machine:

```
Input state
IF state = 1 then
    Turn machine on
    state = 2
ELSIF state = 2 then
    Put machine into standby mode
    state = 3
```

```
    ELSIF state = 3 then
        Turn machine off
        IF user flips switch
            state = 1
    ELSE
        Input not valid
        state = 3
```

- CASE: For this example, a simple CASE statement is used to implement a state machine:

```
Input state
CASE 1:
    Turn machine on
    state = 2
CASE 2:
    Put machine into standby mode
    state = 3
CASE 3:
    Turn machine off
    IF user flips switch
        state = 1

Default:
    Input not valid
    state = 3
```

Whether you opt for the IF-ELSIF-ELSE or the CASE example will mostly depend on your coding style and the project. Both will do the same job; however, the CASE commands will typically streamline the code and arguably make it easier to follow. Regardless, both iterations will often be used. To round out this chapter, we're going to create a working state machine. However, for now, we're going to move on and create something that will in many respects resemble a state machine. This concept that we will explore is known as an expert system.

Expert systems

In the early days of **artificial intelligence** (**AI**), when computer and data science weren't nearly as fleshed out, there was an AI technique called **expert systems** or **expert machines**. Essentially, these systems were a series of IF-THEN statements that, when executed correctly, could provide a level of expertise in a given system. Nowadays, these systems aren't used as much in the traditional programming landscape due to the advances of actual AI systems, such as deep learning and the like. Nonetheless, expert systems can still have a great effect in the automation industry.

An expert system is a program or functionality of a program that is used to simulate human-like judgment and decision-making. Essentially, you can think of these systems as simulated experts for a domain. For example, suppose you have a cutting machine; you can use an expert system to automatically adjust the saw blade in relation to the consistency of the material.

Though traditional AI rules the modern world, expert systems are relatively easy to implement and lightweight compared to many modern AI systems. Though these systems can easily incorporate various technologies, you can get away with simply using control statements to implement basic expert systems. In terms of automation, expert systems are used in many different automation projects. One notable application for an expert system is **fault diagnosis**.

Before we can demonstrate the expert system, we first need to understand what rules and facts are:

- **Facts:** Things the system currently knows or believes to be true
- **Rules:** The IF–THEN logic that uses facts to infer new facts or actions

An example of rules and facts would be the following pseudocode:

```
If car does not start then
    Check lights
    If lights are dim then
            Check battery
            If battery is less than 12 volts
                    Change battery
```

In this case, the car not starting, the lights being dim, and the battery being less than 12 volts are examples of facts. The logic they are embedded in (that is, the If statements) represents the rules that govern the system's behavior.

Knowledge base

An expert system is built around what is called a **knowledge base**. In a very brief sense, a knowledge base is a set of facts and rules that serve as the system's brain. The more complex the knowledge base is, the more knowledgeable the system will be. To see an expert system in action, let's explore an example.

Expert system example

To begin, let's create an expert machine that can troubleshoot an error with an industrial saw machine.

To implement this program, we can use the following variables:

```
VAR
    cutsNotStraight : BOOL;
    sawBladeHours   : INT;
    changeBlade     : BOOL;
END_VAR
```

In this case, these variables will simulate a system that checks whether cuts are straight and the number of hours on a saw blade. If the hours on the blade are greater than 20, the system will direct the user to change the saw blade. To do this, the following logic can be implemented:

```
IF cutsNotStraight = TRUE THEN
    IF sawBladeHours > 20 THEN
        changeBlade := TRUE;
    ELSE
        changeBlade := FALSE;
    END_IF
ELSE
    changeBlade := FALSE;
END_IF;
```

Much like with the pseudocode, the system will do a series of IF checks to evaluate the rules. Once you implement the code, set the cutsNotStraight variable to TRUE and sawBladeHours to 50.

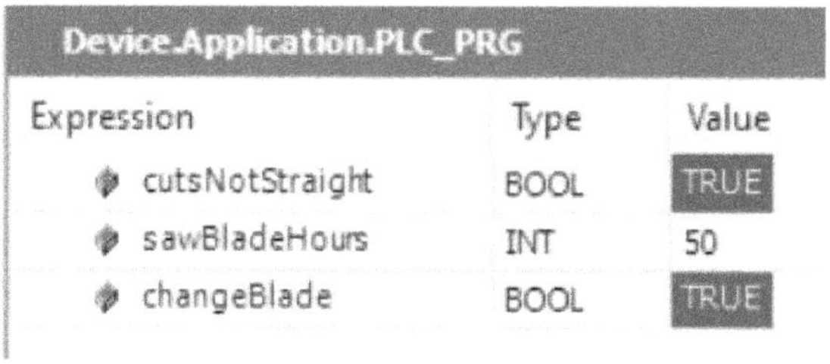

Device.Application.PLC_PRG		
Expression	Type	Value
◈ cutsNotStraight	BOOL	TRUE
◈ sawBladeHours	INT	50
◈ changeBlade	BOOL	TRUE

Figure 1.16: Expert system output

As can be seen, based on the current facts, the expert system suggests changing the blade out by setting the changeBlade variable to TRUE. Now, either change the cutsNotStraight variable to FALSE or the sawBladeHours variable to something less than 20. When you do that, you should be met with the result in *Figure 1.17.*

Expression	Type	Value
cutsNotStraight	BOOL	FALSE
sawBladeHours	INT	10
changeBlade	BOOL	FALSE

Figure 1.17: Expert system second output

In this case, the changeBlade variable is set to FALSE, as one would expect. In other words, we now have a simple expert system that can determine whether a saw needs to be changed!

Now that we have a solid understanding of expert systems, we can move on to our final project!

Final project: Making a simple state machine

For our final project, we're going to make a simple FSM that will control the state of a machine that is used to produce a certain number of parts. The machine will have multiple states, such as an on, off, and error state. To begin, let's design our state machine!

State machine design

To begin, let's jot out some pseudocode to get a basic feel for the general skeleton of the system:

```
Case state
1:
    Machine off
2:
    Production run (on)
    If error detected
        state = 3
3:
    Error state
    state = 1
```

As we can see, all we have are three states: on, off, and error. If an error is detected during a production run, the machine will transition to an error state. With the general logic in place, let's implement the code.

Variables for the state machine

These are the variables that will be used for the state machine:

```
PROGRAM PLC_PRG
VAR
    machineState      : INT := 1;
    motorSpeedCutOff  : INT := 10000;
    runTime           : INT := 2;
    setSpeed          : REAL;
    numOfParts        : REAL := 8;
    motorOff          : BOOL;
    exc               : __SYSTEM.ExceptionCode;
END_VAR
```

This program will have a number of preset values. The machineState variable is preset to 1, which means that the machine will automatically start in the off state. The setSpeed variable is the quotient of numOfParts divided by the runTime value. The setSpeed variable is a simulated motor speed value that will set the speed of a theoretical motor. These values simulate the number of parts that a line should produce in a given amount of time. If the operator accidentally inputs a 0 value for runTime, the line will transition to an error state, which will reset everything.

Now that the variables for the state machine have been established, we can explore the logic that drives the state machine. As can be seen in the following section, the general structure for our state machine is very simple.

Exploring state machine logic

This is the logic for the state machine:

```
CASE machineState OF
1:
    //machine off state
    motorOff := TRUE;
2:
    //machine run state
    __TRY
        //set motor speed
        setSpeed := numOfParts / runTime;
        IF setSpeed >= motorSpeedCutOff THEN
```

```
            motorOff := TRUE;
            machineState := 1;
        ELSIF setSpeed < motorSpeedCutOff THEN
            motorOff := FALSE;
        END_IF
    __CATCH(exc)
        //throw machine into error state
        machineState := 3;
    __ENDTRY
3:
    //error state
    runTime     := 0;
    setSpeed    := 0;
    machineState := 1;

END_CASE
```

As is the case with many state machines, the machine is built around a CASE statement. For this machine, *Case 1* is the off state. In other words, when in *Case 1*, the machine is turned off. *Case 2* is the machine running state. *Case 2* computes and controls the motor speed. Since this is wrapped in a TRY-CATCH block, if there is any error, the machine will go into *Case 3*, which is an error state, and will then immediately transition into an off state.

Case 1 — non-running state machine

This is the state machine in what is considered an off state:

Device.Application.PLC_PRG		
Expression	Type	Value
◈ machineState	INT	1
◈ motorSpeedCutOff	INT	10000
◈ runTime	INT	2
◈ setSpeed	REAL	0
◈ numOfParts	REAL	8
◈ motorOff	BOOL	TRUE
◈ exc	EXCEPT...	RTSEXCPT_NOEXCEPTION

Figure 1.18: State machine in an off state

The state machine is set to *Case 1* by default, which means that the machine is in an off state. This can be seen by examining the setSpeed variable being set to 0 and the motorOff variable being in a TRUE state, which, in this case, means the motor is off.

Case 2 – running state machine

These are the variable outputs when the machine is running:

Device.Application.PLC_PRG		
Expression	Type	Value
⬦ machineState	INT	2
⬦ motorSpeedCutOff	INT	10000
⬦ runTime	INT	2
⬦ setSpeed	REAL	4
⬦ numOfParts	REAL	8
⬦ motorOff	BOOL	FALSE
⬦ exc	EXCEPT...	RTSEXCPT_NOEXCEPTION

Figure 1.19: Running state machine

When the machineState variable is set to 2, the machine will go into a running state, in which setSpeed is computed to 4 and the motorOff variable is set to FALSE. This means that the motor is running.

Case 3 – state machine exception thrown

To trigger an error, set machineState to 2 and set runTime to 0.

These are the variables when an error is thrown:

Device.Application.PLC_PRG		
Expression	Type	Value
⬦ machineState	INT	1
⬦ motorSpeedCutOff	INT	10000
⬦ runTime	INT	0
⬦ setSpeed	REAL	0
⬦ numOfParts	REAL	8
⬦ motorOff	BOOL	TRUE
⬦ exc	EXCEPT...	RTSEXCPT_FPU_DIVIDEBYZERO

Figure 1.20: Exception thrown

If you look at the value of the `machineState` variable, it is set to 1, which means off; however, the `runTime` and `numOfParts` variables are zeroed out, which only happens when the machine passes through the exception state. Essentially, the state machine transitioned states faster than you could notice it.

Compared to the other concepts we explored, state machines are an amalgamation of different concepts. A state machine is a pattern and, as such, how you implement the code will vary. Overall, you should now have a decent understanding of state machines and the core concepts explored.

Chapter challenge

As a chapter challenge, integrate an expert system into the state machine. Add an extra state that will house a simple knowledge base that can determine whether the user inputted an invalid value when an error occurred. Be sure to add variables as needed to support the expert system!

Summary

This chapter has been an introduction to some of the more complex topics that you may encounter as a PLC programmer. We explored some advanced features of ST, such as pointers and error handling, as well as some complex topics such as state machines and expert systems. So far, we have only touched on the basics of what ST in the IEC 61131-3 standard can do. In the next chapter, we're going to explore how to get the most out of variables!

Questions

1. What are three keywords that can be used with a TRY-CATCH block?
2. What is an expert system?
3. What is the difference between a fact and a rule?
4. What is a state machine transition?
5. What is an FSM?
6. What is the main purpose of a TRY-CATCH block?
7. What are some applications for an expert system?

Further reading

Have a look at the following resources to further your knowledge:

- *CODESYS pointer*: `https://content.helpme-codesys.com/en/CODESYS%20` `Development%20System/_cds_datatype_pointer.html`

- *CODESYS TRY-CATCH-FINALY*: `https://content.helpme-codesys.com/en/CODESYS%20` `Development%20System/_cds_operator_try_catch_finally_endtry.html`

2

Complex Variable Declaration: Using Variables to Their Fullest

Every programmer who has ever written a program more complex than the famous *Hello World!* program has used variables in some manner. Variables and programming go hand in hand. A programmer can in no way avoid using variables. Variables and the way we group them are a very rich topic and can extend well beyond simply naming a memory address and storing a single value in it.

If you have ever programmed in a language such as C++, you may be familiar with how variables can be used and organized. Many of the advanced features that are often associated with complex variables are actually complex data structures. Practically, though, these principles can be conceptually linked to variable usage and can drastically improve and enhance your code. This chapter is going to demonstrate how to get the most out of variables by exploring the following topics:

- Constants
- Arrays
- Global variable lists
- Structs
- Enums
- Persistent variables
- Final project: motor control program

Technical requirements

As usual, to follow along with this chapter, all you will need is CODESYS installed and working on your machine. The source for the examples presented can be found at the following URL: https://github.com/PacktPublishing/Mastering-PLC-Programming-Second-Edition/tree/main/Chapter%202.

Understanding constants

The values that are stored in variables often change. This behavior is quite useful when it comes to things like taking inputs from an operator, such as the number of parts that a machine needs to make. However, there are times when we need to store a value that doesn't change. This may seem counterintuitive, but it's nonetheless very common. To accomplish this, we use what are called **constants**.

To understand constants and how they differ from a typical variable, you need to first understand what the following two terms mean:

- **Mutable:** A mutable variable is a value that can change during the execution of a program. A mutable variable can be initialized (declared and set) when the program is compiled and then overwritten by other data during the program's execution. Thinking back to the aforementioned example, consider a variable that holds the number of parts for a machine to produce. In one run, the operator may want to make 20 parts, and in the next run, they may want to make 100 parts. In terms of the program, the variable will need to be the same; however, the value in the variable must change.

- **Immutable:** An immutable variable is a value that cannot change. In other words, when a value is assigned to a variable, it cannot be changed during the program's execution. At first glance, this may seem useless. What use is a variable that doesn't change? In reality, this is one of the most desired behaviors for a well-written program. Generally, you want a value to change as little as possible. Every time a value changes, it will change the behavior of a program in a positive or negative way. A few common reasons for why you want a variable to be immutable are as follows:

  - Makes code easier to test and understand
  - Reduces the amount of erroneous data being introduced into the program
  - Improves traceability in the program
  - Prevents values from being changed accidentally

In terms of applications, some common use cases for constants are:

- Mathematical constants such as Pi
- Setting machine values that should not be changed
- A variable that will be used throughout the program and should not change

This list is by no means comprehensive, nor will you always need to declare constants for the preceding bullet items. Whether or not you declare a constant is up to you and the application that you're developing. In short, you will declare a constant when you want to add a level of protection to ensure that the variable's value does not change.

Declaring a constant is very simple; all you need to do to declare an immutable variable is employ the following syntax:

```
VAR CONSTANT
    const: INT := 23;
END_VAR
```

This code will declare a variable called const with an initial value of 23. Essentially, the syntax is the same as declaring a regular variable with the additional keyword CONSTANT after the VAR keyword. Much like regular variable blocks, you can declare as many constants in the variable block as you want.

The CONSTANT block will go in the same tab that regular variables are declared in. For example, the following code is valid:

```
PROGRAM PLC_PRG
VAR
    //This is a test variable
    test: INT;
END_VAR

VAR CONSTANT
    const : INT := 23;
    PI    : REAL := 3.14;
END_VAR
```

Let's say you attempt to change the value of the const variable to 6, as in the following code:

```
const := 6;
```

When you run this code, you will get a compile error. *Figure 2.1* shows what will happen when you attempt to change the value of a CONSTANT variable:

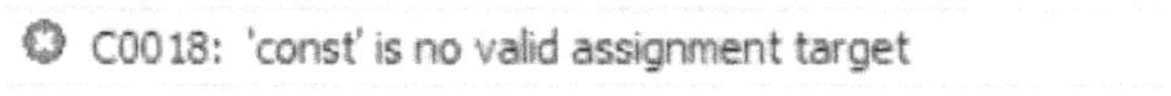

⚙ C0018: 'const' is no valid assignment target

Figure 2.1: Compile error

Constants are one of the most important concepts to understand when properly implementing variables. No matter what you're programming, a general rule of thumb is to have your variables change as little as possible. As we have seen thus far and will see throughout the rest of the book, changing the value of a variable is necessary; however, it is a good practice to declare whatever you can as a constant.

Constants are just one concept in the grand scheme of variables. The next concept we'll look at is arrays. Much like constants, **arrays** are another vital concept that every software engineer must understand.

Investigating arrays

Arrays are one of the most common types of variables that you will use as a programmer. All programming languages utilize arrays in some form; Structured Text is no different. Much like C++, C#, or Java, arrays can be used in PLC programming as well. In fact, most PLC programs of significant functionality will utilize arrays in some way. Before we dive into the depths of arrays, we're going to explore what they are.

A quick review of arrays

Arrays can be thought of as logically related values. By definition, an array is a data structure that holds multiple values that are identified by an index. In other words, an array is a variable on steroids. Where a variable can hold one value, an array can hold many. Consider a college professor who has a class of 20 students. Also, suppose the professor must write a program that can track each student's GPA. For a situation like this, the professor may write a program such as the following:

```
Student1 = 3.00
Student2 = 2.98
Student3 = 4.0
StudentN = 1.0
```

This program would serve the professor well for exactly one term. As soon as a new term starts, the professor will need a new program that reflects the new batch of students. This will put the professor in a bit of a dilemma. On the one hand, the professor needs a piece of software that can keep track of GPAs, but on the other, they shouldn't need to write a program each term. They need a program to be robust enough to accommodate the influx of new students.

For a PLC programming system that supports the IEC 61131-3 standard, such as CODESYS, the syntax for declaring an array is as follows:

```
ArrayName: ARRAY[<Start_Element>..<End_Element] OF <Type>
```

Unlike arrays in a traditional programming language such as C# or Java, which always start at 0, arrays in IEC 61131-3 can start at any value. This aspect of arrays in automation programming can cause a bit of confusion for programmers who are classically trained developers. However, with a little practice, it's easy to get used to working with arrays that do not start with 0. To explore this, let's look at an example.

Array declaration

In the following example, we're going to rewrite the GPA program with an array:

```
PROGRAM PLC_PRG
VAR
    gpa : ARRAY[1..10] OF REAL;
END_VAR
```

In this case, the only variable we're going to declare is the array itself.

Array logic

In terms of the logic, we will use the following code to load data into the array:

```
//Load a value into the first five students
gpa[1] := 3.98;
gpa[2] := 2.80;
gpa[3] := 4.00;
gpa[4] := 1.99;
gpa[5] := 3.45;
```

When this code is run, it will produce the output in *Figure 2.2*:

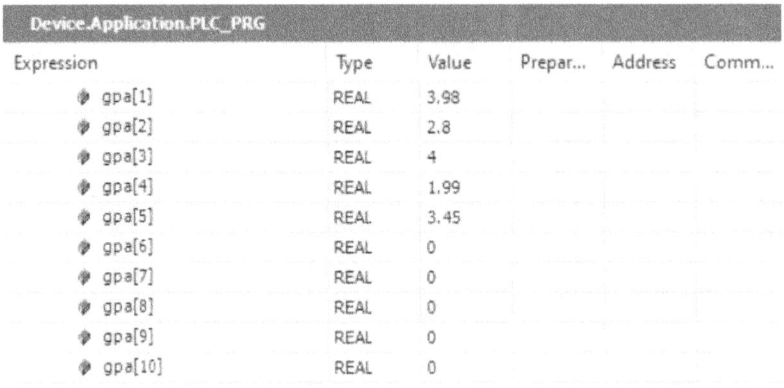

Device.Application.PLC_PRG					
Expression	Type	Value	Prepar...	Address	Comm...
● gpa[1]	REAL	3.98			
● gpa[2]	REAL	2.8			
● gpa[3]	REAL	4			
● gpa[4]	REAL	1.99			
● gpa[5]	REAL	3.45			
● gpa[6]	REAL	0			
● gpa[7]	REAL	0			
● gpa[8]	REAL	0			
● gpa[9]	REAL	0			
● gpa[10]	REAL	0			

Figure 2.2: Array output

Notice that in the screenshot, there are 10 elements, but only 5 are showing an active value. This is because in the source code, we only loaded data for the first five elements! This means that, like with other languages, you can load as many or as few of the elements as you need.

In terms of automation, it is common to use arrays to store values such as weights or sizes for parts produced during a production run or other data, such as settings for the machine. One common application for an array is to store motor positions for a machine. For example, suppose you have a robotic arm with three axes. A common mechanism to store the positions would be a code snippet such as the following:

```
PROGRAM PLC_PRG
VAR
    axes : ARRAY[1..3] OF REAL;
END_VAR
```

In terms of the logic, the code would look something like this:

```
axes[1] := <value>
axes[2] := <value>
axes[3] := <value>
```

From here, there would be other core logic that will process and move the arm into the specified positions.

The arrays explored here are examples of what are called **1D arrays**. Essentially, the arrays hold one set of values. This is sufficient for most applications; however, arrays in IEC 61131-3-based systems are capable of much more. In the next section, we're going to explore multidimensional arrays.

Multidimensional arrays

Multidimensional arrays can best be thought of as arrays embedded in arrays. In terms of automation, these are very handy. One application for multidimensional arrays is to control devices in clusters. For example, you may have three banks of motors, with each bank containing five motors. By using a multidimensional array, we can access the motor bank using an array index and then specify the target motor with another index.

Before we can explore how to utilize the array, we need to first establish the pattern for declaring a multi- or *n*-dimensional array!

Multidimensional array pattern

The syntax for declaring a multidimensional array is as follows:

```
multidimensionalArray: ARRAY[<outer>, <inner>] OF INT;
```

In terms of a working array, the code would look like the following:

```
multidimensionalArray: ARRAY[1..10, 1..5] OF INT;
```

This example will declare an array of 10 elements. Each element, in this array, is an array composed of five elements. A more common way of expressing this is a 10 x 5 array.

As of now, this 10 x 5 array does nothing. To make this array do something meaningful and explore how it works, we're going to load some values into it.

Working with n-dimensional arrays

The general pattern for accessing a value in an *n*-dimensional array is with the following:

```
multidimensionalArray[<outer>, <inner>]
```

To explore this, we are going to implement a 10 x 5 array as we did previously, and we're going to load a value of 21 into the 2 x 3 element position.

The first step to implementing this program is declaring the array, which can be done with the following code:

```
PROGRAM PLC_PRG
VAR
    multidimensionalArray: ARRAY[1..10, 1..5] OF INT;
END_VAR
```

The logic for loading the number 23 into the target elements is as follows:

```
multidimensionalArray[2,3] := 23;
```

This essentially follows the pseudocode pattern. In this snippet, we're accessing the third element of the second array. When the code is executed, you should be met with *Figure 2.3*:

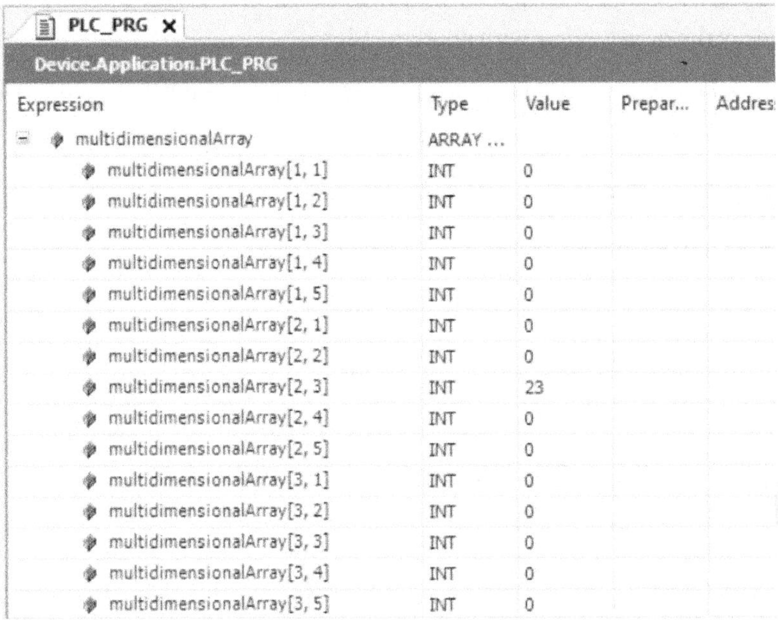

Expression	Type	Value	Prepar...	Addres:
⊟ multidimensionalArray	ARRAY ...			
multidimensionalArray[1, 1]	INT	0		
multidimensionalArray[1, 2]	INT	0		
multidimensionalArray[1, 3]	INT	0		
multidimensionalArray[1, 4]	INT	0		
multidimensionalArray[1, 5]	INT	0		
multidimensionalArray[2, 1]	INT	0		
multidimensionalArray[2, 2]	INT	0		
multidimensionalArray[2, 3]	INT	23		
multidimensionalArray[2, 4]	INT	0		
multidimensionalArray[2, 5]	INT	0		
multidimensionalArray[3, 1]	INT	0		
multidimensionalArray[3, 2]	INT	0		
multidimensionalArray[3, 3]	INT	0		
multidimensionalArray[3, 4]	INT	0		
multidimensionalArray[3, 5]	INT	0		

PLC_PRG — Device.Application.PLC_PRG

Figure 2.3: N-dimensional array output

As can be seen in *Figure 2.3*, the number 23 is loaded into the correct location.

Multidimensional arrays are very powerful, but as can be seen, they can convolute code. You generally want to use the fewest dimensions possible. The more dimensions an array has, the harder it is to do the following:

- Troubleshoot the array
- Load values into the correct coordinates (location)

Most textbooks will describe a multidimensional array as a series of rows and columns. The first array can be thought of as rows, while the second array can be thought of as columns. I don't like this explanation because when you end up working with a more complex array, such as a four- or five-dimensional array, it can be nearly impossible to conceptualize it. In my opinion, it is better to think of *n*-dimensional arrays as arrays embedded in arrays.

Typically, loading data into an array requires some type of **loop**. When working with *n*-dimensional arrays, you will need a loop that has *n* embedded loops. In the next example, we're going to explore how to loop through a 2 x 5 array!

Looping through an n-dimensional array

One of the most common operations that a developer will perform on an array, whether it's multidimensional or not, is to loop through it. To loop through a multidimensional array, you will need to implement a series of embedded FOR loops. For each dimension an array has, you will need one FOR loop. So, if you have a 1D array, you will need one FOR loop. If you have a 5D array, you will need five FOR loops. To demonstrate this, let's loop through the following array:

```
PROGRAM PLC_PRG
VAR
    multidimensionalArray: ARRAY[1..2, 1..5] OF INT;
    i, j : INT;
END_VAR
```

This is a 2 x 5 array. In other words, this is an array that has two elements, and each element is an array with five elements. This is also a 2D array, which means that we will need two FOR loops to process each element.

In this example, we're going to load values into each element in the array. To accomplish this, we can use the following code:

```
FOR i := 1 TO 2 DO
    FOR j := 1 TO 5 DO
        multidimensionalArray[i, j] := j;
    END_FOR
END_FOR
```

In this case, the i and j variables are counter variables. We're using the counter variables to access the target element's coordinates. When the code is run, we should get the following output:

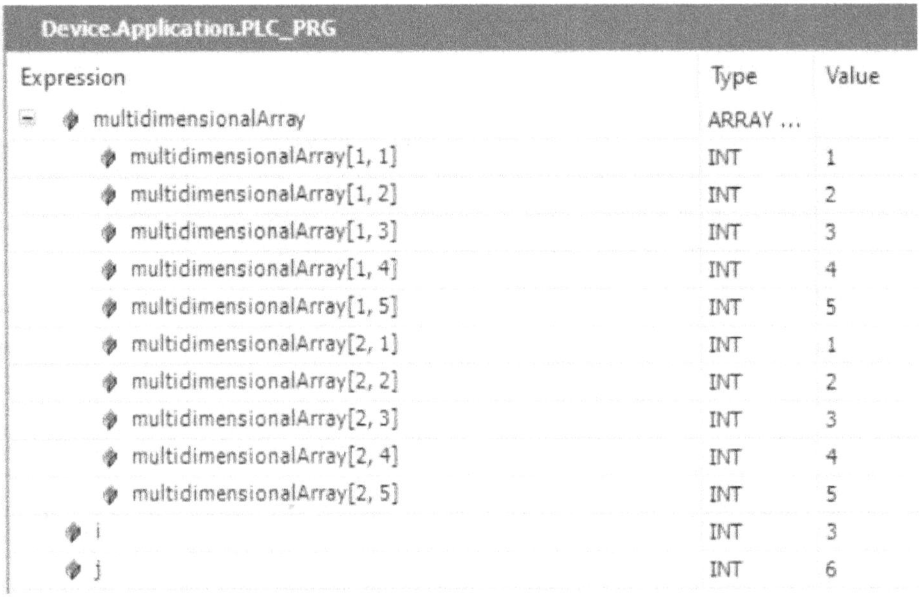

Device.Application.PLC_PRG		
Expression	Type	Value
⊟ ● multidimensionalArray	ARRAY ...	
● multidimensionalArray[1, 1]	INT	1
● multidimensionalArray[1, 2]	INT	2
● multidimensionalArray[1, 3]	INT	3
● multidimensionalArray[1, 4]	INT	4
● multidimensionalArray[1, 5]	INT	5
● multidimensionalArray[2, 1]	INT	1
● multidimensionalArray[2, 2]	INT	2
● multidimensionalArray[2, 3]	INT	3
● multidimensionalArray[2, 4]	INT	4
● multidimensionalArray[2, 5]	INT	5
● i	INT	3
● j	INT	6

Figure 2.4: Looping through a multidimensional array

As can be seen, we have values of 1–5 repeated throughout the array.

As stated before, as a PLC developer, you need to be very careful when implementing large multidimensional arrays. The more dimensions an array has, the harder it will be to work with. Though there is no set rule for the maximum number of dimensions in an array, you typically want to keep the dimension count to a minimum.

With arrays fully explored, we can move on to a more automation-centric topic: **global variable lists (GVLs)**.

Exploring global variable lists

Contrary to other forms of programming, **global variables** are quite common in automation programming. This stems from different subsystems needing to know the state of a certain variable. This section is dedicated to understanding what GVLs are and how to properly implement and use them.

What is a global variable?

Before we dive into what a GVL is, we need to understand what a global variable is. Typically, when a variable is declared in one **program organization unit (POU)**, it is not accessible from another. This can cause a problem, as many machines must know the state of certain values, such as emergency stops, to properly function. This is where global variables can come in handy. A global variable is a variable whose scope is not bound to one POU. A global variable can be read or modified from anywhere in the PLC project. This makes a global variable a convenient way to keep track of critical states, such as whether certain subsystems are on or off, whether an emergency stop is engaged, and many other things! They do have some downsides, though. In the next section, we're going to explore some of the dangers of global variables.

Dangers of global variables

In traditional programming, global variables are frowned upon, and their use is strongly discouraged. They can easily be havens for bugs and cause erroneous behavior in a program. Since a global variable can be accessed by anything in a program, it is very easy to accidentally alter a value without realizing it. On top of that, when a program consists of hundreds or even thousands of POUs, it can be nearly impossible to track down where the value is being erroneously altered.

Manually changing a global variable can also cause issues. This goes back to the potential of a program being hundreds or thousands of POUs in size. If you manually alter a value, it can ricochet across the program and cause the machine to behave in unexpected and potentially dangerous ways. Overall, you typically only want to use global variables sparingly. Though they are convenient to use, they can be very, very dangerous.

When to use a global variable

There are no hard and fast rules for when to use a global variable; however, there are applications that are more appropriate than others. The following is a short and by no means exhaustive list of when a programmer would use a global variable:

- **States:** A common use case for global variables is to track the global state of the machine. For example, most industrial machines utilize a state machine to operate. It is not uncommon to use a global variable to set the machine's state. This will allow subsystems that are not necessarily linked to the main system to keep track of the machine's overall state.

- **Logging:** Often, log data needs to be shared across multiple files and systems. Typically, the easiest way to do this in a PLC program is to use a global variable.

- **Emergency stops**: Though it can be argued that this would be considered a state, it is not uncommon for PLC programmers to attach an emergency stop switch to a global variable. The logic behind this is that each POU that needs to know about the state of the emergency stop can easily read the state without having to pass the variables between POUs.

Up until now, all our projects have been composed of one POU file (PLC_PRG); however, programs of significant functionality are usually composed of different POUs (files). One such common file is a GVL. When it comes to IEC 61131-3 based programming systems such as CODESYS and similar systems, global variables are declared in unique ways compared to more traditional programming languages such as Java or C#. In the next section, we're going to explore declaring and implementing global variables.

Creating a GVL

Compared to other programming systems, such as Java or C#, global variables in an IEC 61131-3 style system are typically declared in a GVL. GVLs typically must be manually added to a project. The process for adding a GVL in CODESYS is as follows:

1. Right-click **Application** in the PLC project tree.
2. Hover over **Add Object**.
3. Click **Global Variable List**.

Once those steps have been completed, you should see the following window appear:

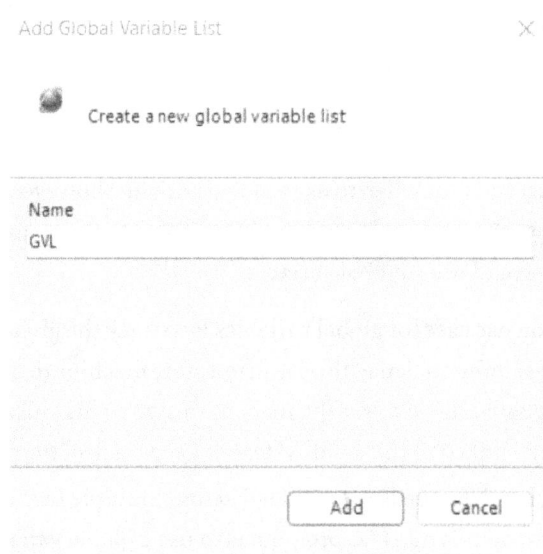

Figure 2.5: Global variable list wizard

This is the wizard that will generate the GVL. In the case of the GVL file, all you have to do is input a unique name in the **Name** field and press **Add**. For this example, input testGVL into the **Name** field. Once you press **Add**, testGVL should be added to the project tree. The code in the file should resemble the following code snippet:

```
{attribute 'qualified_only'}
VAR_GLOBAL
END_VAR
```

This is the code that powers a GVL. Like the way we have been declaring variables before, a variable is declared between the VAR_GLOBAL and END_VAR blocks.

> **Note**
>
> Every programming system is different. This procedure works for CODESYS; however, if you are using another IEC 61131-3 programming system or a system that is not IEC 61131-3 compliant, the procedure may vary.

It is a common practice to have the term GVL as a suffix for the GVL's name. This is a way for you to clearly signal to other developers that the file is indeed a GVL, so they can know how to treat it. For example, suppose you have a GVL for a water pump, you may opt to name it something like waterPumpGVL.

Now that we know how to make a GVL, we need to see one in action!

Demonstrating a GVL

The first step in demonstrating a GVL is to add a variable to the GVL file that we just created. In this case, add a variable called gvlVar to the GVL file and give it an int data type. When you are done, your GVL file should look like the following:

```
{attribute 'qualified_only'}
VAR_GLOBAL
    gvlVar : INT;
END_VAR
```

This code creates a single global variable. In other words, this code snippet creates a variable that can be accessed by any POU in the PLC project. To access the global variable, we can implement the following code in the PLC_PRG file:

```
testGVL.gvlVar := 32;
```

Notice that the pattern to access the global variable is as follows:

```
GVL_Name.Variable := <value>
```

When the code is executed, double-click the GVL in the project tree, and you should be met with *Figure 2.6*:

Expression	Type	Value
gvlVar	INT	32

Figure 2.6: GVL output

When it comes to global variables, organization matters. In the next section, we're going to explore some tips to organize your GVLs!

Organizing GVLs

Global variables and, by extension, GVLs, are dangerous as it is; couple that with an unorganized program structure and you have a recipe for utter disaster. The key to properly using GVLs is organizing them into cohesive units that are clearly and logically labeled.

A key consideration is that you can create more than one GVL in a project. When it comes to organizing your code, you typically want more than one GVL. Each variable list should be responsible for the global variables for a specific part of the system. For example, suppose you have a saw machine. For this device, suppose there is a saw system, and there is a blower system that cleans the saw blade. These are two distinct systems that compose the totality of the machine. This means that if you opted to use global variables for the overall system, you should have one GVL for the saw and one for the blower, as in *Figure 2.7*:

Figure 2.7: GVL files

To explore using these two GVLs, let's look at an example. First, let's implement an on variable in sawGVL with the following code:

```
{attribute 'qualified_only'}
VAR_GLOBAL
    sawState : BOOL;
END_VAR
```

In this example program, the sawState variable will dictate whether the saw is off or on.

In the next code snippet, we're going to create a variable called blowerState that will serve the same purpose with the exception of it controlling the blower:

```
{attribute 'qualified_only'}
VAR_GLOBAL
    blowerState : BOOL;
END_VAR
```

To use the variables in the PLC_PRG POU, implement the following code:

```
sawGVL.sawState        := TRUE;
blowerGVL.blowerState := TRUE;
```

When you run the program, and you click on sawGVL, you should be met with *Figure 2.8*:

Expression	Type	Value
sawState	BOOL	TRUE

Figure 2.8: sawGVL state

If you see the result in *Figure 2.8*, double-click on blowerGVL, and it should match *Figure 2.9*:

Expression	Type	Value
blowerState	BOOL	TRUE

Figure 2.9: blowerGVL state

In the next section, we're going to look at some safety considerations to consider when using GVLs.

Safety considerations for global variables

As alluded to before, global variables can cause the machine to behave unexpectedly. When troubleshooting, it is not uncommon to manipulate variables to trigger certain behaviors. Though it can be dangerous to manipulate any variable, global variables are particularly dangerous to manually manipulate since they are often used to control many different aspects of the machine. When manipulating global variables, you need to be extremely cautious, as triggering a variable can induce the following:

- Damage to property, such as the machine or the surrounding area
- Injury to people around the machine
- Death in extreme cases

Typically, it is best to avoid manually manipulating global variables, but at times, it is necessary. If you do find yourself in a situation where you need to manually manipulate a global variable, do not input random values. For example, if you try to manually change the position of a machine, you should input a value that is realistic and is commonly used for that machine. Using junk data can cause a ricochet effect across the machine that can put it into a dangerous state.

Though dangerous and often frowned upon, GVLs offer a way to organize variables that can be easily used across multiple parts of the overall PLC program. A safer way to organize variables is to use what is known as a **struct**. Therefore, in the next section, we're going to explore what structs are and how to use them!

Understanding structs

Structs are a way to organize related variables. Conceptually, structs are similar to a GVL in the sense that when properly implemented, a struct will contain all the necessary variables to accurately describe a process; however, a struct differs in that, unlike a GVL, a struct is not globally accessible. In other words, the values are not global and cannot be altered by any given part of the overall program. In this section, we're going to look at how to properly declare and implement a struct.

Declaring a struct

The key to declaring a struct ironically starts with a good name. Fundamentally, a struct describes the attributes of a thing, so the name of a struct should be a noun. This is a pitfall that many inexperienced programmers will fall into. Many inexperienced developers who don't fundamentally understand what a struct is can often give it a name that is a verb or a name that is not fully descriptive of the struct's responsibility. The end result will be a struct that doesn't fully describe anything and has an ambiguous functionality.

A quality struct name might be one of the following:

- `Car`
- `Motorcycle`
- `Saw`
- `Power_Supply`

Where a poorly named struct might be one of the following:

- `Run`
- `Dodge`
- `Utility`

As can be seen, the second batch of names is verbs. They do not accurately describe an object and, as a result, will produce a poor struct with an ambiguous purpose and poorly related values.

Normally, you will end the name of the struct with the word `struct`. This will signal to other developers the exact nature of the data structure and eliminate confusion about how to use it. For example, a common name for a struct might be something like the following: `carStruct`.

> **Note**
>
> Ending a struct name with the word `struct` is ideal; however, many may not follow that rule.

Let's now apply these rules and make a struct!

Implementing a struct

Creating a struct is very similar to creating a GVL. Similar to a GVL, you create a struct in CODESYS or a similar system with the following steps:

1. Right-click **Application**.
2. Hover over **Add Object**.
3. Click **DUT**.

When you finish these steps, you should see a wizard that is very similar to the wizard used to create a GVL, except that it has a few more options. The wizard can be viewed in *Figure 2.10*. For now, the only thing that you will need to do is change **DUT** in the **Name** field to motorStruct and click **Add**. Once you click **Add**, a new file will appear in the application tree under **Application**.

DUT wizard

The following screenshot is the **data unit type (DUT)** wizard:

Figure 2.10: DUT wizard

As stated earlier, this wizard will be used to create many different types of data structures and will determine whether a struct will inherit from another struct. For now, do not check the **Extends** box.

After you change the name and press **Add**, the struct that is added will be similar to the code in the following code snippet:

```
TYPE motorStruct :
STRUCT
END_STRUCT
END_TYPE
```

Similar to the way we add variables to a GVL, we add variables to a struct by simply declaring them.

Consider this example: Suppose we have a motor, and we want a specific struct to manage the motor's current speed, maximum RPMs, and minimum RPMs. We could use the following code in the struct:

```
TYPE motorStruct   :
STRUCT
     motorSpeed : INT;
     maxRPM      : INT;
     minRPM      : INT;
END_STRUCT
END_TYPE
```

Unlike GVLs, you have to explicitly state whether a POU can use a struct or not, which is why they are safer to use than GVLs. To do this, you create a variable in the variables list of the file that needs to manipulate the struct:

```
PROGRAM PLC_PRG
VAR
     motor1 : motorStruct;
END_VAR
```

In this case, motor1 is a reference to the motorStruct data type. In short, motor1 will have attributes called motorSpeed, maxRPM, and minRPM that you can manipulate in a similar fashion to GVL variables

Multiple objects

A struct can best be thought of as an object's attributes. In the preceding example, we have motor1, which may need its own operational parameters to properly do whatever it is that it needs to do in the system. However, motor2 may need a different set of values. In this case, motor2 will have all the attributes that motor1 has, but the operating values will differ. The beauty part of a struct is that we can recycle the motorStruct attributes and give motor2 different values. Consider the following code:

```
PROGRAM PLC_PRG
VAR
     motor1 : motorStruct;
     motor2 : motorStruct;
END_VAR
```

In this case, motor1 will reference motorStruct and be able to use all of its attributes, and motor2 will be able to do the same. It is important to remember that though motor1 and motor2 reference the same struct, they are independent of each other. This means we can set the values for each motor as needed without affecting the other motor. To demonstrate this, consider the following code

The values for motor1 and motor2 can be set with the following logic in the PLC_PRG file's logic section:

```
motor1.maxRPM    := 3000;
motor1.minRPM    := 1000;
motor1.motorSpeed := 1500;

motor2.maxRPM    := 4000;
motor2.minRPM    := 1000;
motor2.motorSpeed := 2000;
```

These are the outputs when the code is run:

Expression	Type	Value
⊟ ◈ motor1	motorStruct	
◈ motorSpeed	INT	1500
◈ maxRPM	INT	3000
◈ minRPM	INT	1000
⊟ ◈ motor2	motorStruct	
◈ motorSpeed	INT	2000
◈ maxRPM	INT	4000
◈ minRPM	INT	1000

Figure 2.11: Motor struct output

As can be seen, though both motors are referencing the same struct, the values are clearly independent of each other.

All things considered, the preceding example is a way of recycling code; however, there is another, more powerful trick to recycling code in a struct!

Inheriting with structs

If you think about objects, you may have a base object that is a vehicle, and another that represents a car. When you consider a vehicle, it will typically have an engine and wheels. Now, if you think about a car, it has an engine and wheels, but it also has a steering wheel. In the modern programming landscape, where we don't want redundant code, it doesn't make much sense to have to redeclare the engine and wheels variables. When it comes to structs, there is a shortcut to getting the use of the vehicle variables in the car struct. To do this, we can use what is called **inheritance**.

Inheritance is an object-oriented principle, and we're going to dive deeper into it in the following chapters. For now, think of inheritance as a way of copying and pasting variables from one struct to another without physically doing it. To demonstrate inheritance, create two structs, and name one vehicleStruct and the other carStruct. When you create carStruct, ensure that you check the **Extends** box and enter vehicleStruct in the box next to it, as in *Figure 2.12*:

Figure 2.12: The carStruct setup

In vehicleStruct, set up the following variables:

```
TYPE vehicleStruct :
STRUCT
    wheels  : INT;
    engine  : BOOL;
END_STRUCT
END_TYPE
```

In carStruct, set up the following variable:

```
TYPE carStruct EXTENDS vehicleStruct :
STRUCT
    steeringWheel : BOOL;
END_STRUCT
END_TYPE
```

Notice that carStruct has the EXTENDS keyword. This keyword is used to signal to the programming system to use the variables of vehicleStruct in carStruct. If the steps in *Figure 2.12* are followed properly, this code will be automatically generated. Once the variables are implemented, you can move on to implementing the logic in the PLC_PRG POU file.

In the variable section of the PLC_PRG POU, implement the following:

```
PROGRAM PLC_PRG
VAR
    car1 : carStruct;
END_VAR
```

Next, implement the following logic:

```
car1.engine             := TRUE;
car1.steeringWheel      := TRUE;
car1.wheels             := 4;
```

When the code is executed, you should be met with the following:

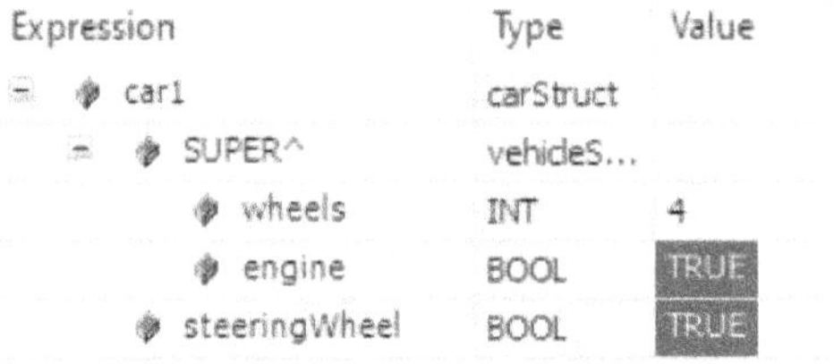

Expression			Type	Value
⊟ ⬦ car1			carStruct	
	⊟ ⬦ SUPER^		vehicleS...	
		⬦ wheels	INT	4
		⬦ engine	BOOL	TRUE
	⬦ steeringWheel		BOOL	TRUE

Figure 2.13: Struct inheritance

As can be seen, though the car1 variable is referencing carStruct, we are able to access the variables in vehicleStruct!

In all, inheritance is a way of recycling code. As we will see in future chapters, there are certain relationships that need to be followed to get the most out of extending a struct. For now, we're going to explore another important data structure known as an **enum**.

Getting to know enums

Like a struct, an **enumeration (enum)** is also a user-defined data type composed of comma-separated named values. The enumeration's named values are **constants**; they and their numeric assignments are fixed at design time. Enumerations are excellent tools for defining threshold limits, motor speeds, temperature limits, and more. You declare an enumeration with the same wizard that was used to declare a struct, so be sure to view *Figure 2.10*.

For this example, create an enum named `motorSpeeds`, using the same DUT wizard as before, but by checking **Enumeration** as opposed to **Structure**, and leaving **Textlistsupport** unchecked. For simplicity, implement `motorStruct` from the first struct example. Once the code is generated, you can remove the enum_member attribute that is auto-generated. Once that is done, modify the code to match the following:

```
{attribute 'qualified_only'}
{attribute 'strict'}
TYPE motorSpeeds :
(
    maxSpeed := 2000,
    minSpeed := 500

);
END_TYPE
```

Notice that the values in the enum end with a comma, except for the last entry. This is because values in an enum are separated with a comma; this is how the system knows when one value ends and the next begins.

The `PLC_PRG` variable should match the following:

```
PROGRAM PLC_PRG
VAR
    motor1 : motorStruct;
END_VAR
```

For this project, we only need the reference for `motorStruct`.

After adding the reference variable, modify the code in the `PLC_PRG` POU to match the following:

```
motor1.maxRPM     := motorSpeeds.maxSpeed;
motor1.minRPM     := motorSpeeds.minSpeed;
motor1.motorSpeed := 1500;
```

When you run this program, you should get the following output, which is the result of setting the motor's speeds with an enum:

Expression	Type	Value
⊟ ✦ motor1	motorStruct	
✦ motorSpeed	INT	1500
✦ maxRPM	INT	2000
✦ minRPM	INT	500

Figure 2.14: Motor speeds set with an enum

Notice how the maxRPM and minRPM fields now reflect the values set in the enum.

If you opt not to include a value for an enum entry, remember that the first value declared will be set to 0 by default, and each subsequent entry will be set as the previous value plus 1. In our case, if we didn't have the values set, maxSpeed would be 0, and minSpeed would be 1. Now that we have a grasp on enums, we need to shift our attention to our final project.

Final project: Motor control program

To demonstrate all the concepts we have covered so far, we're going to build a motor control program. The program will simulate five motors. The motors will be in an array, and the program will set the speed of the motors based on values in a GVL. To begin, let us create a motor structure:

```
TYPE motorStruct :
STRUCT
    motorStateMsg : STRING[20];
    motorState    : BOOL;
    motorSpeed    : INT;
END_STRUCT
END_TYPE
```

This code will create a struct that will dictate whether the motor is on with a Boolean variable, motorSpeed (which will be set with an enum value), and a STRING that will tell which state the motor is in. After the structure is created, add an enum named motorSpeeds. Once you create the enum, add the code to match the snippet:

```
{attribute 'qualified_only'}
{attribute 'strict'}
TYPE motorSpeeds :
(
    maxSpeed := 4000,
    avgSpeed := 3000,
    minSpeed := 2000,
    offSpeed := 0
);
END_TYPE
```

Next, we need to add a GVL called `motorStateGVL`. After you create the list, add the variables to match the code, as shown in the following snippet:

```
{attribute 'qualified_only'}
VAR_GLOBAL
    maxSpeed : BOOL;
    minSpeed : BOOL;
    avgSpeed : BOOL;
END_VAR
```

Now, in the `PLC_PRG` POU file, we will create an array of `motorStruct` references and a counter variable, like so:

```
PROGRAM PLC_PRG
VAR
    motors: ARRAY[1..5] OF motorStruct;
    count : INT;
END_VAR
```

What this code will do is create an array of five motors of the `motorStruct` type. The main logic of the file should match the following:

```
IF motorStateGVL.avgSpeed = TRUE THEN
    FOR count := 1 TO 5 DO
        motors[count].motorStateMsg := 'avg speed';
        motors[count].motorState := TRUE;
        motors[count].motorSpeed := motorSpeeds.avgSpeed;
        motorStateGVL.minSpeed := FALSE;
        motorStateGVL.maxSpeed := FALSE;
    END_FOR
ELSIF motorStateGVL.maxSpeed = TRUE THEN
    FOR count := 1 TO 5 DO
        motors[count].motorStateMsg := 'max speed';
        motors[count].motorState := TRUE;
        motors[count].motorSpeed := motorSpeeds.maxSpeed;
        motorStateGVL.avgSpeed := FALSE;
        motorStateGVL.minSpeed := FALSE;
    END_FOR
ELSIF motorStateGVL.minSpeed = TRUE THEN
    FOR count := 1 TO 5 DO
```

```
        motors[count].motorStateMsg := 'min speed';
        motors[count].motorState := TRUE;
        motors[count].motorSpeed := motorSpeeds.minSpeed;
        motorStateGVL.avgSpeed := FALSE;
        motorStateGVL.maxSpeed := FALSE;
    END_FOR
END_IF
```

This code will check to see which setting the motors are set to and loop through the array to determine which speed from the enum to set them to. For example, set avgSpeed in the motorState GVL to TRUE, and you should be met with the output in *Figure 2.15*:

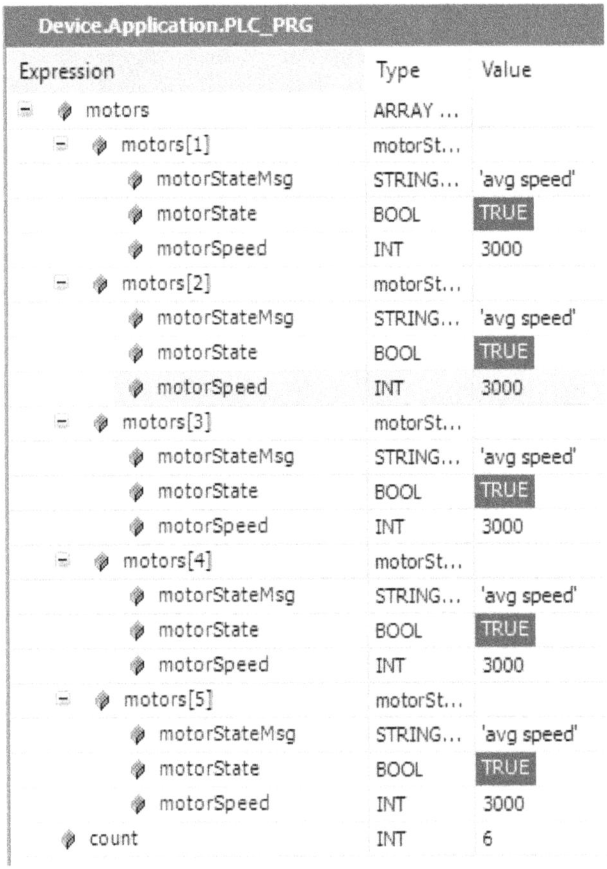

Device.Application.PLC_PRG		
Expression	Type	Value
⊟ ◆ motors	ARRAY ...	
⊟ ◆ motors[1]	motorSt...	
◆ motorStateMsg	STRING...	'avg speed'
◆ motorState	BOOL	TRUE
◆ motorSpeed	INT	3000
⊟ ◆ motors[2]	motorSt...	
◆ motorStateMsg	STRING...	'avg speed'
◆ motorState	BOOL	TRUE
◆ motorSpeed	INT	3000
⊟ ◆ motors[3]	motorSt...	
◆ motorStateMsg	STRING...	'avg speed'
◆ motorState	BOOL	TRUE
◆ motorSpeed	INT	3000
⊟ ◆ motors[4]	motorSt...	
◆ motorStateMsg	STRING...	'avg speed'
◆ motorState	BOOL	TRUE
◆ motorSpeed	INT	3000
⊟ ◆ motors[5]	motorSt...	
◆ motorStateMsg	STRING...	'avg speed'
◆ motorState	BOOL	TRUE
◆ motorSpeed	INT	3000
◆ count	INT	6

Figure 2.15: Motor array state

Now, turn whichever variable you set to TRUE back to FALSE, set another variable (such as minSpeed) to TRUE, and view your output.

Summary

In this chapter, we explored many types of variables, such as GVLs, enums, constants, structs, and more. In traditional PLC programming, concepts such as these are rarely used. However, as technology advances, these concepts are going to become more ingrained in the development of automation equipment. The concepts we have explored in this chapter will allow you to better organize your code. They will also serve as a way to better encapsulate data. Not every PLC system will support these advanced concepts; however, when they do, it is recommended that you use them where applicable because they can greatly improve the quality and reusability of your code.

As we continue our journey into more advanced PLC programming, variables will play a vital role. In the next chapter, we are going to explore functions, and understanding variables will be pivotal in getting the most from them.

Questions

1. What is a GVL?
2. What suffix should a GVL have at the end of its name?
3. Suppose there is a 3D array. How many FOR loops are needed to loop through each element?
4. What is a constant?
5. How does a constant differ from a normal variable?
6. What is a struct?
7. What is a key difference between a struct and a GVL?
8. What suffix should the struct file have?
9. What are some issues that global variables cause?

Further reading

- *CODESYS enumerations*: https://content.helpme-codesys.com/en/CODESYS%20 Development%20System/_cds_datatype_enum.html

- *CODESYS Structs*: https://content.helpme-codesys.com/en/CODESYS%20 Development%20System/_cds_datatype_structure.html

Join our community on Discord

Join our community's Discord space for discussions with the authors and other readers: https://packt.link/embeddedsystems

3

Functions: Making Code Modular and Maintainable

As a college-level programming instructor, the first thing I like to teach after teaching the basics, such as flow control, is functions. For many new and non-classically trained programmers, the purpose of functions often makes little sense. They tend to see functions as a useless code organization technique that convolutes their project. I usually counter this logic by stating that programmers should be like sewists. When a sewist creates a quilt, they take individual patches and sew them together. When it's time to create the quilt, there is little concern about creating a patch. The only thing they are worried about is integrating the patch into the quilt as a whole.

For the most part, a programmer should consider themselves to be a sewist of software, and the patch of choice should be functions. As we will explore in this chapter, codebases should be as modular as possible. In a well-written, modular program, adding or removing functionality will be as simple as adding or removing function calls. We will explore functions through the following concepts:

- What modular code is and the reasons for using it
- Functions
- Return types
- Arguments

To combine everything, we are going to build a temperature conversion function. The temperature converter is a common implementation of a method because the code will need to be run many times and it will need be in a singular location.

Technical requirements

The code for the examples can be found here: `https://github.com/PacktPublishing/Mastering-PLC-Programming-Second-Edition/tree/main/Chapter%203`.

What is modular code?

A fallacy many inexperienced programmers make is to dump all their programming logic into a central location. The fallacy stems from developers thinking that by having all their code in a central location, their program will be easier to maintain. This is a major design flaw that can easily kill a program. Therefore, to effectively combat this issue, programmers use modular code.

A definition of modular code

Modular code can best be described as code that is broken out into logical units of functionality. That is, a well-written program will consist of multiple logically organized files (modules), and each file will consist of some type of related functionality. For example, a code module might be the functionality for a motor drive, a series of actuators, or unit of logically related values.

In essence, the easiest way to think of a modular program is to think of the human body. Your body is composed of many different organs that all perform a certain task. For example, your heart pumps blood throughout your body while your lungs process air. A program should be thought of in the same regard. A program should be broken into multiple parts, and each one of those parts should perform a single task and be optimized to perform it well.

A close example of modular code could be the GVLs and structs we explored in *Chapter 2*. Essentially, the way we organize GVLs and structs can be thought of as a way to modularize code. In IEC 61131-3, there are other and arguably better ways to organize code. In fact, there are many ways to easily architect a modular program.

How code is organized

There are many ways to create modular code. To create modular code, we use special files called **Program Organizational Units (POUs)** and **Data Unit Types (DUTs)**. Some common constructs that can be used to organize code are as follows:

- Functions
- Function blocks
- Structs
- Enums
- GVLs

Note

POUs and DUTs are not technically files, such as files on a disk. However, many programming systems organize these constructs in a file-like container in the PLC project tree. As a result, many developers will usually refer to and treat these constructs as files. Therefore, the word file will be used as a catch-all term to refer to one of these constructs.

Each of these files will have a different purpose. For example, the struct type that we explored in the previous chapter is designed to provide a logical unit to declare related variables, while GVLs are used to declare global variables that can be used throughout a project.

This chapter is concerned with a coding unit called a function. A function is one of the fundamental building blocks for any program. Essentially, this code unit is a callable block of code that does one thing very well. Before we get into the mechanics of using functions, we first need to explore some modulation strategies!

Strategies for creating modular code

Creating modular code requires experience. As you progress as a developer, you will naturally migrate to creating modular code via trial and error. However, to jump-start the process, there are a few strategies you can keep in mind to help flesh out a modular program.

Limiting the amount of code in the PLC_PRG file

The PLC_PRG file is an entry point for a PLC program. This file should only contain the logic necessary to start the machine and coordinate its overall operations. In other words, you should not overburden this file with too much logic. The one and only responsibility for this file should be to call the necessary modules to start and possibly coordinate the operation of the machine.

The PLC_PRG file should be thought of as a conductor of an orchestra, in that this file should only control the execution and coordination of modules that compose the PLC program, without handling specific functionality. This is where many developers can make a mistake. Packing too much logic in this file can lead to a very complex entry point for the PLC program, and the file can become overly convoluted. This can harm maintainability and cause misunderstandings about how the machine will operate. It will also typically harm modularity as packing the PLC_PRG file with logic can burden the file with too much responsibility.

There is no silver bullet for creating an effective PLC_PRG file. Typically, you want to only do things such as setting variables or calling functionality from this file; in other words, as we will see in this and subsequent chapters, call functions! Since the PLC_PRG file is best thought of as the conductor of the program, it is ideal to use this file to house the program's main state machine.

Creating a quality PLC_PRG file is of paramount importance for a well-crafted modular program; however, there are other ways to ensure modularity. In the next section, we're going to explore the separation of responsibilities.

Separation of responsibilities

The key to creating modular code resides in what's called the separation of responsibilities. Essentially, when developing a PLC program for a machine, a solid understanding of how the machine works and the individual components of the machine is necessary. For example, suppose you have an integrated robot, that is, a machine composed of a robotic arm that can home itself, paint a part, and start an oven. For a system such as this, a module could be moving the robot to a spray-paint position, turning on pumps to start the flow of paint to the robot, starting the oven, or homing itself. The key here is that one module is responsible for one thing.

Understanding the separation of responsibilities will be explored throughout the rest of this book. Mastering this skill is as much an art as it is a science. However, for now, we're going to switch gears and explore why we should use modular code.

Why use modular code?

There are many reasons to use modular code; however, the following is a list of some very common ones:

- **Reusability**: Arguably, the most important reason to use modular code is reusability. Much like the parts of a car, a code module can be used in multiple projects and in multiple parts of the project, which cuts down on development time, unexpected behaviors, and bugs.
- **Code removability**: Much like with a mechanical machine that has been upgraded to use fewer parts, a modular program should allow for the easy removal of all individual modules.
- **Code upgrades**: When a program is modular, it will be easy to upgrade. If a behavior must be changed, all you need to do is find the code module that handles that behavior and change it without needing to break other, unrelated code. Upgrading modular code is like upgrading a car part, you only need to upgrade that part and the necessary related parts for it to work.
- **Code scalability**: Modular code is usually scalable, meaning that codebases can be more easily maintained and modified as they grow.

Now that we understand what modular code is and why we should use it, we're going to switch gears and actually implement a code module with functions!

Exploring functions

On the surface, functions may seem like we are merely splitting up a program into multiple files; however, this is a vast oversimplification and misunderstanding. At their core, functions are callable blocks of code that are designed to perform a task. Before we can start coding a function, we need to first understand how to name one.

The art of naming functions

Function naming is a bit of an art. Functions are actions that a program will perform. For example, a function could be tapping the brakes on a car, calculating the sum of some numbers, flying to Mars, or anything of the like. Since functions are actions, we typically give them a verb name. For example, if we were creating a method to fly to Mars, we might name it something such as `flyToMars`. Here, we have a name that is indicative of an action.

Though you should lean on using a verb for a name, there are many functions named simply `length` or `size` that are floating around different languages and plugins. A verb should be used whenever possible, but from a practical point of view, context and situations matter. There are also times when using a verb can bloat the name of a function and add no real value. For example, if we have a program called `convertSize`, and in the program we have a function called `squareLength`, it's obvious that the function's job is to convert the length of a square. It's okay to use a non-verb in the following scenarios:

- Math operations, such as creating a function for a mathematical operation: for example, calculating the Fibonacci sequence.
- Effectively acting as a function that gets something.
- Verb names are reserved words in the language, convoluted, or redundant.
- When a noun or non-verb is better suited. For example, names such as `LedOn`, `LedOff`, `on`, and `off` are common in many systems, especially embedded ones such as PLCs.
- Documentation and learning material will sometimes use simplistic nouns to reduce mental complexity. This is a teaching strategy that is often used in the classroom and will be used from time to time in this book but should be avoided where possible in production code.

Realistically, every codebase of significant size will use nouns as names, and you will also find yourself using nouns. These facts of life do not negate the verb rule or relegate it in any way. You should use verb-based names wherever possible!

In all, it is best to use a verb or verb-type name for a function; however, context and, more importantly, the standards of your team matter. There is one naming rule that should never be broken, which is including the word *and* in the name. Names with the word *and* indicate more than one, which is very bad for a function. In the next section, we're going to explore why we should never see the word *and* in a function name!

What goes into a function?

Architecting good functions is an art. Even traditional programmers often struggle with writing quality functions that are easy to use. A major reason for this is that they often pack too much responsibility into a given function. There is one law that should always be followed when writing a function, and it goes as follows:

A function should do one thing only and do it well!

Having a function that does more than one thing will break modularity and risk the introduction of bugs, unscalable code, and, above all else, redundant code.

A good test to determine whether your function is only doing one thing and one thing well is to summarize it in a sentence. If the word *and* appears in the sentence, then the function is doing too much. To fix this issue, break the statement after the word *and* into a function of its own. For example, consider these two sentences:

- This function turns on the assembly line
- This function turns on the assembly line and hopper

The definition of the first function is correct. This sentence describes a function that does one thing (turns on the assembly line). On the other hand, the second function does way too much. The second function turns on both the assembly line and the hopper. If a modification ever must be made to either one of those operations, or a situation occurs where only one of the operations needs to be turned on, you're going to risk either introducing defects into the other process or having to create redundant code to control the targeted operation.

Another good rule to follow is to break out redundant code into a separate function. If you see two or more lines of code constantly appearing throughout your program or function file, consider placing that code into a function of its own. Try your best to determine what that code is doing and put it in a function with a name that describes its purpose.

Now that we have explored the fundamental concepts of a function, such as why we use them, when we use them, and what should go into a function, we can move on and explore how to create one. Compared to other languages, creating a function in many IEC 61131-3 systems and, more specifically, CODESYS, is a bit more in-depth.

Creating a function

Unlike in many traditional languages, a function in a system such as CODESYS or TwinCAT lives in a file of its own in the project tree. If you're coming from a background in a general-purpose programming language such as Java or C#, functions are split across different files/POUs in the project tree, like the way classes are broken out in those languages. This is where the misconception of equating splitting files with functions comes from. If you are a traditional programmer, this may throw you off, as you will be used to implementing multiple functions or methods in a single file construct. However, this feature helps promote modularity. In short, one file equals one function, which equals one responsibility for the program!

To create a function, the first thing you should do is create a new Structured Text program, then right-click **Application**, navigate to **Add Object**, and click **POU**. When you do this, you should be met with the following popup:

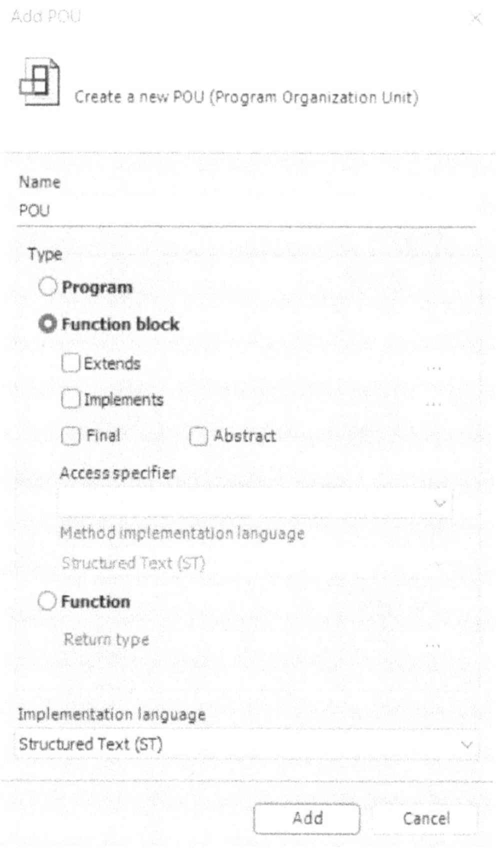

Figure 3.1: POU wizard

In my opinion, *Figure 3.1* shows one of the most used windows for any object-oriented programmer developing with CODESYS or a similar system. At first glance, we see options called **Function** and **Function block**. There is a major difference between these two options, and they should not be confused. A function block is akin to a class in C++, Java, or other traditional programming language. For this chapter, we will only be interested in the **Function** option. The POU that the **Function** option will create is a callable block of code. Essentially, these are the sewist patches that were alluded to earlier.

For our example, we are going to select the **Function** option and input Addition into the **Name** field. For **Return type**, click the button with the three dots, and select **INT**. The return type is very important for functions. It specifies the data type of the value that the function will ultimately output. For now, just ensure that your POU creation wizard matches *Figure 3.2*:

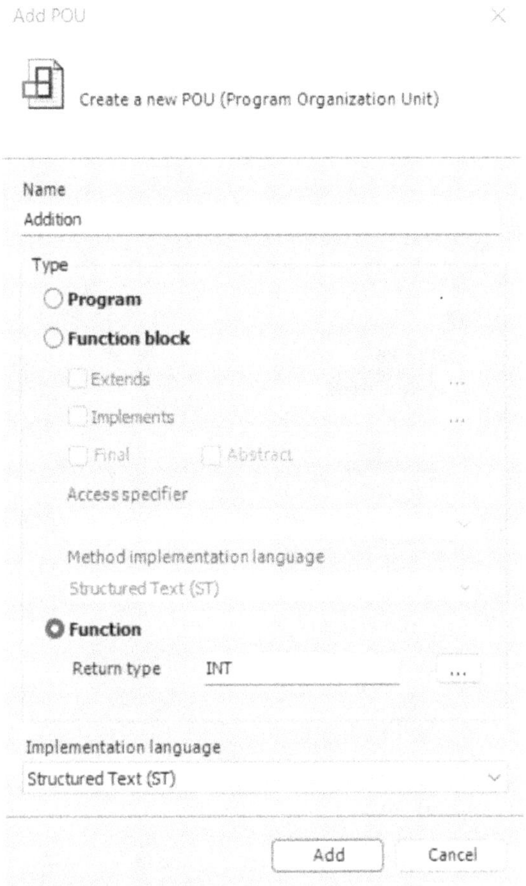

Add POU   ×

Create a new POU (Program Organization Unit)

Name
Addition

Type
○ **Program**
○ **Function block**
  ☐ Extends   ...
  ☐ Implements   ...
  ☐ Final   ☐ Abstract
  Access specifier

  Method implementation language
  Structured Text (ST)

◉ **Function**
  Return type   INT   ...

Implementation language
Structured Text (ST)

Add   Cancel

Figure 3.2: POU setup for the Addition function

> **Note**
>
> We use the word `Addition` for the function name despite it not being a verb because the word `Add` is an illegal name for a custom function.

Notice in the preceding figure that an **Implementation language** option can also be selected. This is very important to remember, as each function can be written in a programming interface that best suits it. For example, for simple programs, Ladder Logic might be more appropriate for simple operations such as flipping a bit. Regardless, for this example, we're going to keep the language as **Structured Text (ST)**.

The function we are going to make will add two hardcoded numbers and return the value. If you're not sure what return values are, we're going to explore that in the *Examining return types* section of this chapter. For now, your main focus should be on simply understanding how a function operates.

After filling out the wizard and clicking the **Add** button, a file with the name `Addition` should be generated in the file tree under `Application`. Navigate to the file and open it. You should see the following code:

```
FUNCTION Addition : INT
VAR_INPUT
END_VAR
VAR
END_VAR
```

Notice that there are two variable sections. The `VAR_INPUT` section is used for variables that will be used for what are called arguments or parameters. This is a concept that we will explore in the *Understanding arguments* section. The `VAR` section is used to declare variables that are internal to the function. This means that the variable cannot be accessed from outside the function and cannot be used for arguments. For this example, let's add two variables, a and b, of type `INT` to the `VAR` section. We're going to assign the values 3 and 4 to the variables, respectively. In other words, your code should match the following snippet:

```
FUNCTION Addition : INT
VAR_INPUT
END_VAR
VAR
    a : INT := 3;
    b : INT := 4;
END_VAR
```

In the logic section of the `Application` file, input the following code:

```
Addition := a + b;
```

Essentially, this line of code means that the output of the function is the sum of the a and b variables.

The code in a function block will not execute until it is called. Generally, this is the purpose of the `PLC_PRG` file. Remember – in most well-written programs, this file is equivalent to the main function or entry point for a program. Its main job is to only invoke functions that are needed to kickstart and run the PLC program. For the most part, you want this file to be as short as possible.

With that in mind, a function is invoked by calling its name and passing in the necessary arguments. You can call a function from another function or any other file that is allowed to call functions. For our example, we're going to call the function from the `PLC_PRG` file. Therefore, navigate to that file, open it, and add the following code:

```
PROGRAM PLC_PRG
VAR
    sum : INT;
END_VAR
```

This variable will be used to hold the return value from the function. Since our function's return value is of the integer type, it is important to declare the sum variable as an integer as well.

This code snippet shows how a function is invoked:

```
sum := Addition();
```

The code boils down to invoking the `Addition` function and assigning the return value to the sum variable. When the code is run, you should see an output that is congruent to the screenshot in *Figure 3.3*.

Expression	Type	Value	Prepar...	Address	Comm...
🔶 sum	INT	7			

Figure 3.3: Output from the Addition function

In the real world, a function will often need to return some type of data to the line that invoked it, similar to what we did here. This may be a calculation, a status, or anything else. To master this concept, we're going to explore how to return data from a function!

Examining return types

Return types can often be very confusing for new programmers. The main hang-up for many of the students I have taught is that they often have a difficult time understanding what a return type is. As we have seen, a return type is simply the kind of value that a function returns. Expanding this, the actual value that a function returns can be thought of as the output of the function.

Each function must be declared with a return type. This return type can be any data type that is supported by IEC 61131-3; for example, the integer data type from the Addition function. In all, a function can return exactly one value of the type the function was declared with. So, if you declared a function with a return type of INT, you could only return an integer, similar to what we did with the Addition function. As we saw, returning a value is as simple as assigning the function name to a statement, as we did in the preceding code snippet. Overall, this is a simplistic definition of return types. If this concept is still unclear, you will understand it as the book progresses, as this will be a concept that will be used from here on out.

Key takeaways about return types are as follows:

- A return type is a function's output
- There can be exactly one return type per function
- One function can return exactly one value
- A return type can be any supported data type

> **Note**
>
> In CODESYS, it is possible to have more than one output value from a function by declaring additional variables in the VAR_OUTPUT block; however, only the value assigned to the function's name is considered the true return value. The others are output parameters.

Even though a function does return a value, sometimes it shouldn't. Sometimes we simply want to terminate the function before the value can be returned. For this, there is a special command known as RETURN.

The RETURN statement

A function does not always have to return a value. In certain cases, it might be more appropriate to simply terminate a function as opposed to returning a value. For example, an otherwise fatal error that won't return a valid value or something along those lines will usually benefit from using a RETURN statement.

Compared to traditional languages such as Java or C++, the RETURN statement does not return a value. However, as stated before, it does terminate a function. This means that it is very common to use the RETURN statement in some type of control statement, such as in an IF statement or inside a CATCH block. Consider the following scenario. Suppose we have a system where the RPMs of a motor are input as a multiple of 1,000. If the operator wanted to program the machine for 4,000 RPMs, they would enter the value as 4, and the function would multiply the value by 1,000.

The function should return the converted RPM values; however, if an invalid entry is input, such as a negative number, the function will simply terminate. For this example, let's create a new function called RPMs with a return value of type INT, and once the file is generated, add a variable called rpmsInput in the VAR section. Once done, your code should look like the following:

```
FUNCTION RPMs : INT
VAR_INPUT
END_VAR
VAR
    rpmsInput : INT := 4;
END_VAR
```

The following code snippet is the function's main logic, which is responsible for converting the user input into the proper RPM value or executing a RETURN command for an invalid value.

This is the logic that represents the RPM function:

```
IF rpmsInput < 1 THEN
    RETURN;
ELSE
    RPMs := rpmsInput * 1000;
END_IF
```

This code will return the converted RPMs as long as rpmsInput is greater than or equal to 1. If the value is less than 1, then the function will simply terminate. To run this code, the PLC_PRG file will also need to be modified to include a variable named x of type INT and the following logic, which will invoke the RPM function:

```
x := RPMs();
```

When the code is run with all the values as shown, you should see the output presented in *Figure 3.4.*

| ◆ x | | INT | 4000 |

Figure 3.4: Successful RPM conversion

Now, if rpmsInput is changed to -2, you should be met with an output similar to *Figure 3.5.*

| ◆ x | | INT | 0 |

Figure 3.5: Invalid RPM conversion

For *Figure 3.5* the RETURN defaulted the value to 0. In other words, the value we got was not from the RETURN statement but from the function itself. Overall, the RETURN statement resulted in the function terminating before a value was returned.

There are four important takeaways regarding the RETURN statement. They are as follows:

- The RETURN statement will terminate a function, usually before a value is returned. This means that the RETURN statement is the last command executed in the function.
- A function can contain many different RETURN statements when they are used in a control statement, but only one will execute per function call.
- The RETURN statement is usually wrapped in some type of control statement, such as an IF statement or a CATCH block.
- Unlike many traditional programming languages, the RETURN statement does not return a value; it simply terminates the function.

Our Addition function is great as long as we want to add 4 and 3. However, outside of that very specific use case, the function is useless. In the real world, we would want our functions to be more generic. In the case of our example function, it should be able to add more than two numbers. Therefore, it is time to explore function arguments!

Understanding arguments

Where return types are a function's output, arguments are a function's input(s). Arguments are optional, as we saw with the `Addition` function, where no inputs were used. For many functions, especially for functions that do math, arguments are usually necessary to ensure the reusability of the function. For example, our `Addition` function only summed two hardcoded values. Unless our PLC program needed to add 3 and 4 together each time the function was called, the function serves no purpose. To make this function usable, we need to provide some inputs to make it more generic and usable in different circumstances.

The first step in creating functions with arguments is declaring variables in the `VAR_INPUT` section of the function file. For our modified `Addition` function, we are going to have the function take two inputs, a and b. As such, we're going to modify that section of code to match the following:

```
FUNCTION Addition : INT
VAR_INPUT
    a : INT;
    b : INT;
END_VAR
VAR
END_VAR
```

As can be seen, we have two variables labeled a and b. These variables will hold the values we input into the function. For this example, we are going to keep the same addition logic that we used before, so that logic should match the following:

```
Addition := a + b;
```

The key takeaway here is that no actual values are being assigned to any of the variables in the function. In this case, all values are supplied when the function is called. To demonstrate this, we are going to modify our code in the `PLC_PRG` file. The first thing we will do is modify the variable list to match the following:

```
PROGRAM PLC_PRG
VAR
    sum : INT;
    input1 : INT;
    input2 : INT;
END_VAR
```

In this case, we still have our sum variable, which will hold the sum of the input1 and input2 variables, which will serve as our function inputs. To run the function, we use the following code:

```
sum := Addition(input1, input2);
```

Notice that we have input1 and input2 separated by a comma in the parentheses. This is how we supply the function with the arguments. In this case, input1 will be assigned to variable a in the Addition function, and input2 will be assigned to variable b. For this technique of passing in variables, the first argument is mapped to the first variable declared in the VAR_INPUT block, and the second is mapped to the second declared variable.

To test the code, run the program and write the values 2 to input1 and 3 to input2. When the code is executed, you should be met with *Figure 3.6*.

Expression	Type	Value
sum	INT	5
input1	INT	2
input2	INT	3

Figure 3.6: Argument passing

Though this is a very common technique for passing arguments to a function, it is not the only way. Depending on what you're trying to accomplish, it is sometimes better to explicitly state which variables get which value. This technique is known as **named parameters**.

Named parameters

In computer science, the concept of named parameters allows developers to explicitly state which variable gets which value. By using named parameters, we are not bound to the traditional one-to-one mapping approach of argument assignment that we explored in the last example. Named parameters allow us to assign a value to a specific variable by assigning it in the argument list. To demonstrate this, create a function named Subtraction with a return type of INT. Once the file is generated, add a and b variables to the VAR_INPUT list as in the following code block:

```
FUNCTION Subtraction : INT
VAR_INPUT
    a : INT;
    b : INT;
END_VAR
VAR
END_VAR
```

As with the `Addition` function, the logic for the `Subtraction` function is as follows:

```
Subtraction := a - b;
```

Once the function is set up, we can work on invoking the code in the `PLC_PRG` construct. Notice in the second bullet, there are two different ways of passing arguments to the `Subtraction` function:

- **Variables for function call**: These variables will hold the values of the different subtraction calls:

```
PROGRAM PLC_PRG
VAR
    diff1 : INT;
    diff2 : INT;
END_VAR
```

- **Logic to invoke the function**: The following code snippet represents two different ways to pass arguments to the subtraction function:

```
diff1 := Subtraction(3, 2);
diff2 := Subtraction(b:=3, a:=2);
```

When this code is run, you will get an output similar to *Figure 3.7*.

Expression	Type	Value
◈ diff1	INT	1
◈ diff2	INT	-1

Figure 3.7: Different argument order

As can be seen, we passed variables to the function using different techniques and got two different results. On the first line, we traditionally passed in the variables, and a got assigned 3 and b got assigned 2. Using this technique, the difference is 1. In the second line, we explicitly stated that b will be set to 3 and a to 2, which, in this case, negates the traditional order of the arguments:

```
diff2 := Subtraction(b:=3, a:=2);
```

In this code snippet, we are manually assigning values to the variables. When we use this methodology, we can pass the arguments in any order we want.

Passing arguments, though common, can be very cumbersome. Suppose you had many arguments, but for the most part, the values never changed. It would be very inefficient to have to pass those values every time the function was called. Much like in many other languages, there is a solution for this. This special technique of argument assignment is often referred to as **default arguments**.

Default arguments

Default arguments are a way of pre-setting parameter values for a function. This technique allows us to preset arguments that often won't change but will give us the ability to set them if necessary. Sometimes the term *default initialization* is used to describe the behavior of default arguments in formal documentation. The two terms describe what are usually equated to as the same thing in practice, with the only real difference occurring under the hood. At the surface level, default initialization and default arguments are essentially the same thing in terms of everyday development; that is, presetting function arguments. The term *default arguments* is more commonly used in daily language across many programming contexts, and more generic documentation exists that can be used as a resource. In practice, I strongly recommend using the term *default arguments* due to its common usage.

I often get asked by my entry-level students why we should use default arguments. A common usage is with arguments that do not change. If we have arguments that hardly change, it is often better to simply assign them a default value. Default parameters provide an overridable value for the function to use if a value is not explicitly assigned when the function is invoked. In other words, a default value is not immutable and can change when necessary.

Now that we have some background information on default parameters, let's look at an example of this concept in action. For this example, let's revisit our RPMs function. As it stands, we use a hardcoded conversion constant of 1,000. For the most part, this is fine; however, for whatever reason, suppose we need to use a different conversion value. If we decide that the value needs to change, we can simply pass in a new value to override 1,000. This is a much more convenient technique, as we will not be forced to constantly supply the same value to the function, which is an easy vector for bugs.

The magic of default arguments resides in the VAR_INPUT section. You create a default input by simply assigning a value in that block. With that being said, create a new project with a RPMs function with the same set up we used before. The VAR_INPUT should be set to the following code snippet, which will set the rpmsConversion variable to 1000 by default:

```
FUNCTION RPMs : INT
VAR_INPUT
    rpmsInput       : INT;
    rpmsConversion  : INT := 1000;
END_VAR
VAR
END_VAR
```

As can be seen in the code snippet, all we did was simply assign the value 1000 to the variable. This means that this value will not necessarily have to have a value assigned to it when we invoke the function.

For this example, we're going to use the following logic for the function:

```
IF rpmsInput < 1 THEN
    RETURN;
ELSE
    RPMs := rpmsInput * rpmsConversion;
END_IF
```

To invoke this function, we're going to use the following logic in the PLC_PRG file.

These are the variables that we will use to demonstrate the RPMs function:

```
PROGRAM PLC_PRG
VAR
    convertedRpms : INT;
    motorRpms     : INT := 4;
END_VAR
```

The core logic for calling the function is as follows:

```
convertedRpms := RPMs(rpmsInput := motorRpms);
```

When the program is run, the following will be output:

| convertedRpms | INT | 4000 |
| motorRpms | INT | 4 |

Figure 3.8: RPMs conversion output

As it stands, we will multiply the input by 1,000; however, for whatever reason, if we need to multiply by 100, it is still possible to do so without changing any code. We can simply pass the extra argument when we call the RPMs function in the PLC_PRG file. To demonstrate this, we can simply modify the RPMs call in the PLC_PRG file in the following manner:

```
convertedRpms := RPMs(rpmsInput := motorRpms, rpmsConversion := 100);
```

When this line of code is run, it will produce output similar to *Figure 3.9*.

```
convertedRpms                                    INT            400
motorRpms                                        INT            4
```

Figure 3.9: Overridden RPMs conversion

Default arguments/initialization is a very powerful concept in PLC programming. As was demonstrated in the preceding code example, default arguments are excellent to use when you have a value that may only need to change sometimes. By simply giving an argument variable a value, you can free yourself and other developers from needlessly passing in values that typically won't change. In turn, this means that your code will become more stable and, as such, will be easier to maintain and modify in the long run. Now that we know about functions, return types, and various forms of passing arguments, let's explore calling functions from other functions!

Calling a function from a function!

As was touched on earlier, a function can call another function; in fact, it is quite common to call a function from another function. Calling a function from another function is like the parts of a car working together to transport you from point A to B. When leveraged correctly, calling a function from another can drastically improve your code and allow you to simplify it. In this section, we're going to look at how this is done with an example.

Simplifying your functions with facades

Automation machinery often requires a series of predefined steps to turn on, do a task, or, in many cases, even shut down. This can be further complicated when each predefined step requires complex logic to operate. In cases such as these, there are typically two courses of action that a programmer can take. The first is to use redundant code. Each time a complex process must be executed, the developer can copy and paste the code into that function. However, based on what we've seen so far, this is a very bad option. For example, if any part of the process must be modified, you'll need to ensure it's changed across the program. This means that bugs are bound to occur if any part of the complex operation needs to be changed.

The other option is to package all the complex logic into functions and use another function to call these composites. This calling function is called a *facade*. The goal of this function is to hide the complexity of a complex process. In all, the goal of this function is to provide a simple interface for the programmer to call instead of having to remember the order and the support logic for a complex operation to succeed.

For this example, suppose we have a welding bot. For the bot to start welding, we need to send a ping signal to the bot and check if an old welding job is done before we can start a new one. For this example, all we're going to do is set a few variables to simulate this. To begin, create a function named weld and implement the following variables:

```
FUNCTION weld : BOOL
VAR_INPUT
END_VAR
VAR
    weldDone : BOOL;
END_VAR
```

In terms of the main logic for the function, implement the following:

```
weldDone := TRUE;
weld := weldDone;
```

When the program is executed, the weldDone variable will simulate the completion of a welding job. In a real-life program, this function would contain complex logic that would control things such as the necessary parts to execute the weld.

Once you have this function implemented, you can move on to implementing the next function, which we will call sendStartUpSignal. This function will have the following variables and logic:

```
FUNCTION sendStartUpSignal : BOOL
VAR_INPUT
END_VAR
VAR
    pingResult : BOOL;
END_VAR
```

The logic for the function is as follows:

```
pingResult := TRUE;
sendStartUpSignal := pingResult;
```

Finally, we're going to create a function that is going to hide the startup complexity for the bot. To do this, simply create a function called start and add the following statements:

```
FUNCTION start : BOOL
VAR_INPUT
END_VAR
VAR
```

```
      pingStatus    : BOOL;
      weldStatus    : BOOL;
      overallStatus : BOOL;
  END_VAR
```

In this case, we have three variables: one holds the status of the welder, the other the status of the ping, and, finally, an overall status that is dependent on the other two variables. If both the ping and weld variables are TRUE, this function will send a TRUE value back to the code line that called it; else, it will send a FALSE value.

The logic for this function can be implemented with the following:

```
  pingStatus := sendStartUpSignal();
  weldStatus := weld();
  IF pingStatus = TRUE AND weldStatus = TRUE THEN
      overallStatus := TRUE;
  ELSE
      overallStatus := FALSE;
  END_IF
  start := overallStatus;
```

This is the complex logic that is meant to be hidden. Without this facade function, any time or place we needed to kickstart the welding process we would have to copy and paste this code. As can be seen, the code has several steps that could be easily mixed up or overlooked; however, with this function, if it ever needs to be changed or reused, we only need to modify it in one place. Also, if the process has to be used more than once, the function prevents us from having to implement the logic in multiple places.

To use this function and kickstart the welding process, all we need to do is implement the following variable in the PLC_PRG file:

```
  PROGRAM PLC_PRG
  VAR
      status : BOOL;
  END_VAR
```

The core logic for kickstarting the weld process is the following line:

```
  status := start();
```

When you have all the code implemented, run the program, and you should be met with the output shown in *Figure 3.10*.

Expression	Type	Value
◈ status	BOOL	TRUE

Figure 3.10: Facade output

The key point to remember is that the goal of a facade is to reduce the burden on the programmer. If you see yourself with a complex series of processes that must be called multiple times in a defined order, you may want to try to implement a facade function. In automation, these functions can be used to great effect as they can greatly reduce the overall complexity of the codebase while simultaneously keeping it nice and tidy. With facade functions out of the way, we can move on to creating our final project.

Final project: Temperature unit converter

Often, as PLC programmers, we are asked to provide software that monitors temperatures. These temperatures could be inside the housing of a control panel, the temperature of a part we are fabricating, or anything of the like. It is quite common to need to be able to convert between temperature units, especially when the program is deployed to places around the world.

Temperature converters are prime examples of functions as they will often need to be used multiple times in a program. As such, we want these conversion calls to be able to be used with minimal effort; that is, we don't want to write the code multiple times. For our function, we are going to create a state machine to trigger our conversion from one unit to another.

Our program will need to perform the following operations:

1. F -> C
2. F -> K
3. C -> F
4. C -> K
5. K -> F
6. K -> C

Our state machine will have six states. Therefore, create a function called tempConverter with a return type of REAL and match the code to the following snippets.

These are the variables that will be used for the temperature conversion:

```
FUNCTION tempConverter : REAL
VAR_INPUT
    state : INT;
    temp  : REAL;
END_VAR
VAR
END_VAR
```

Once you have these variables in place in the function file, add the following logic:

```
CASE state OF
    1:
        // F -> C
        tempConverter := ((temp - 32) * 5) / 9;
    2:
        // F -> K
        tempConverter := (((temp - 32) * 5) / 9) + 273.15;
    3:
        // C -> F
        tempConverter := ((temp * 9) / 5) + 32;
    4:
        // C -> K
        tempConverter := temp + 273.15;
    5:
        // K -> F
        tempConverter := (((temp - 273.15) * 9) / 5) + 32;
    6:
        // K -> C
        tempConverter := temp - 273.15;
    ELSE
        RETURN;
END_CASE;
```

Once you have the function file squared away, modify the PLC_PRG file to match the following variables:

```
PROGRAM PLC_PRG
VAR
```

```
    convertedTemp : REAL;
    state         : INT;
    temperature   : REAL;
END_VAR
```

Once you have the variables in place, you can call the conversion function with the following call:

```
convertedTemp := tempConverter(<state>, <temperature>);
```

For a simple test, we will convert 100°F to Celsius. To do this, we will input 1 for `state` and `100` for `temperature`. When we write the value, we will get the following output:

convertedTemp	REAL	37.77778
state	INT	1
temperature	REAL	100

Figure 3.11: Fahrenheit to Celsius conversion

As we can see, it correctly converted Fahrenheit to Celsius. Now, you can input different values and states to test the code.

Chapter challenge

A function such as `tempConverter` is very common in the real world. However, if you notice, the name is not a verb. For this chapter challenge, figure out a better name for the function, and also apply the one-sentence rule to determine whether the function is doing too much.

At this point, we have had an in-depth look at functions. As the book progresses, we will be using functions and their cousins, known as methods, more and more. This chapter is a foundational chapter for the rest of the book, so if you're not comfortable with the material, it is recommended that you go back and re-read this chapter and play with some of the examples or create a few of your own.

Summary

In this chapter, we explored functions, return types, arguments, facade functions, and more. The goal of this chapter was to demonstrate how to modularize code. Moreover, this chapter was an introduction to code organization. In general, if a program is to survive, it must be organized, and to create an organized program, you need to make it as modular as possible.

The key to a long-lasting codebase is modularity, and one of the core ways we can modularize and organize a program is by breaking it up into functions. Functions are the fundamental backbone of any program; however, there are other, more sophisticated ways to create modular and more organized code. In the next chapter, we're going to take a look at a concept that can be almost thought of as the big brother to functions, object-oriented programming!

Questions

1. What is a function?
2. What are default arguments?
3. What are default arguments sometimes referred to as in formal documentation?
4. What are named parameters?
5. In which order are arguments received in a function?
6. What goes into a function?
7. What is a return type?
8. Can the INT return type be used with a variable of type REAL?
9. What is a facade function?
10. Why do we want modular code?
11. Can you call a function from another function?

Further reading

- *CODESYS Functions*: https://content.helpme-codesys.com/en/CODESYS%20 Development%20System/_cds_obj_function.html

4

Object-Oriented Programming: Reducing, Reusing, and Recycling Code

As we saw in the previous chapter, the key to a healthy codebase is organization and no redundant code. Though functions are one vital way of reducing, organizing, and reusing code, there is a much more effective way. This methodology is called **object-oriented programming (OOP)**.

OOP is the backbone of most modern programming languages. In fact, to effectively use most modern languages such as Java, C++, C#, Python, and many others, a solid understanding of OOP is not only vital but mandatory. OOP is, for the most part, novel to PLC programming. Many PLC programmers are not fully aware OOP is supported, much less how to use it.

OOP offers many capabilities that, when leveraged correctly, can create very organized codebases that have no redundant code. There is a lot to understand about OOP and how to properly leverage it. This chapter is going to introduce the concept by exploring the following topics:

- What is OOP?
- Why OOP should be used
- Exploring function blocks
- Exploring methods
- Getting to know objects
- Function block naming
- Getting to know getters and setters

- Understanding recursion and the THIS keyword
- Using function blocks in Ladder Logic (or using its more formal name, Ladder Diagrams)

To wrap things up, we're going to create a custom function block that will serve as a unit converter.

Technical requirements

The source code for this chapter can be found at the following URL: https://github.com/ PacktPublishing/Mastering-PLC-Programming-Second-Edition/tree/main/Chapter%204.

The key to learning OOP is to practice, practice, and practice. It is strongly recommended that you follow along with the material in the book; that is, type it out by hand. However, if you run into technical issues, you can download the code from the GitHub repository to explore it. It is also highly recommended that you try to expand the functionality of the code in the URL.

What is OOP?

The first step in mastering OOP is understanding what it is. The best analogy, in my opinion, that one can use to conceptualize OOP is digital blueprints. If you think about a car, a company may produce thousands of cars a day based on a single set of blueprints. Each car might be a little different; for example, one car might be green while another red, or one car may have a cloth interior while another has a leather. Regardless, each car model will generally be the same.

OOP is a paradigm. In other words, OOP is a way to structure and conceptualize your codebases. There are many programming paradigms out there; however, OOP is by far the most popular. Almost all modern programming systems support the paradigm, and it is widely used across most applications in most industries.

When it comes to PLC programming, we can apply the same concepts that govern many traditional programs. If you've ever programmed in a language such as C++, C#, Java, Python, or any other modern language, you might be familiar with what's called a **class**. In terms of PLC programming, the closest structure that we have to a class is what's known as a **function block**. A function block can be thought of as a digital blueprint for something. For example, you can create a function block to model a power supply, a motor, or even a car.

Function blocks are not the same as the functions we saw in the last chapter. A function block is a special POU that has many special properties and can support its own self-contained functions and variables. Not only that, but these POUs also have the ability to use code from other function blocks in very special ways. We're going to explore function blocks in detail in the *Exploring function blocks* section; however, for now, just know that function blocks are the backbone of OOP for

PLCs. To expand our understanding of OOP, we need to switch our attention to understanding what OOP is not!

What OOP is not

OOP is a complete mystery to many PLC programmers. This is mostly due to its novelty. OOP is not supported by every PLC brand, and the extent to which it is supported will often vary greatly between brands that do. Typically, OOP is only fully supported by computer-based PLCs such as Beckhoff PLCs. However, with the rise of new technologies that support Industry 4.0, AI, and other cutting-edge technologies, OOP is going to be integrated into more PLC systems to support modern trends.

First and foremost, OOP is not breaking up your program into multiple POU files. Many PLC developers who are not familiar with the paradigm simply assume that OOP is simply splitting logic across multiple POUs. This misconception could not be farther from the truth. Fundamentally, it is a way to organize and reduce redundant code by modeling real world objects.

Another misconception among automation programmers is that OOP is a special language; this is also not true. As stated before, OOP is a programming paradigm. It is a way of structuring a program. Many different languages support OOP and its features. In other words, it is language agnostic, meaning that the principles that govern it are not unique to any given language, and in the case of PLCs any specific brand or programming system.

The final misconception that we're going to explore is that OOP can convolute code. In all fairness, this misconception is only partially false. The reason why it can convolute code has nothing to do with the paradigm itself; instead, it has to do with the skill of the programmer. To properly use the OOP paradigm, a developer must thoroughly understand the rules that govern it and how to properly implement them. As we will explore in the next chapter, OOP works best when a series of relationships are followed. When the rules and relationships are not properly followed using OOP, features can and often will render a codebase unmaintainable.

When leveraged correctly, OOP can greatly improve a codebase. It can keep codebases clean and healthy. In the next section, we're going to take a deep dive into why we should use the paradigm.

Why OOP should be used

OOP is the cornerstone of modern programming. Almost all modern programs utilize it in some way. In fact, for almost all traditional modern programming jobs, a solid foundation in OOP is mandatory. In short, it is the future of all software projects, and automation codebases are no different.

The benefits of OOP

When it comes to OOP, there are technical and non-technical benefits. By that, there are benefits that are directly related to writing code and higher-level benefits that are more concerned with code maintenance and productivity. To begin this exploration, we're going to first look at the non-technical benefits:

- **Reusability**: When implemented properly, OOP allows for code to be implemented in one location with the ability to be used across multiple projects. As a result, it allows developers to write and test one code module that can be used almost anywhere.

- **Code maintenance**: OOP promotes code organization by organizing code into what can be thought of as digital blueprints. This means if you have a module in the form of a function block that handles, for example, a robotic claw, and an issue arises, it can be easily located and fixed.

- **Reduces redundant code**: This ties into reusability and code maintenance. When implemented properly with the correct relationships, OOP can eliminate redundant code because it promotes code reusability. This means that codebases can be smaller and more manageable.

- **Reduces the storage usage on the controller**: Smaller codebases can translate into less required storage for the codebase. This means that, though relatively minor, the overall footprint of the codebase will be reduced. In turn, the smaller footprint equates to more space on the device for things such as logging, more sophisticated HMIs (assuming the HMI is housed on the control unit), and many other things as well.

- **Leverages design**: OOP allows developers to implement more sophisticated program designs. As stated before, it is governed by a set of rules that, when implemented properly, can allow for well-crafted programs that are smaller and cleaner than their non-OOP counterparts.

- **Increases productivity**: Object-oriented code generally produces an overall better-quality product faster and cheaper than a non-object-oriented software system. When designed properly, the modules can be ported to another project (as will be explored later), bugs will be easier to find and fix, and code is less likely to be accidentally broken.

To appreciate the benefits that directly relate to writing code, we need to explore some of the technical benefits of the paradigm. To do this, we are going to explore what are commonly called the four pillars of OOP!

The four pillars: a preview

In terms of writing code, there are four main technical benefits to using OOP, that are as follows:

- Encapsulation
- Abstraction
- Inheritance
- Polymorphism

These four principles have a lot of complexity behind them. For now, it is only necessary to be aware of what the four pillars are, as we will focus on them in the next chapter. For this chapter, we are not going to focus on the pillars in practice. For now, we will use public access specifiers, which will allow us to access function block attributes from anywhere in the program.

> In some OOP contexts, the word "attribute" is often used in a stricter sense and is used to refer to a variable or field of a class or function block. In IEC 61131-3, attributes are keywords or properties that modify programming elements (for example, to control storage or behavior), particularly variables. These terms are often confusing for beginners. This book will use the word "attribute" in a more general sense to refer to any component of a function block.

In all, if a PLC system does support the *IEC 61131-3* standard for OOP, there is no reason not to adopt it. Even if there is an initial investment to get a team up to date with OOP, it will ultimately be worth it in the long run. Without implementing OOP, an organization will lose the ability to easily modify, debug, produce, and, ultimately, reuse code.

As we will explore, the function block POU is the backbone of OOP for many PLC systems; however, we have not explored what function blocks are in any real depth. Therefore, the next section will explore the intricacies of function blocks!

Exploring function blocks

A function block is, for all intents and purposes, a class in any other programming language. Chances are, if you're reading this book, you've programmed a PLC at some point, and you've probably used a function block at some point as well. Considering the name *function block*, it is forgivable to confuse a function block with a standard function; however, a function block is radically different from a function. Many inexperienced PLC programmers will often confuse a function with a function block. However, it is important to understand that a function is simply a callable block of code that will carry out a specific task. In contrast, a function block is a blueprint for something.

A simple way to think of a function block is as a container. Function blocks hold programming code such as unique functions, called methods, and variables.

> **Note**
>
> A **method** is a function that lives in a function block. Methods will be explored deeper in the *Exploring methods* section. However, for now, it is only important to know that a method is a special type of function.

The first step in understanding a function block is learning to declare and use one.

Function blocks are generated via the same wizard we used to create a function. If you are unfamiliar with that process, see *Chapter 3*. To create a function block, right-click on **Application**, then navigate to **Add Object**, select **POU**, check **Function block**, and set the fields to match the following screenshot:

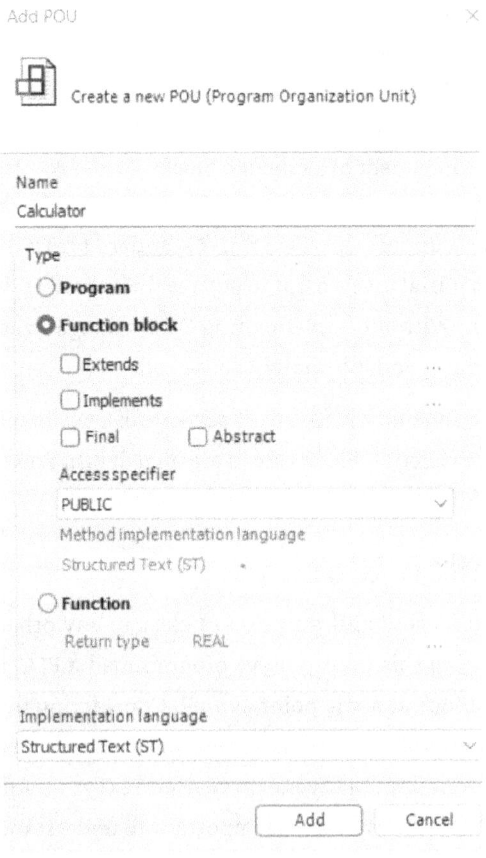

Figure 4.1: Calculator function block

The function block we are creating is going to be a `Calculator` function block. For this example, all we're going to do is perform a basic hookup for the function block to explore its basic functionality. That is, we're only going to implement a line that says, `Hello world`. However, be sure to follow all the steps because after we get the function block hooked up, we're going to add more functionality to it so it can actually be a calculator.

Once you create the function block, you should be met with a POU file in the project tree called `Calculator`. In the file, you should see the following code in the variable block:

```
FUNCTION_BLOCK PUBLIC Calculator
VAR_INPUT
END_VAR


VAR_OUTPUT
END_VAR


VAR
END_VAR
```

The logic in the file that was generated when you created the function block is analogous to a constructor method in a language such as C++, C#, or Java when invoked. However, unlike a traditional constructor, the code in the function block will run only when you explicitly call it. It's not uncommon to use the logic in the function block to initialize things such as setting values, turning things on, and so on, to get the overall function block ready to use. Conceptually, the only real difference is that a function block's code will need to be invoked, while a true constructor is a class method that runs when the class is initialized. For the sake of practicality, I like to think of the code in a function block as constructor logic. As such, I will normally architect the base function block's logic to do tasks that would be found in a traditional constructor.

Depending on what you're working on, it is not uncommon to have only a function block with no methods. However, it is typically a good idea to break up the functionality of a function block into methods to help other developers and your future self better understand and use it.

To demonstrate the initialization, first, match the code in your function block to the following snippet:

```
FUNCTION_BLOCK PUBLIC Calculator
VAR_INPUT
END_VAR
```

```
VAR_OUTPUT
    msg : STRING(20);
END_VAR

VAR
END_VAR
```

Once you have completed this, insert the following code:

```
msg := 'Hello world!';
```

If you have any experience with OOP, you will know that we now need to create a variable to reference the class—or in this case, the function block.

The reference variable is declared in the variable block of the PLC_PRG file, like so:

```
PROGRAM PLC_PRG
VAR
    c1 : Calculator;
END_VAR
```

In this case, the c1 variable is an object reference, or in IEC 61131-3 lingo, an instance of the Calculator function block. In other words, c1 is an object that is based on the blueprint provided by Calculator. We're going to explore the concept of objects in the next section, but for now, the code is set up in such a way that only one calculator is produced based on the blueprint.

Now, we have essentially created a calculator. To use it and have it display our message, all you need to do is implement the following code in the PLC_PRG file:

```
c1();
```

When we run the code, we should get the following output:

Expression	Type	Value
⊟ ● c1	Calculator	
⌐● msg	STRING(20)	'Hello world!'

Figure 4.2: Calculator function block output

Essentially, what *Figure 4.2* represents is a function block being run. At first glance, this looks like we called a function in a very convoluted way. In all fairness, for this example, we did. An application such as this is almost a waste of a function block. In other words, it's overkill to use a function block in this manner; it would typically be more appropriate to use a standard function. To get the most out of a function block, we need to add methods to it to give it unique functionality!

Exploring methods

To get the most out of a function block, you should add methods to it. Methods are a special type of function that are declared in and must be invoked through a function block. Where a standard method is globally accessible in a PLC program, a method can only be invoked in a file that references the function block, as we saw in the previous section. Another unique feature of a method is that when inheritance is invoked, function blocks can share certain methods with other function blocks. To start exploring methods, we're going to add four methods to the Calculator function block we created in the last example.

Adding methods

Adding a method is relatively simple. To add a method, all you have to do is right-click the Calculator function block, hover over **Add Object**, and click **Method...**, similar to what is shown in *Figure 4.3*.

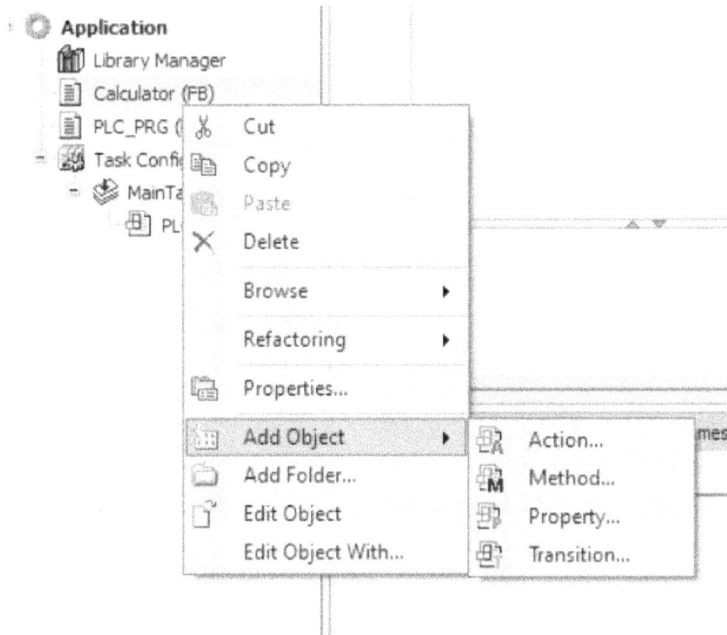

Figure 4.3: Adding a method

When you follow these steps, you will be met with a wizard that will generate a method similar to what is shown in *Figure 4.4*. For this particular method, we're going to name it addNumbers, set the **Return type** value to REAL, and the **Access specifier** value to **PUBLIC**:

Add Method   ✕

⬚M   Create a new method

Name
addNumbers   ⌄

Return type
REAL   ...

Implementation language
Structured Text (ST)   ⌄

Access specifier
PUBLIC   ⌄   ☐ Abstract

                                              Add   Cancel

Figure 4.4: Method wizard

It is very important that your inputs match those in *Figure 4.4*. There are a couple of fields that need to be explained. First, the access specifier ties in with the concepts of abstraction and encapsulation. There are two main access specifiers that you will use on a daily basis: private and public. It is very important that you select the right one for your method. A public access specifier will allow any file that references the function block to use that method. On the other hand, a private access specifier will only allow a method to be accessed from within the function block itself. For example, you will only be able to call a private method from a method that lives in the same function block or from logic in the function block itself.

The second gotcha is the language selection. Most systems, especially CODESYS, will allow you to implement your methods in different IEC languages. For example, you can implement a method in **Structured Text (ST)**, **Ladder Diagram (LD)**, or any of the other languages. Though you can mix and match languages, it is typically a good idea to implement the whole function block in a single language.

Challenge

Now that you have created the addNumbers method, create methods called subNumbers, mulNumbers, and divNumbers with the same parameters as addNumbers. These will be the four functions of the calculator. When you are done, your function block should look like the one shown in *Figure 4.5*.

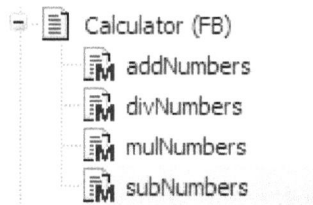

Figure 4.5: Function block methods

After you complete the challenge, you can start implementing the code. In each of the methods, you should see a variable block as in the following code snippet:

```
METHOD PUBLIC addNumbers : REAL
VAR_INPUT
END_VAR
```

Much like traditional functions, a method can take arguments. The arguments will behave in the same manner as they would with a traditional function. For this example, the methods will take two arguments, which we will call val1 and val2. These inputs will be the specific numbers that the methods will perform the mathematical operation on.

Similar to how we used inputs with functions, we will declare these variables in the VAR_INPUT block. With that being said, the variable block of each of the methods should resemble the following:

```
VAR_INPUT
    val1 : REAL;
    val2 : REAL;
END_VAR
```

Note

When working with mathematical operations, ensure that you pay attention to the data types. For this example, we're going to use REAL as that type is robust enough to handle any calculation.

Once you have all the input variables set up, you can move on to implementing the logic for the methods. To do this, add the following logic to each of the methods.

Add this for the addNumbers method:

```
addNumbers := val1 + val2;
```

Add this for the subNumbers method:

```
subNumbers := val1 - val2;
```

Add this for the mulNumbers method:

```
mulNumbers := val1 * val2;
```

Add this for the divNumbers method:

```
IF val2 <> 0 THEN
    divNumbers := val1 / val2;
ELSE
    divNumbers := 0;
END_IF
```

Note

Notice that the divNumbers method has a little extra logic compared to the other methods. This is to ensure that if a division-by-zero situation occurs, the code will not crash. Essentially, the method will return 0 if the bottom number is input as 0.

Once you have the method code assembled, you can move on to preparing the PLC_PRG file. These are the variables that will be needed for the Calculator program:

```
PROGRAM PLC_PRG
VAR
    calculator : Calculator;
```

```
    sum         : REAL;
    dif         : REAL;
    pro         : REAL;
    rat         : REAL;
END_VAR
```

Here, the `calculate` variable is a reference to the `Calculator` function block. The other variables that are of the `REAL` type are the holders for the return values.

The following `PLC_PRG` file code will invoke the methods:

```
sum := calculator.addNumbers(1, 3);
dif := calculator.subNumbers(3, 2);
pro := calculator.mulNumbers(5, 5);
rat := calculator.divNumbers(8, 2);
```

Essentially, we use the same syntax as we did when we were accessing variable values from the core function block. We also pass arguments the same way we did when we were calling functions.

Once you have the `PLC_PRG` file set up and you run the code, you should be met with the following output:

Expression	Type	Value
+ ★ calculator	Calculator	
★ sum	REAL	4
★ dif	REAL	1
★ pro	REAL	25
★ rat	REAL	4

Figure 4.6: Calculator output

As can be seen, all the methods were correctly called and are computing the correct values.

Getting to know objects

So far, we have demonstrated how to create function blocks and how to add methods to them. However, in terms of actual implementation, we haven't fully explored how to leverage these to do anything productive. A key use case for function blocks is reducing code for things that are similar. In other words, suppose we have a blueprint for a car. From that blueprint, we can create an unlimited number of vehicles from that single set of prints; however, each vehicle will be unique regarding certain values such as car color, VIN, or gas mileage.

In a more technical sense, an object is an instance of a function block. That is, each object you create will consist of all the attributes of the function block, such as the variables and methods, but they are independent of each other. For example, if a function block has a variable called MPG in it and has two objects that derive from it, one object can have an MPG value of 31 while another will have a value of 22. Essentially, an object is a copy of all the function block's code. This means that although the objects are copies of one another, they will operate and evolve separately throughout the program's run cycle. To fully grasp this concept, let's look at an example!

Suppose we're working on a car wash. A car wash will have many different motors that each do their own thing to clean a car. Depending on the way the car wash is designed, it will most likely use the same motor type throughout the system. For this example, assume our car wash is composed of three motors that can be put into three modes that we'll call on, off, and pause.

For this project, create a function block called motors and add three functions named motorOn, motorOff, and motorPause to it. For all three methods, set the access specifier to PUBLIC, and though it technically won't matter for this example, set the return type to WSTRING. When you're finished, your function block should look like *Figure 4.7*.

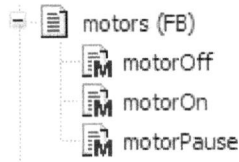

Figure 4.7: The motors function block

The motors function block will have what other OOP-based programming languages call a class-level variable in it. This is a variable that will be accessible throughout the function block. The code for the motors function block POU file should look like the following:

```
FUNCTION_BLOCK motors
VAR_INPUT
END_VAR

VAR_OUTPUT
    state : WSTRING;
END_VAR

VAR
END_VAR
```

The code for the motorOn method should only consist of the following:

```
state := "on";
```

The code for the motorOff and motorPause methods should be the following, respectively:

```
state := "off";
state := "pause";
```

The true magic for this program will be in the PLC_PRG file. The code for this file should match the following:

```
PROGRAM PLC_PRG
VAR
    brushMotor          : motors;
    wheelShiner         : motors;
    roller              : motors;

    brushState          : WSTRING;
    rollerState         : WSTRING;
    wheelShinerState : WSTRING;
END_VAR
```

In this code, brushMotor, wheelShiner, and roller are the objects. They are instances of the motors function block and are sometimes referred to as reference variables in traditional OOP terminology. They have all the attributes that are in the motors function block, but as stated before, they will operate independently. This will be demonstrated with the state variables which will simply hold the state for the individual objects.

The main logic for this program will be as follows:

```
brushMotor.motorOn();
brushState := brushMotor.state;

wheelShiner.motorPause();
wheelShinerState := wheelShiner.state;

roller.motorOff();
rollerState := roller.state;
```

Once the code is implemented, your output should resemble *Figure 4.8*.

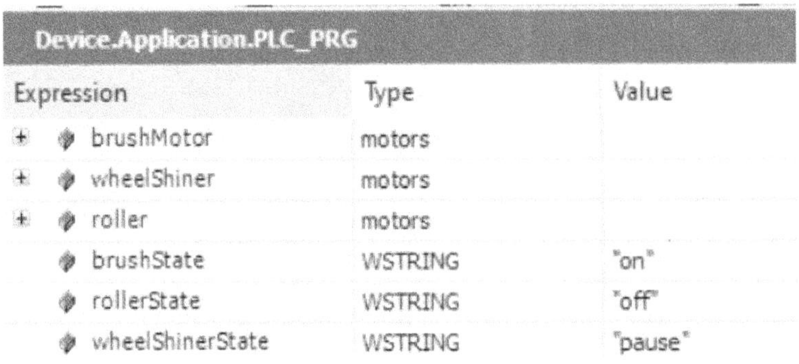

Device.Application.PLC_PRG		
Expression	Type	Value
+ ⚙ brushMotor	motors	
+ ⚙ wheelShiner	motors	
+ ⚙ roller	motors	
⚙ brushState	WSTRING	"on"
⚙ rollerState	WSTRING	"off"
⚙ wheelShinerState	WSTRING	"pause"

Figure 4.8: Motor objects

If you have never programmed in OOP before, this code may not make a lot of sense. For instance, why does state have three unique values all at once? In short, objects such as roller, wheelShiner, and brushMotor all get a copy of the function block and its attributes. This means that each object variable will get its own state variable, its own pause method, and so on. Therefore, each one of these objects can mutate the state variable without affecting the other objects! Function blocks can easily become a mess. The key to a well-crafted function block starts with a solid name. In the next section, we're going to explore function block naming and how it can ensure you have a quality function block.

Function block naming

The key to writing a quality function is understanding that it represents a thing. This is vital to understand for programmers. Many inexperienced programmers will often use function blocks or classes in general-purpose programming languages as dumping grounds for random methods and variables. This is a terrible practice, as a function block is supposed to model a thing. If you think back to school, you may remember that a noun represents a thing. This means that, much like a struct, a function block should also use a noun as a name.

As we saw with structs, a quality name can easily lay the foundation of a quality function block. A function block should represent one thing and one thing only. Therefore, having a quality noun name will enforce that principle. If you think about it, if you have a function block called television, adding methods to support the operations of a car won't make any logical sense. Also, much like with structs, you want to avoid using a verb as a name. Again, if you think back to school, a verb is an action word; it does not describe a thing, so it won't make logical sense to name a function block a verb since it's supposed to represent a thing.

The following are some basic tips to help you pick an appropriate name for a function block:

- Keep the name short and to the point
- Ensure that the name is a noun
- Avoid using verbs
- Avoid using ambiguous names that can be interpreted differently by different people, especially if your team natively speaks other languages or stems from different cultures

More on function block best practices will be explored in *Chapter 6*; however, now that we have some background on using function blocks and methods, we can move on to using a special method type known as getters and setters.

Getting to know getters and setters

Until this point, all of our methods and variables have been set to PUBLIC. This means that any file that has a reference to the function block can use its attributes. Believe it or not, this is considered bad programming. When you're using the OOP paradigm, you want only the minimum attributes exposed to any file other than the files in the function block.

The golden rule of OOP is that if a programmer does not need to use an attribute, they should not be able to see it. This is the core of encapsulation and abstraction. However, there is a caveat. Sometimes we need to use the encapsulated variables. So, how can we use a variable that should not be accessible? The answer is properties.

A **property** is broken into two components that are placed in two different files. One file is called get, and the other is called set. Properties help you keep data encapsulated (that is, hidden). For example, consider the following code for our motors function block:

```
FUNCTION_BLOCK motors
VAR_INPUT
    state : WSTRING;
END_VAR


VAR_OUTPUT
END_VAR


VAR
    motorSpeed : REAL := 1000;
END_VAR
```

In this case, motorSpeed would be a function block-wide configuration. If you try to access this variable and change it from another file, as in *Figure 4.9*, you will get an error.

```
m1.motorSpeed := 3000;
```

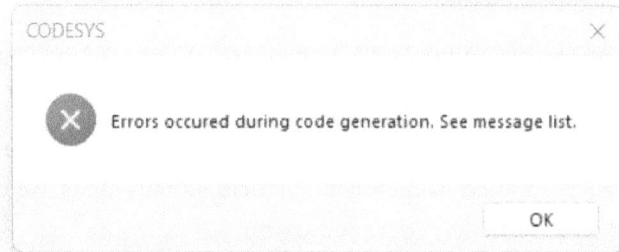

Figure 4.9: Error message

Essentially, this message is saying that you cannot directly access the function block variable because it is not an input variable. However, we can set the value with a method. Though you can (and many times, will) mutate these variables with a PUBLIC method, the optimal way of mutating the value is with the Get and Set values of a property.

The overall goal for a property is to ensure that the program is reading or writing to a value on your terms. Properties allow us to set validation logic to ensure that we're setting a valid value to the target variable or ensure that we can read the value under the correct circumstances. The first step in this process is setting up a property.

Getting to know properties

Adding a property is a lot like adding a method, with the only exception being that you will select **Property...** instead of **Method...**. To follow along, add the property in *Figure 4.10* to the motors function block:

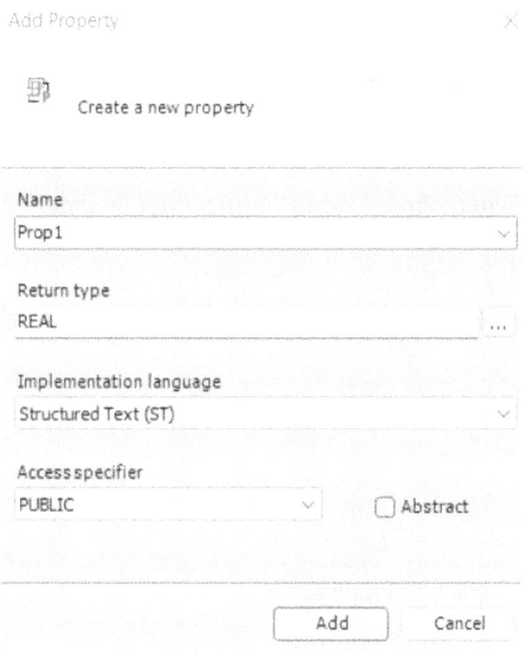

Figure 4.10: Property creation wizard

For this example, we're going to name the property Prop1 and give it a return type of REAL. As usual with this chapter, we will give the property an **Access specifier** value of **PUBLIC**. Once you click the **Add** button, you should see a property component generated with two methods, similar to what is seen in *Figure 4.11*:

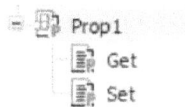

Figure 4.11: Property-generated getter and setter

Now that we have our property in place, let's use it!

Using the Get method

The easiest of the two types of methods to use is the Get method. Though it can support logic, it is very common for it to just read a value, which is why it is typically the easiest to use. For this example, we're going to read the motorSpeed variable in the motors function block. To do this, all we need to do is implement the following line in the Get method:

```
Prop1 := motorSpeed;
```

Once you do that, you can implement the following variables in the PLC_PRG file:

```
PROGRAM PLC_PRG
VAR
    m1    : motors;
    speed : REAL;
END_VAR
```

The core logic for the example is as follows:

```
speed := m1.Prop1;
```

When this example is executed, you should see the same output as in *Figure 4.12*.

Device.Application.PLC_PRG		
Expression	Type	Value
+ 🔷 m1	motors	
🔷 speed	REAL	1000

Figure 4.12: Getter output

As can be seen, the program reads the motorSpeed variable and assigns it to the speed variable. In other words, we bypassed the error message in *Figure 4.9*. Now that we can read a value, we need to learn how to set a value.

Using the Set method

The Set method typically takes more complex logic for validation purposes. For example, if your device is using a motor and the motor has a maximum speed of 2,000 RPMs, you don't want to accidentally set the motor speed to 6000. This is where the validation logic comes into play. To explore this, implement the following logic in the Set method:

```
motorSpeed := Prop1;
```

```
IF motorSpeed > 2000 THEN
    motorSpeed := 1111;
END_IF


Prop1 := motorSpeed;
```

In this case, we're going to read in a value and then perform a check on it. If the value we set is greater than 2000, we're going to default the motor speed to 1111.

To use this code, we need to modify the PLC_PRG POU to have the following variables:

```
PROGRAM PLC_PRG
VAR
    m1    : motors;
    speed : REAL;
END_VAR
```

The logic for this example should be as follows:

```
m1.Prop1 := 6000; //set
speed := m1.Prop1; //get
```

When the code is ran you should be met with *Figure 4.13*:

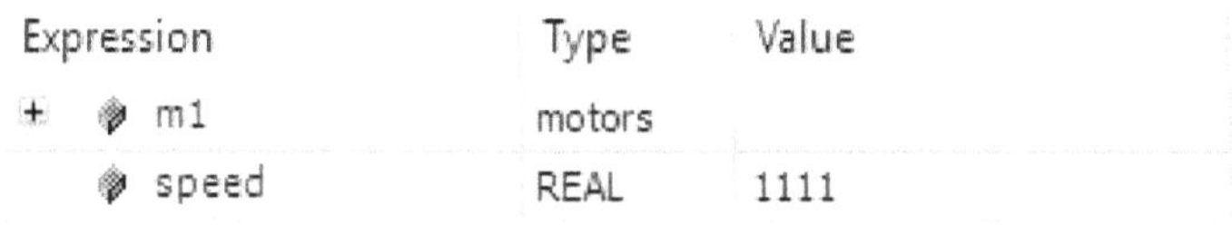

Expression	Type	Value
+ ◈ m1	motors	
◈ speed	REAL	1111

Figure 4.13: Getter output

Whether you're using the getter or the setter will depend on which side the property is being invoked on. If the property reference is on the left of : =, it will invoke the setter method, and if it is on the right, it will invoke the getter.

The rules of properties

To properly leverage properties, you need to follow a few rules:

- **Responsibility**: When it comes to a property, it is important to limit what it can read or write to. More specifically, a property should only read or write to a single variable. If you have a property that is reading or writing to more than one function block variable, you should create another property to interact with that variable. In other words, you can have as many properties as you need!

- **Use responsibly**: A major objective of OOP is to hide attributes such as methods and variables whenever you can. In other words, you should only expose attributes that are needed for the function block to work properly. This means that you do not want to create a property for every function block variable. In fact, it is best to only create a property for a variable that absolutely has to be interacted with. In short, these are values such as configuration values (for example, motor speeds, temperature ranges, and so on).

- **Naming**: Since a property should only manipulate a single variable, it is best to include the name of that variable in the property. Our example did not follow this rule; however, for production code, this is vital as it will make troubleshooting and modifications much easier.

As stated earlier in this section, setters and getters are simply special methods and can have complex logic to properly vet values that are being assigned to a variable. The past getter and setter examples are simply the bare bones of how to use them. It is recommended that you play around with getters and setters. Now that getter and setter properties have been explored, it is time to move on to another concept, known as **recursion**.

Understanding recursion and the THIS keyword

Recursion is a looping concept that isn't used much in today's world. However, it is a concept that often pops up in interviews and is something that all developers should understand. Simply put, recursion is where a method calls itself to solve a smaller part of a problem. Recursion is a valid concept that is important to know; however, for many applications, some type of loop will be more appropriate.

If you do opt to use recursion, exercise great caution. Recursion is generally considered resource-heavy, and in the automation world, where many PLCs have traditionally limited computing resources, it can become a heavy burden on the PLC. Recursion is also somewhat dangerous, as it is easy to create what is known as an infinite recursive loop, which is a loop that will call itself forever. Many modern compilers do check for this and will usually throw a compile error before the code is run. However, you should be aware of this and need to look out for the issue.

The THIS keyword

To understand recursion, you must first understand the THIS keyword. The CODESYS documentation states that the THIS keyword is a function block pointer to its own function block instance. In other words, THIS is a keyword for a function block pointer that points to itself. The general syntax for the THIS keyword is as follows:

```
THIS^.method()
```

Recursion in action

To demonstrate recursion, let's implement a very common recursive function that calculates the factorial of a number. To do this, add a new method to the `Calculator` function block, name it `factorial`, and give it the return type of `INT`:

```
METHOD factorial : INT
VAR_INPUT
    x : INT;
END_VAR
```

This logic will calculate the factorial:

```
IF x <= 1 THEN
    factorial := 1;
ELSE
    factorial := x * THIS^.factorial(x - 1);
END_IF
```

The line in the `IF` statement is what calls the method. The method takes an argument by default, called x. When the method is called, it is supplied with an initial value. That value has 1 subtracted during each iteration, and the value is multiplied by the current value of x. For example, if the initial value supplied is 4, the method will compute the following: $4*3*2*1=24$.

To demonstrate the code in action, modify the `PLC_PRG` file to match the following:

```
PROGRAM PLC_PRG
VAR
    calculator : Calculator;
    sum        : REAL;
    dif        : REAL;
    pro        : REAL;
    rat        : REAL;
    fac        : INT;
END_VAR
```

In the case of the variables, all we did was add `fac`:

```
sum := calculator.AddNumbers(1, 3);
dif := calculator.SubNumbers(3, 2);
pro := calculator.MulNumbers(5, 5);
rat := calculator.DivNumbers(8, 2);
fac := calculator.Factorial(4); //Factorial method
```

When the code is run, you should see the following output:

Expression			Type	Value
+	🔷	calculator	Calculator	
	🔷	sum	REAL	4
	🔷	dif	REAL	1
	🔷	pro	REAL	25
	🔷	rat	REAL	4
	🔷	fac	INT	24

Figure 4.14: The factorial output

As can be seen, `fac` is showing 24, which means the `factorial` method works!

At this point, you should have a decent understanding of methods and function blocks. Methods and function blocks are the backbone of any well-written object-oriented program.

It is no secret that LL is a very common PLC programming language. A logical question is, can we integrate ST and, more importantly, these techniques with LL? The answer to this question is a resounding yes! In the next section, we're going to explore how to integrate ST code with LL.

Using function blocks in LL

It's no secret that LL still rules the automation world; however, ST shines for modern, complex applications. Regardless, for many applications, the simplicity of LL is much preferred. So, for certain applications, it would be ideal to be able to mix the programming languages. Luckily, many IEC 61131-3-based systems accommodate this with the use of function blocks. Function blocks implemented in one of the IEC 61131-3 programming languages can typically be utilized in another. For example, if we write a function block in ST, we can use it in LL or any of the other IEC languages.

> **Note**
>
> Whether or not you are allowed to implement a function block in another programming language will depend on the programming system that you are using. Many modern systems will support the mix-and-match approach; however, it's important to be mindful of the limitations of your programming environment.

To grasp this concept, let's explore an example!

Exploring the power of ST in LL

To explore how ST can be used to power an LL program, let's create a new project and set the language to **Ladder Diagram**. This will allow us to implement the main PLC program in LL as opposed to ST. After you do that, you need to create a function block named `Calculator` and set the language to **Structured Text**. Once you create the function block, add a method called add_nums to it. When you're done, your function block should resemble *Figure 4.15*.

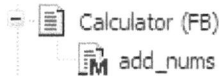

Figure 4.15: Calculator function block

The method's variable code section should resemble the following:

```
METHOD PUBLIC add_nums : REAL
VAR_INPUT
    a : REAL;
    b : REAL;
END_VAR
```

The logic should match the following:

```
add_nums := a + b;
```

Once you complete these steps and implement the code, navigate to the PLC_PRG file, and add a **Box with EN/ENO** control from the toolbox as in *Figure 4.16*.

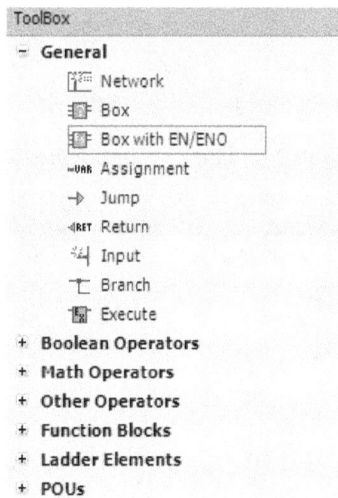

Figure 4.16: Toolbox

To utilize the method, implement the following variables in the PLC_PRG file:

```
PROGRAM PLC_PRG
VAR
    en1 : BOOL := TRUE;
    x   : REAL := 2;
    y   : REAL := 3;
    sum : REAL;
    c   : Calculator;
END_VAR
```

In this case, en1 is a variable that will enable the function block. The x and y variables are the values that will be added together, sum will hold the sum of the two numbers, and finally, c is a reference to the Calculator function block.

Once you have the variables in place, you can move on to adding the method to the PLC program. To do this, simply select the **Box with EN/ENO** control and drag it over to the logic section of the PLC_PRG POU file. Once you add the method, assign the variables to the box so that it resembles *Figure 4.17*.

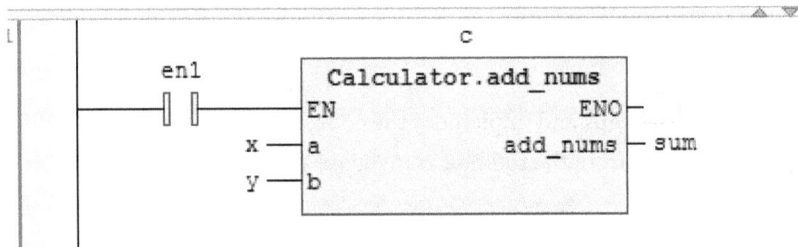

Figure 4.17: LL PLC program

When the program is executed, you should be met with *Figure 4.18*.

Figure 4.18: LL program output

As can be seen, the program produces the sum of the two numbers and stores it in the sum variable.

Challenge

Add a method that can multiply, divide, and subtract two numbers in ST and create an LL program that can use each of the functions. Once you complete the challenge, you can move on to the final project.

In all, this technique allows you to implement a complex algorithm that would otherwise be difficult to write in LL in ST while still being able to use the simplicity of LL. Typically, you want to keep your whole codebase in a single language; however, there are exceptions to this rule, mainly in the form of libraries, which we'll explore in subsequent chapters. For now, though, we're going to focus on our final project!

Final project: Creating a unit converter

In automation programming, it is very common to have to convert between different units of measurement to support clients around the world. This is especially true if you have a single codebase that supports a specific machine that is deployed to many different regions. To accommodate the different units of measurement, we're going to create a function block.

You might be asking yourself, Why a function block? Why not just use simple functions? In practice, we could get away with a simple function; however, in software engineering, it is important to try and think ahead. In the future, we may want to have more units to convert between. If we were to use pure functions, we could end up with a bunch of conversion functions hanging out in the project tree. However, if we use a function block, we can add a level of organization to the project and create different objects that can share the code.

For our final project, we are going to create a very simple function block that can convert the following units:

- lbs to kgs and kgs to lbs
- Feet to meters and meters to feet

Depending on what you're working on, there will probably be many more units; however, it's important to remember that this is just an example.

The first thing we need to do is create a function block called UnitConverter and add two methods called convertWeight and convertLength to it. To implement these methods, both should have a **Return type** value of REAL and an **Access specifier** value of **PUBLIC**. When you're done, your function block should look like the following:

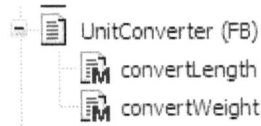

Figure 4.19: Unit converter with methods

For this project, we do not have to make any changes to the UnitConverter function block. The only changes made will be to the convertLength and convertWeight methods, which will be as follows.

These are the convertLength method variables:

```
METHOD PUBLIC convertLength : REAL
VAR_INPUT
    lengthInput : REAL;
    metric      : BOOL;
END_VAR


VAR
    conversionFactor : REAL := 3.281;
END_VAR
```

This is the convertLength method logic:

```
IF metric = TRUE THEN
    // feet to meters
    convertLength := lengthInput / conversionFactor;
ELSE
    // meters to feet
    convertLength := lengthInput * conversionFactor;
END_IF
```

These are the convertWeight method variables:

```
METHOD PUBLIC convertWeight : REAL
VAR_INPUT
```

```
    weightInput : REAL;
    metric      : BOOL;
END_VAR

VAR
    conversionFactor : REAL := 0.4536;
END_VAR
```

This is the `convertWeight` method logic:

```
IF metric = TRUE THEN
    // lb to kg
    convertWeight := weightInput * conversionFactor;
ELSE
    // kg to lb
    convertWeight := weightInput / conversionFactor;
END_IF
```

Essentially, both methods will work off a Boolean value. If the value is TRUE, it will convert the numerical argument to its metric counterpart; if it is FALSE, it will convert to a standard value.

To call these methods, we will add the following lines of code to the `PLC_PRG` file:

```
PROGRAM PLC_PRG
VAR
    unitConverter : UnitConverter;
    meters        : REAL;
    feet          : REAL;
    pounds        : REAL;
    grams         : REAL;
END_VAR
```

The logic for the `PLC_PRG` file is as follows:

```
meters := unitConverter.convertLength(32, TRUE);
feet   := unitConverter.convertLength(32, FALSE);
pounds := unitConverter.convertWeight(100, TRUE);
grams  := unitConverter.convertWeight(100, FALSE);
```

When the code is run, you should get the following output:

Expression	Type	Value
⊞ 🖋 unitConverter	UnitCon...	
🖋 meters	REAL	9.753124
🖋 feet	REAL	104.992
🖋 pounds	REAL	45.36
🖋 grams	REAL	220.458557

Figure 4.20: Unit conversion

This is an example of a real-world function block that can be put into a production machine. However, if you do decide to put a unit converter into your machine, you may want to add some more conversion methods.

Summary

OOP is the backbone of all modern programs. OOP is so ingrained in the IT world that you can't function as a programmer without an in-depth knowledge of the concept. The days of being able to get away with simply programming machines, in a procedural sense, with LL are quickly fading.

This chapter was simply a soft introduction to OOP. When creating a program, it is usually considered wise to approach the program from an OOP point of view. This means that instead of just jumping to using simpler structures, such as functions (though there is a time and a place for functions and other simple structures), from this point forward in your programming journey, you're going to want to think about how things relate to each other and can be condensed into logical units. With that being said, OOP is way more than just organizing your code into function blocks. Now that we have a grasp on function blocks, methods, properties, recursion, and implementing LL programs with components written in ST, we can learn how to leverage these to reduce redundant code, create cleaner code, and apply actual architecture to programs. In the next chapter, we're going to explore some common OOP techniques to get the most out of the paradigm!

Questions

1. What is a function block called in traditional programming languages?

2. What is recursion?

3. What is the purpose of the THIS keyword?

4. What are the two methods that make up a property?

5. What is the difference between a getter and a setter?

6. What is a method?

Further reading

*   *CODESYS function blocks*: https://content.helpme-codesys.com/en/CODESYS%20 Development%20System/_cds_obj_function_block.html

*   *CODESYS "This" keyword*: https://content.helpme-codesys.com/en/CODESYS%20 Development%20System/_cds_pointer_this.html

*   *Integrating Structured Text With Ladder Logic*: https://www.youtube.com/ watch?v=BQLvaiX66N4

Join our community on Discord

Join our community's Discord space for discussions with the authors and other readers: https:// packt.link/embeddedsystems

5

OOP: The Power of Objects

Object-oriented programming (OOP) is much more than just programming with function blocks. Assuming that OOP is simply programming with function blocks has led to the downfall of many projects. In reality, OOP can best be thought of as a way to model real-world objects and concepts using function blocks as their digital blueprints. The power of OOP comes from the function blocks themselves being governed by a series of pillars or rules that can reduce code, create clear relationships, and allow for logical cohesion. The ideal result of a well-crafted OOP program is a codebase that is slimmer, easier to maintain, and makes more sense in general.

OOP can be a very abstract concept to those who have never used it before. At times, the techniques explored in this chapter may seem counterintuitive, convoluted, or flat-out useless. However, the key to a well-written, modern, and quality codebase that will last the test of time is understanding how to utilize these concepts in a practical way. This chapter will explore OOP in a more nuanced manner and will delve into how to form relationships between function blocks. To explore the nature and concepts of OOP, we will explore the following topics:

- Understanding access specifiers
- Exploring the pillars of OOP
- Exploring the PROTECTED access specifier
- Inheritance versus composition
- Examining interfaces

To round out the chapter, we will create a simulated assembly line using the concepts that we explore in this chapter.

Technical requirements

Though we are dealing with complex programming, there are no extra plugins needed to follow along with the examples. As usual, to follow along, you will need a copy of CODESYS installed on your computer. The code for the examples can be found at the following GitHub URL: `https://github.com/PacktPublishing/Mastering-PLC-Programming-Second-Edition/tree/main/Chapter%205`

Understanding access specifiers

Before we begin our deep dive into OOP, we need to first understand the mechanics of hiding data. That is, we need to understand how to hide attributes. To do this, we need to understand the concept of **access specifiers**. As we explored in *Chapter 4*, we typically want to restrict or hide as many attributes as we can from other POUs. We do this to prevent potential bugs from sneaking into our program. The key to hiding attributes is access specifiers. In short, an access specifier is a way of signaling to the PLC program whether another POU outside of a function block is allowed to manipulate or use an attribute, mainly a method. In the context of IEC 61131-3, an access specifier will typically only refer to a method; however, you can think of internal function block variables in a similar manner. Therefore, let's take a deeper look at access specifiers!

Exploring the different types of access specifiers

Different programming languages and PLC programming systems often support a number of different access specifiers that do different things. For our purposes, we're going to explore what I like to call **the big three**. The big three are arguably the most common access specifiers that are found in most OOP languages. The big three are as follows:

- **Private**: Private methods cannot be accessed outside the function block they are declared and implemented in. In other words, these methods are restricted and can only be used internally. Private methods are best utilized for operations that are orchestrated by other methods inside the function block they are in. Ultimately, private functions should be the worker bees of a function block. When implemented correctly, most of the methods in a function block will usually be private.

- **Public**: This access specifier will grant any POU file that has a reference to the function block the ability to use the method. Generally, you want to limit the use of public methods in production code; however, they are necessary at times. You will usually use a public method as a way to kick-start a process or invoke a specific task from outside the function block. When implemented correctly, a public method either does a task or invokes a series of private methods. A pitfall that many inexperienced programmers will fall into is that they will make all their methods public. This is asking for trouble and will cause issues later down the road, as methods that are not supposed to be directly invoked will, which in turn, will cause erroneous behavior. A general rule I like to follow is to use public attributes sparingly.

- **Protected**: Of the big three access specifiers, protected is used the least. Protected signals that only a function block and its children can invoke a method. This is a concept we will explore in the next section.

> Note
>
> Arguably, the two most used access specifiers are PUBLIC and PRIVATE.

When I was first learning about access specifiers, I used a simple trick to help me conceptualize which one I should use. For the trick, I would think of each access specifier as a security level:

- **Level 1** - public: Use when unrestricted access is allowed and needed
- **Level 2** - protected: Use when only certain files need access to the method
- **Level 3** - private: Use when components should only be used by other internal attributes

Now that we have a basic idea of what access specifiers are and when to use them, let's see them in action! In the next section, we're going to explore the PRIVATE access specifier.

PRIVATE access specifier in action

To explore the PRIVATE access specifier, we're going to make some coffee! This program will be composed of a function block, CoffeePotFB, and the following three methods:

- startHeater (private method with a return type of WSTRING)
- startWaterPump (private method with a return type of WSTRING)
- makeCoffee (public method with a return type of BOOL)

To implement the methods, simply click on the function block, select **Add Object**, and then **Method**. Use the information in the bullets to fill out the wizard. When you're done creating the files, you should have the following tree:

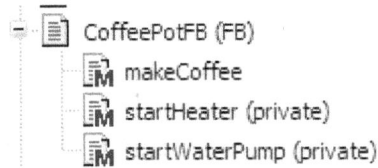

Figure 5.1: Coffee tree

In *Figure 5.1*, both the private methods are marked as such. The three methods will not have any variables, but the CoffeePotFB function block will. The variables for CoffeePotFB will be as follows:

```
FUNCTION_BLOCK PUBLIC CoffeePotFB
VAR_INPUT
END_VAR
VAR_OUTPUT
    heater : WSTRING;
    water  : WSTRING;
END_VAR
VAR
END_VAR
```

Essentially, the function block will have only two variables that will hold a message string to simulate the internal workings of the coffee pot. The logic for these methods will be as follows:

- startHeater method logic:

    ```
    startHeater := "heater started";
    ```

- startWaterPump method logic:

    ```
    startWaterPump := "water pump on";
    ```

- makeCoffee method logic:

    ```
    heater := startHeater();
    water := startWaterPump();
    ```

In this case, we have a heater and a water pump method that simulates the internal components of the device. In real life, we wouldn't want our users messing with these dangerous components. The heater could burn them, and the pump could possibly hurt them under the right circumstances. If the user had to manually work these components, at the very least, they would get a lousy cup of coffee. To be responsible, we need to hide these methods from our users and give them an easy-to-use interface to start the coffee-making process.

In this case, we're hiding away the internal complexity (needing to manually work the heater and water pump) by using the PRIVATE access specifier. At the same time, we're going to provide an easy-to-use interface to start the coffee-making process – that is, the makeCoffee method.

The final part of the program is the code that will go into the PLC_PRG POU file. The variables for this POU are as follows:

```
PROGRAM PLC_PRG
VAR
    coffeeMaker : CoffeePotFB;
    heater      : WSTRING;
    water       : WSTRING;
END_VAR
```

Here we have a simple object variable and two variables to hold our output from the methods. The program's main logic should match the following:

```
coffeeMaker.makeCoffee();
heater := coffeeMaker.heater;
water  := coffeeMaker.water;
```

When the program is run, you should be met with *Figure 5.2*:

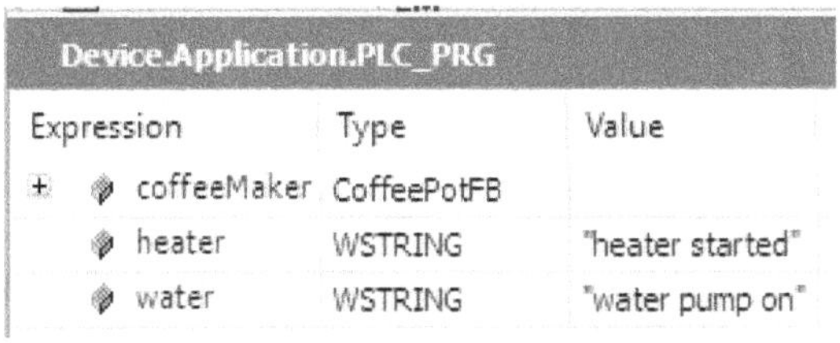

Device.Application.PLC_PRG		
Expression	Type	Value
+ ⬦ coffeeMaker	CoffeePotFB	
⬦ heater	WSTRING	"heater started"
⬦ water	WSTRING	"water pump on"

Figure 5.2: Program output

What this program did was use one `PUBLIC` method to call two `PRIVATE` ones. The `startHeater` and the `startWaterPump` methods could not be called from outside the function block. If you try to call the methods from, say the `PLC_PRG` POU file, you'll find that you can't. In other words, we hid some of the internal complexity and delegated all the complex tasks, such as starting the water pump and heater at the correct time, to the `coffeeMaker` method.

The coffee pot example used two OOP pillars, encapsulation and abstraction. In the next section, we're going to take a deeper look at these two concepts, along with the inheritance and polymorphism principles.

Exploring the pillars of OOP

The core of OOP is composed of four pillars. When implemented correctly, these pillars can ensure that you can easily scale, modify, and troubleshoot your codebase. In the automation industry, where money is directly related to time, having a flexible and easy-to-troubleshoot codebase is vital not only to the success of a machine but to an organization in general.

Depending on who is asked, OOP is governed by four pillars: *encapsulation, abstraction, inheritance,* and *polymorphism*. Some sources, especially automation sources, will cite only three pillars due to some developers grouping abstraction and encapsulation together and essentially classifying them as two sides of the same pillar. Academia usually teaches that there are four pillars, and it is more common to hear about four pillars as opposed to three in traditional programming circles. For this book, we're going to treat encapsulation and abstraction as two distinct concepts.

Encapsulation versus abstraction

As a programming instructor, two of the most basic OOP principles, **encapsulation** and **abstraction**, are two of the hardest to teach. This is because formal textbook definitions are a bit too abstract to be applied by novice programmers. Formally, abstraction is described as *the process of exposing only the essential features of an object while hiding the implementation details*, while **encapsulation** is defined as *the bundling of data (variables) and behavior (methods) into a single unit (such as a class or function block) and restricting access to the internal details using visibility controls*.

In my opinion, the most productive way to approach these concepts is to think of abstraction as a design choice and encapsulation as the mechanism to implement abstraction. What this boils down to is the following:

- **Abstraction:** In most cases, the fewer elements that can access a variable or method, the better. The concept of hiding these components and only having the necessary ones visible to other files is what is known as abstraction. Essentially, abstraction is the concept of hiding components/data from other POU files. By the formal definition, this may seem more like encapsulation, but I like to group this principle with abstraction because you have to choose what will be hidden and what will be exposed to outside attributes. In the case of the coffee pot, we hid the internal workings of the heater and the water pump, but we gave the user an easy-to-use interface (the `coffeeMaker` method) to do the hard work for them. What we did in that example was essentially a practical form of abstraction.

- **Encapsulation**: Encapsulation is a concept that deals with binding data into logical units. Function blocks are the normal modular unit in which logically related attributes are bound. The goal of encapsulation is to group related data and ultimately hide the complexity of the unit. Again, think of the heater and water pump as this was the coffee pot's internal complexity that we hid away.

Some traditional sources may say these definitions are blurry or inverted; however, there is a reason why both of these concepts were originally lumped together as a single pillar and still are in many PLC documents. Put simply, both concepts walk together. In practice, you can't use one without the other and you're not going to cleanly separate the two concepts when you're on the shop floor programming a machine. When you're designing your program, you will often think of abstraction as what you need to hide and what you need to expose – in other words, what will be PRIVATE, PUBLIC, or PROTECTED. On the other hand, you'll typically think of encapsulation as the primary mechanism to hide complexity. In other words, you want to think of abstraction as a design choice and encapsulation as the mechanism to implement the design.

These concepts are often confusing, especially when one views them through a formal definition or a purely conceptual lens. The simplest way to think of abstraction and encapsulation is through the lens of your computer. Your computer has millions of electrical components. These components are grouped together to form a module, and the modules are wired together to form a working computer. When considering the average user, they don't, and in many cases shouldn't, know how computer electronics work to operate the machine. For the typical user to operate a computer, all they need to know how to do is turn the computer on, use a mouse, and use the keyboard. Realistically, the average person poking around the electronics will be detrimental in many cases. An untrained person would probably damage the computer, and in some cases, themselves!

This example is essentially the basis for encapsulation and abstraction. The complex electronics that power a computer are hidden from the user. Computer manufacturers go to great lengths to encapsulate electrical components so the average user can't get to them. This hiding is analogous to the encapsulation of software attributes. On the other hand, the user needs to know how to press the power button to turn the computer on, but they don't need to know what order the chips need to be powered on in. Attributes such as the power button, keyboard, and mouse can be thought of as abstractions. The user only needs a high-level understanding of how to press a button or move a mouse to make the computer work. In other words, the complexity of the computer is abstracted out!

When designing a program, you want to keep the same mentality. You only want to expose attributes that are absolutely necessary to other POUs while hiding the rest. You need to think of other POU files as the computer user; knowing or being able to do too much could easily lead to damage or erroneous behavior. A good rule of thumb is to only allow other files to see attributes – in this case, those easy-to-use interfaces that are necessary for a given operation to be carried out.

For languages that support it, abstraction and encapsulation are arguably the two most important topics in OOP. In all, encapsulation and abstraction are accomplished with access specifiers. For this book, the visibility of attributes such as methods will be denoted with the PUBLIC or PRIVATE access specifier.

Though encapsulation and abstraction are very powerful concepts, an equally powerful OOP principle that we're going to explore in the next section is inheritance.

Inheritance

Inheritance mostly conjures up thoughts of receiving material possessions from the dearly departed. In terms of programming, the concept is similar without anyone, or in the case of PLCs, anything needing to pass away. In programming, special relationships can be formed between function blocks. These relationships allow one function block to use certain attributes of another. In other words, inheritance allows you to cut down on redundant code. Inheritance will let you reuse reliable code that exists in one function block inside a different one. As such, inheritance will help you write code that can be used in many different places but will only require you to keep the code in one central location.

In all types of programming, the concept of inheritance is often abused, especially among inexperienced programmers. Inheritance is not meant to be a way of circumventing access specifiers or using attributes from unrelated function blocks! Instead, you use inheritance when there exists an "is-a" relationship between two function blocks. To demonstrate this, think of a vehicle.

A vehicle will have an engine and wheels; however, there are different types of vehicles. For example, one could have a car or a truck. If we were to convert this concept to a PLC program, we would have a car function block and a truck function block. However, since both of these are vehicles, we can cut down on some coding and add a vehicle function block as well. In this case, the vehicle function block would serve as a generic template known as a base or parent function block. Since we can safely say that a car *is a* vehicle and a truck *is a* vehicle as well, we can inherit the visible attributes from the vehicle function block and cut down on some coding. In this case, since the car and truck function blocks are inheriting from the vehicle block, we can call these child or derived function blocks.

To code this up, we are first going to create a standard function block called VehicleFB. After creating the function block, add two PUBLIC methods called revEngine and spinWheels with a return type of WSTRING. When completed, your function block tree should look like *Figure 5.3*:

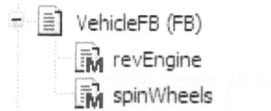

Figure 5.3: VehicleFB tree

For revEngine, add the following line of code:

```
revEngine := "rev";
```

Add the following for spinWheels:

```
spinWheels := "spin";
```

Next, add a new function block named CarFB, but ensure the **Extends** button is checked and VehicleFB is in the box next to it, as in *Figure 5.4*:

Figure 5.4: Inheritance setup

Repeat the same process with a function block named `TruckFB`. After generating the function blocks, enter their code area, and you should see something akin to *Figure 5.5*:

```
FUNCTION_BLOCK TruckFB EXTENDS VehicleFB
VAR_INPUT
END_VAR
VAR_OUTPUT
END_VAR
VAR
END_VAR
```

Figure 5.5: EXTENDS code

The key here is the `EXTENDS` keyword. The `EXTENDS` keyword signals inheritance. In the case of *Figure 5.5*, the code is telling `TruckFB` to inherit from `VehicleFB`.

As counterintuitive as it may seem, we are not going to add any methods to the truck or car function blocks. Instead, we'll add the following variables to the `PLC_PRG` POU file.

```
PROGRAM PLC_PRG
VAR
        car       : CarFB;
        truck     : truckFB;

        carRev    : WSTRING;
        carSpin   : WSTRING;

        truckRev  : WSTRING;
        truckSpin : WSTRING;

END_VAR
```

The following code is for the `PLC_PRG` logic:

```
carRev     := car.revEngine();
carSpin    := car.spinWheels();

truckRev   := truck.revEngine();
truckSpin  := truck.spinWheels();
```

When the code is run, you should get the output in *Figure 5.6*:

Device.Application.PLC_PRG		
Expression	Type	Value
+ ⊘ car	CarFB	
+ ⊘ truck	truckFB	
⊘ carRev	WSTRING	"rev"
⊘ carSpin	WSTRING	"spin"
⊘ truckRev	WSTRING	"rev"
⊘ truckSpin	WSTRING	"spin"

Figure 5.6: Inheritance output

Notice that we are calling the revEngine and spinWheels methods and getting results without implementing them in the car or truck function block. A keen-eyed observer may wonder what's going on. Essentially, since we are inheriting from the vehicle function block, CarFB and TruckFB now have access to those methods. In other words, CarFB and TruckFB inherited their current functionality!

> **Note**
>
> You can add unique methods to the child function blocks. If you do so, these methods will only be visible to that function block and any possible children that it could have. The parent function block will not be able to access PRIVATE or PROTECTED attributes in the child.

Most modern programming languages, including the many PLC programming systems that support the pillar, restrict inheritance to one function block. In other words, you can't inherit from more than one function block at a time. Though you can only extend one function block, you can inherit properties from other function blocks through what is known as the **inheritance chain**.

Essentially, the inheritance chain is like a flow-down system. If you extend a function block that extends another function block, you will be able to access the properties from both. To demonstrate this, add a PUBLIC method to TruckFB called truckBed, set the return type to WSTRING, and add the following code to it:

```
truckBed := "put stuff here";
```

Once you do that, add another function block called BigWheeler and have it inherit from TruckFB. As we did with the last example, we are not going to add any methods to BigWheelerFB.

For this example, we're going to have the following variables in the PLC_PRG POU file:

```
PROGRAM PLC_PRG
VAR
    bigWheel : BigWheelerFB;
    spin     : WSTRING;
    rev      : WSTRING;
    bed      : WSTRING;
END_VAR
```

We're also going to add the following logic to PLC_PRG:

```
spin := bigWheel.spinWheels();
rev  := bigWheel.revEngine();
bed  := bigWheel.truckBed();
```

When the code is executed, you should see what's shown in *Figure 5.7*:

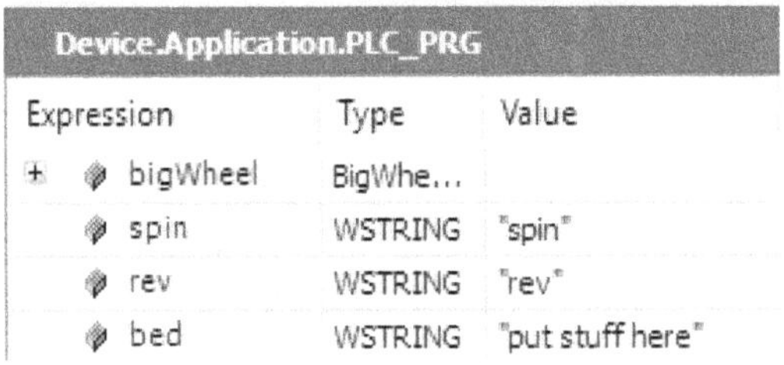

Device.Application.PLC_PRG		
Expression	Type	Value
+ 🖉 bigWheel	BigWhe...	
🖉 spin	WSTRING	"spin"
🖉 rev	WSTRING	"rev"
🖉 bed	WSTRING	"put stuff here"

Figure 5.7: Inheritance chain

Notice that BigWheelerFB was able to call methods from both VehicleFB and the TruckFB. This is the chain at work! In short, *Figure 5.7* shows that if you inherit from a function block that inherits from another, you will be able to use available attributes from both!

> **Note**
>
> In this case, BigWheelerFB is known as a grandchild function block.

Inheritance and the inheritance chain are excellent tools, but they only allow using what was already defined in another function block. However, what if we run into a situation where we need to change the functionality of a method? Is it possible to morph the behavior of the attribute? The answer for some PLC systems is yes!

Polymorphism

Polymorphism is the concept of changing an attribute from one function block to another. For example, the truck we created in the last example is probably going to have more traction when it spins its wheels than the car CarFB models. To give our truck more traction, we're going to change the way the spinWheels method behaves, at least for TruckFB.

There are many ways to implement polymorphism; however, the easiest and probably the most common way to implement the concept is by simply redefining the attribute in the child function block. In this case, all we're going to do is create a spinWheels method in the TruckFB function block. We're going to give this method all the same properties and the exact same name as in *Figure 5.8*:

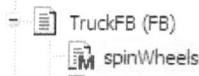

Figure 5.8: TruckFB tree

The only code for this method will be as follows:

```
spinWheels := "lots of traction";
```

For this example, we're going to use the following variables in the PLC_PRG POU file:

```
PROGRAM PLC_PRG
VAR
    truck : TruckFB;
    spin  : WSTRING;
    rev   : WSTRING;
    bed   : WSTRING;
END_VAR
```

As well as the following logic:

```
spin := truck.spinWheels();
rev  := truck.revEngine();
bed  := truck.truckBed();
```

When the code is run, you should see *Figure 5.9*:

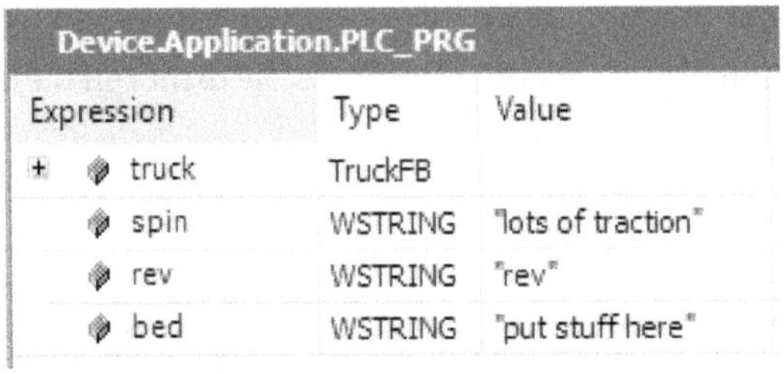

Device.Application.PLC_PRG		
Expression	Type	Value
+ ⬦ truck	TruckFB	
⬦ spin	WSTRING	"lots of traction"
⬦ rev	WSTRING	"rev"
⬦ bed	WSTRING	"put stuff here"

Figure 5.9: Polymorphism in action

Notice that instead of the method returning simply spin, we got lots of traction! Essentially, we overrode the VehicleFB version of the method and gave TruckFB its own unique implementation. You will often use this technique, when available, when the base function block's version of the method is too generic or doesn't reflect the child function block's needs.

> **Note**
>
> This form of polymorphism is called method overriding.

The best way to think of polymorphism is as a means of changing a behavior. In our TruckFB example, we changed our spinWheel method to display the proper traction reading for the function block. In other words, we morphed the behavior of the object.

> **Note**
>
> Some PLC implementations will use a keyword similar to SUPER that will allow you to call the base function block's implementation. That is, in supported systems, you can use a keyword such as SUPER to call the VehicleFB version of spinWheel in the TruckFB function block.

By this point, you should have a solid understanding of the four pillars of OOP. However, one key aspect of OOP that we have yet to explore is the PROTECTED access specifier. As we explored, this is a unique and sometimes confusing access specifier. In the next section, we're going to clear some of the confusion by seeing it in action.

Exploring the **PROTECTED** access specifier

The two main access specifiers that we explored are PUBLIC and PRIVATE. As we explored, the PUBLIC access specifier allows access to any attribute from any POU. On the other hand, the PRIVATE access specifier will restrict a component's visibility and usability to only attributes in the same function block. The PROTECTED access specifier is in between these. If a method is declared as protected, any file in the function block and child function blocks can access it. To demonstrate this, make two function blocks, one called BaseFB and one called ChildFB. When setting these function blocks up, ensure that ChildFB EXTENDS BaseFB.

Once you have the function blocks set up, add a method called testMethod to BaseFB. Be sure to set the access specifier to PROTECTED and the return type to INT, as in *Figure 5.10*:

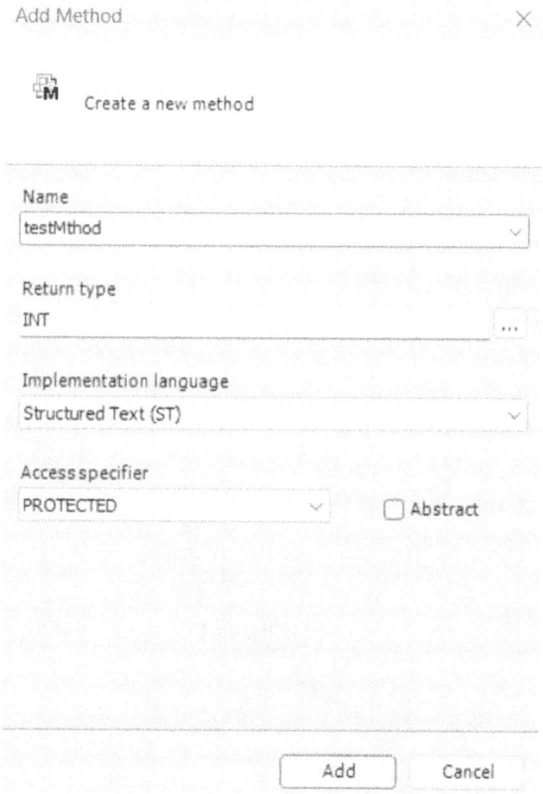

Add Method   ✕

M   Create a new method

Name

testMthod   ⌄

Return type

INT   ...

Implementation language

Structured Text (ST)   ⌄

Access specifier

PROTECTED   ⌄   ☐ Abstract

Add   Cancel

Figure 5.10: The testMethod setup

The only code for this function block will be in `testMethod`, and it should match the following:

```
testMethod := 10;
```

In `ChildFB`, add a `PUBLIC` method with an `INT` return type called `getVal`, and add the following code to it:

```
getVal := testMethod();
```

Finally, in the `PLC_PRG` file, set up the following variables and code:

```
PROGRAM PLC_PRG
VAR
    BF    : BaseFB;
    CF    : ChildFB;
    val1 : INT;
    val2 : INT;
END_VAR
```

In this case, `BF` and `CF` are references to `BaseFB` and `ChildFB`, respectively, while `val1` and `val2` will hold return values.

The core logic for this program is as follows:

```
val1 := BF.testMethod();
val2 := CF.getVal();
```

Once you have all this set up, run the code. You should be met with what's shown in *Figure 5.11*:

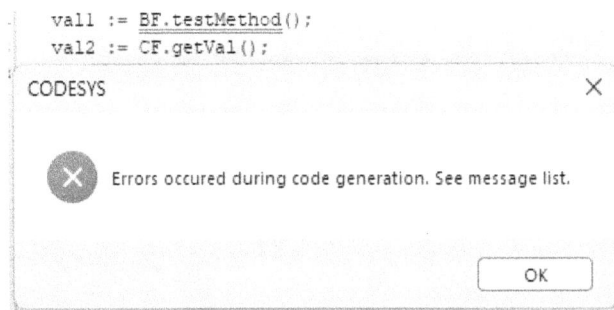

Figure 5.11: Error

What the error is essentially saying is that you're trying to access a PROTECTED attribute from a file that does not inherit from BaseFB. To use testMethod, comment out the first line of the core logic in the PLC_PRG file and rerun the program.

Device.Application.PLC_PRG					
Expression	Type	Value	Prepar...	Address	Comm...
+ ⬦ BF	BaseFB				
+ ⬦ CF	ChildFB				
⬦ val1	INT	0			
⬦ val2	INT	10			

Figure 5.12: Output

In this case, we called a PUBLIC method in the derived function block to invoke the PROTECTED method in the parent block, and we got the expected value.

So far, we have touched upon all four of the pillars of OOP, as well as the PROTECTED access speci-fier. As you might guess, inheritance and polymorphism are very rich and complex topics, and we have merely touched the surface. For now, we are going to start exploring other concepts that are not necessarily a part of the four pillars but are still powerful OOP concepts. One very common principle that needs to be explored is composition and how it contrasts with inheritance.

Inheritance versus composition

Inheritance is a very important concept and is, without a doubt, a great way to recycle code under the right circumstances. However, many new or inexperienced programmers will often use in-heritance as a means of importing code. This is a bad practice because instead of producing clean, organized code, they produce jumbled-up code that has no true relationships between function blocks. As we saw with inheritance, when developing object-oriented code, it is very important to consider the relationships between function blocks. One very common way to implement object-oriented relationships is with a concept known as **composition**.

When to use composition

For many inexperienced, traditional programmers, composition is often an ill-understood but, ironically, often-used concept. The concept of composition can be summarized as assembling things. In other words, composition is where you include object references from one function block in another to essentially build something. Where inheritance utilizes an "is-a" relationship between function blocks, composition uses a *"has-a"* relationship. In other words, we can summarize when to use composition or inheritance with the following:

- **Composition**: If something *has* something else
- **Inheritance**: If something *is* something else

Essentially, the best way to choose which technique to use is to ask yourself whether the function block you're working on *is something* (such as *if a cat is a feline*) or *if something has something* (such as *does a car have an engine?*).

The core idea behind composition is that we are building function blocks from other function blocks. Think of our vehicle example again. If we were programming a car, as we did, we would need an engine, wheels, and so on. If we were using inheritance and wanted to change the engine of our car or truck, we would have to change what we were inheriting from. That is, in our old program, we would have to create a new VehicleFB to inherit from or use some other more complicated technique, such as polymorphism. However, with composition, we can simply change the reference to the engine, and we would be good to go.

Many developers will often favor composition over inheritance for many reasons. One of the core reasons developers usually opt to use a *"has-a"* relationship and composition over an *"is-a"* relationship and inheritance is that inheritance produces more tightly coupled code. This means that a change in the base function block can have a ripple effect that unintentionally changes the behavior of the child function blocks. Since composition is assembling complex objects from other objects, a change to one of the component classes does not necessarily mean that the changes will have the same ripple effect.

Composition in practice

To demonstrate composition, we're going to create a Car function block. To do this, let's analyze some components of a car. A car is composed of many parts that have functionality, such as the following:

- Engine: rev
- Transmission: shift
- Brakes: stop

For this tutorial, we are going to create a Car function block. By the end of the tutorial, this block is going to be composed of an Engine, Transmission, and Brakes function block. Therefore, the project tree for this tutorial should look like *Figure 5.13*:

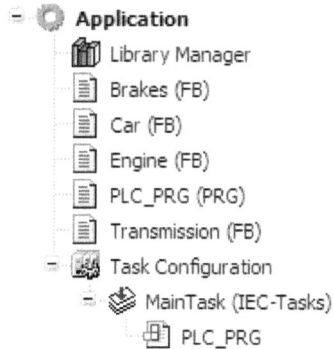

Figure 5.13: Car project structure

The function blocks will be straightforward, as each block will have one method. Each method will only have one line of code and no variables. All the methods will have a return type of WSTRING, and they will be set to PUBLIC. The code for each function block's method along with correct name are as follows:

- The Brakes function block's stop method:

```
stop := "stop";
```

- The Transmission function block's shift method:

```
shift := "shift";
```

- The Engine function block's rev method:

```
rev := "rev";
```

The Car function block will be a bit different. This function block will not have any methods, and the only code will be variables that reference the other function blocks, like so:

```
FUNCTION_BLOCK Car
VAR_INPUT
END_VAR
VAR_OUTPUT
END_VAR
```

```
VAR
    brakes          : Brakes;
    engine          : Engine;
    transmission : Transmission;
END_VAR
```

The final piece of the tutorial that needs to be implemented is the code in the PLC_PRG file, which will look like the following:

```
PROGRAM PLC_PRG
VAR
    car    : Car;
    drive : WSTRING;
    stop   : WSTRING;
    shift : WSTRING;
END_VAR
```

The logic will look like the following:

```
drive := car.engine.rev();
shift := car.transmission.shift();
stop   := car.brakes.stop();
```

When the code is run, you should be met with the output in the following screenshot:

Expression	Type	Value
+ ● car	Car	
● drive	WSTRING	"rev"
● stop	WSTRING	"stop"
● shift	WSTRING	"shift"

Figure 5.14: Car output

As you can see, we accessed the methods from the three component blocks with the Car function block. Essentially, what we did was encapsulate three reference variables in the Car block. If you look at the lines of code, we referenced the Car function block variable, which allowed us to access the internal references to the Engine, Brakes, and Transmission blocks.

In this example, we essentially built a car. A car is not an engine, transmission, or brakes; as such, inheritance was not appropriate to use here. However, a car *has* an engine, transmission, and brakes. This means that to create a car as we did, it was more appropriate to use composition. In this case, we are still able to recycle our function blocks without becoming totally dependent on any given one. In essence, we can remove the old engine and replace it with a high-performance one without completely overhauling the code as we would with inheritance.

Composition is probably one of the most important concepts in OOP. Though the composition technique is not an official pillar of OOP, you will find yourself using it much more often than inheritance, which is why it is sometimes referred to as the fifth pillar.

In PLC programming, sometimes using a function block can actually hamper us. Sometimes, when we model an object, all we need are a few very broad definitions. In the following section, we're going to explore how to leverage generic implementations using interfaces.

Examining interfaces

In the real world, we often work from templates. For example, engineers will not overhaul the design of a car. In reality, certain patterns are always followed. For example, all cars have four wheels, brakes, an engine, a steering wheel, and so on. However, what will change between cars is the way the parts work. In other words, the overall functionality is the same, but the way in which the components operate will vary. For example, a car will always have four wheels, but the size of the wheels and the type of rims will vary from car to car. When sketching out a car, an engineer may draw a few circles as a placeholder to represent wheels but won't decide on a type until the car moves into production. In programming, we can do something similar using what's called an interface.

If you read a textbook on traditional, general-purpose programming languages such as Java or C#, you will see that interfaces are often referred to as contracts. When you opt to use an interface, you are telling the programming system that you agree to, at the very least, implement all the methods prototyped in the interface. However, in my opinion, this is a little confusing.

Generally, when I describe an interface to a new programmer, I usually describe it as a model for something. For example, if we are building an airplane, we will need certain things, such as wings, an engine, and a cockpit, regardless of whether we are building a prop plane or an F-35 fighter jet. Obviously, for each type of plane, these parts are going to be different. When we implement an interface, we are telling our function block that we are going to use the methods that are declared in the interface, but those methods may have different implementations.

To demonstrate how an interface works, let's implement one. For this example, let's pretend we are making an airplane. As was stated before, regardless of the type of plane, each will have a cockpit, engine, and wings. To create an interface, we are going to right-click **Application** in the project tree, click **Add Object,** and then select **Interface**. For this project, name the interface plane.

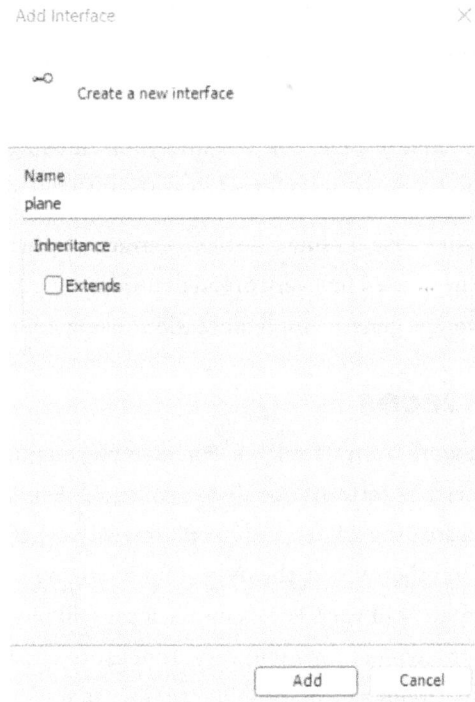

Figure 5.15: Interface creation wizard

The preceding screenshot is the wizard window that you will see when you follow the steps correctly. After creating the interface, we can add methods to it. Adding methods to an interface is the same procedure as adding methods to a function block. Therefore, add an engine, cockpit, and wings method to the interface. For this example, the interfaces will all have a return type of WSTRING, with the exception of engine, which will have a return type of INT.

The cockpit and engine method can be left as is; however, add an argument to the engine method, as in the following snippet:

```
METHOD engine : INT
VAR_INPUT
    rpms : INT;
END_VAR
```

Next, we are going to implement the interface in two different function blocks called F35 and Prop. While spinning up the POUs, click the **Implements** box and select the plane interface, as in *Figure 5.16*:

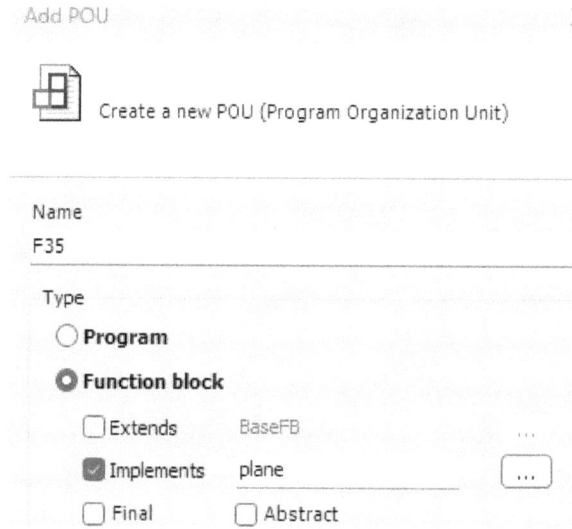

Figure 5.16: Implement interface

After you've completed setting everything up, you should have a structure similar to the following screenshot:

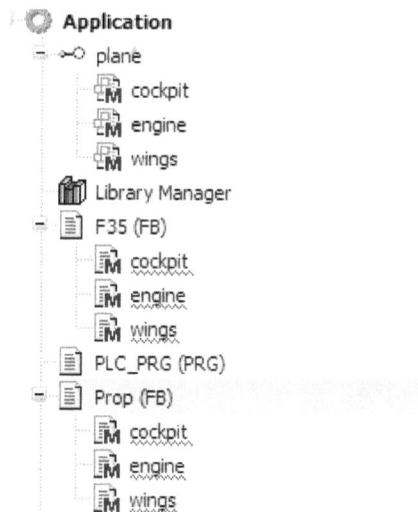

Figure 5.17: Completed structure

Note

Notice that the methods are automatically generated under the function blocks. Since an interface is a contract or model, you will need to have, at a minimum, the method defined in the function block that implements the interface, even if it is empty. Luckily, CODESYS does this for us automatically; however, not every programming environment will.

The F35 block methods will consist of the following code:

- F35 cockpit method:

  ```
  cockpit := "1 seat";
  ```

- F35 engine method:

  ```
  engine := rpms * 1000;
  ```

- F35 wings method:

  ```
  wings := "2";
  ```

Important note

For the following and the preceding examples, write this code in the declared function block methods.

Once the F35 block is squared away, set the prop methods to the following:

- Prop cockpit method:

  ```
  cockpit := "2 seats";
  ```

- Prop engine method:

  ```
  engine := rpms * 100;
  ```

- Prop wings method:

  ```
  wings := "4" ;
  ```

As usual, once those are completed, we're going to make two reference variables and a series of variables to hold the output in the PLC_PRG POU file:

```
PROGRAM PLC_PRG
VAR
    f35         : F35;
    prop_plane : prop;

    f35_cockpit : WSTRING;
    f35_engine  : INT;
    f35_wings   : WSTRING;

    prop_cockpit : WSTRING;
    prop_engine  : INT;
    prop_wings   : WSTRING;
END_VAR
```

This is the logic that will call the functions:

```
f35_cockpit := f35.cockpit();
f35_engine  := f35.engine(10);
f35_wings   := f35.wings();

prop_cockpit := prop_plane.cockpit();
prop_engine  := prop_plane.engine(5);
prop_wings   := prop_plane.wings();
```

When the code is run, you should see the following output:

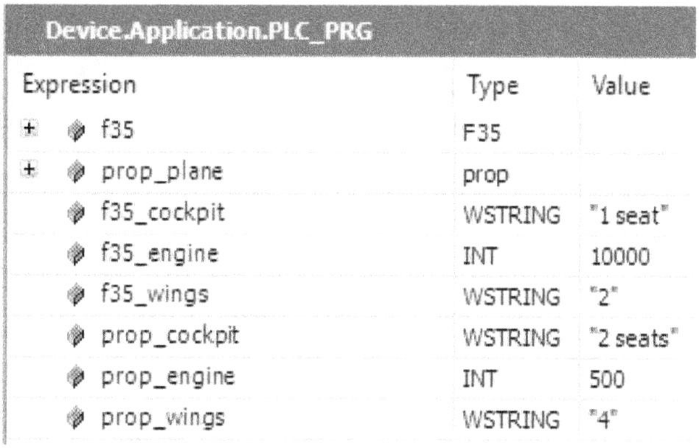

Device.Application.PLC_PRG		
Expression	Type	Value
⊞ ◈ f35	F35	
⊞ ◈ prop_plane	prop	
◈ f35_cockpit	WSTRING	"1 seat"
◈ f35_engine	INT	10000
◈ f35_wings	WSTRING	"2"
◈ prop_cockpit	WSTRING	"2 seats"
◈ prop_engine	INT	500
◈ prop_wings	WSTRING	"4"

Figure 5.18: Interface program output

This example shows that by using an interface, we can automatically import methods and, more importantly, model something. As we can see in the example, the methods have the same name but have different implementations. This means that, though we are building two different types of planes, we are using similar components; however, the ways the components work are different.

> **Note**
>
> You can only implement a method's name, arguments, and return type in an interface; however, you cannot implement executable logic in the interface's method(s). Executable logic can only be implemented in the function block's version of the method.

Another interesting aspect of interfaces is that since they are implemented and not extended, you can use multiple interfaces in a single function block. In other words, there is a difference between implementing an interface and inheriting from another function block. Since you can implement multiple interfaces, you can combine them to model different things.

New or inexperienced programmers usually do not see the benefits of using interfaces. At first glance, what we did may simply seem like a roundabout way of declaring methods. However, a well-written program uses interfaces, and there is an old rule that says that you should "code to an interface."

Coding to an interface will allow you to create more flexible code. So, if you need to add or remove more parts, you can do so without breaking implementation elsewhere. It is also a good way to ensure that projects written by a team are all consistent with method names, arguments, and return types, and are implementing the correct functionality for their sections.

Now that we have a solid foundation in the art of OOP, we can move on to creating our final project.

Final project: Creating a simulated assembly line

Our final project will consist of a production line. The line will consist of a function block called `controller` and another function block called `Line` that will have the following methods:

- `turnMotorsOn`
- `homeMotors`
- `startMotors`

For this project, we're going to assume that a controller has a line; therefore, we'll use composition. In this example, the `Controller` POU is going to act as a façade function block to control the assembly line. The façade nature of this function block will be similar to façade concepts that were explored in *Chapter 3*. In short, the `Controller` function block will provide a level of abstraction for the `Line` block.

The first thing we need to do is create the mentioned methods in the `Line` function block with an access specifier of `PUBLIC` and a return type of `BOOL`.

The `homeMotor` method will consist of the following:

```
outputs.motorState := "motors homed";
```

The `startMotors` method will consist of the following:

```
outputs.startMotors := TRUE;
```

Finally, the `turnMotorsOn` method will consist of the following:

```
outputs.MotorsOn := TRUE;
```

Once that is done, create a GVL called `outputs` and set the following variables:

```
{attribute 'qualified_only'}
VAR_GLOBAL
    motorState  : WSTRING;
    startMotors : BOOL;
    MotorsOn    : BOOL;
END_VAR
```

After the variables are implemented, we can set up the `Controller` function block. The controller function block will, for this example, enforce abstraction by providing a simple, easy-to-use interface for us to start the assembly line.

This function block will consist of only a single method called `start` and a reference variable:

```
METHOD PUBLIC start : BOOL
VAR_INPUT
END_VAR
VAR
    line : Line;
END_VAR
```

The start method's body will be like the following:

```
line.turnMotorsOn();
line.homeMotors();
line.startMotors();
```

Finally, we will start the assembly line in the PLC_PRG file by using the following code:

```
PROGRAM PLC_PRG
VAR
    controller : Controller;
END_VAR
```

The method call to simulate starting the assembly line is achieved with the following code:

```
controller.start();
```

This will result in the following:

Device.Application.outputs		
Expression	Type	Value
motorState	WSTRING	"motors homed"
startMotors	BOOL	TRUE
MotorsOn	BOOL	TRUE

Figure 5.19: Assembly line output

As can be seen in the preceding screenshot of the GVL outputs, we fired three methods with one method call. In reality, you would have multiple methods in the Controller block that would stop the line, pause the line, and so on. It is recommended that you expand the function block using the principles we explored to add the extra functionality.

Summary

In this chapter, we explored the more advanced features of OOP. OOP is a very powerful and new concept in the world of PLC programming. When fully embraced and mastered, your complex project can become greatly simplified. As you become more familiar with OOP and the associated pillars, you will have no redundant code and will have a very maintainable codebase. As you master these concepts, you will be able to do more with less code.

Using OOP is a best practice, but following OOP principles does not guarantee you have quality code. To have quality code, certain rules have to be followed. In the next chapter, we're going to learn some best practices that can be used to produce high-quality codebases!

Questions

1. List the four pillars of OOP.

2. Is there a limit on the number of interfaces you can implement?

3. How many function blocks can you inherit from?

4. What is the difference between `PRIVATE` and `PUBLIC`?

5. What is the `PROTECTED` access specifier?

6. Is multiple inheritance allowed in IEC 61131-3?

7. When would you use inheritance?

8. When would you use composition?

Further reading

Have a look at the following resources to further your knowledge:

*   *CODESYS interface*: `https://content.helpme-codesys.com/en/CODESYS%20 Development%20System/_cds_implementing_interface.html`

*   *CODESYS Object-Oriented Programming*: `https://content.helpme-codesys.com/en/ CODESYS%20Development%20System/_cds_f_object_oriented_programming.html`

6

Best Practices for Writing Incredible Code

Following the theme of this book, there is more to a well-engineered program than simply producing a working codebase. Writing a quality codebase is as much an art as it is a science. New developers often try to impress their bosses and teammates by producing very large and complex codebases, and many non-technical managers will often see this as a sign of in-depth knowledge. However, the exact opposite is true. A quality codebase is easy to follow, as simple as possible, and as well-documented as possible.

There are a lot of gotchas that can bury a codebase. A quality, long-lasting codebase requires much more than working code to last the test of time; however, many codebases suffer from issues that will drastically reduce their lifespan. There are ways to prevent code from being relegated to the cyber-heap before its time. In this chapter, we're going to explore some concepts that can be used to ensure that your codebases last the test of time. This chapter will explore the following topics:

- What is technical debt?
- Understanding naming conventions
- Exploring code documentation
- Understanding and eliminating dead code
- Keeping it simple
- What to look for in a code review
- Things to avoid in software engineering

To wrap things up, we're going to apply our new skills and code review a PLC program!

Technical requirements

This chapter is dedicated to writing quality code and code reviews. The goal of this chapter is to present best practices and ways to eliminate technical debt. The concepts that will be presented here will be generic to all programming languages. In other words, the concepts that we will explore are language-agnostic. These principles can even be applied to graphical languages such as Ladder Logic. So, no technology is needed to follow along.

What is technical debt?

To understand why best practices are important, we first need to understand what technical debt is. This is a concept that is often overlooked by traditional developers and automation programmers alike. Technical debt occurs from quick fixes, shortcuts, documentation gaps, and so on. In automation, a big cause of technical debt is quick fixes. Most quick fixes are not fully fleshed out, and though they will get you through the day, the fix will usually come with consequences.

Technical debt is like financial debt. If you take out a loan to pay for something, you'll end up having to pay the initial cost for the item (the loan) and interest that will usually compound. In the long run, you'll end up paying several times what the item would normally cost. Technical debt is no different; however, the currency of technical debt is time and code complexity. If you're honest with yourself, you'll realize that a vast majority of your quick fixes are sloppy at best and not well thought out, especially those that you implement right before the end of your shift. When the next upgrade comes, those quick fixes can cause a lot of problems, as they can often be hard to understand, and they usually add a lot of complexity because they were designed as a quick workaround for a problem. As a result, you'll end up spending a lot of extra time fixing or upgrading whatever it is you're working on.

The currency of technical debt is time, as mentioned previously. Each bit of technical debt will equate to extra time that needs to be dedicated not only to fixing the original problem but also reverting the patch. To lower the technical debt of a program, you have to repair not only the original problem but the quick fix as well. This means that technical debt will frustrate customers as they will have extra downtime and typically have to spend extra money to fix the problems that were caused by technical debt. It will also cost the organization you work with, as your time will be dedicated to fixing workarounds, and most importantly, it will cost you valuable time that you could dedicate to more productive endeavors.

In terms of automation, the major culprit of technical debt stems from modifying code as a workaround for broken or misbehaving physical components. Using code as a workaround can easily render a codebase unmodifiable, which, in turn, means that it will have to eventually be overhauled when a modification is inevitably needed. It is never okay to modify code as a workaround for physical components. These actions will increase your technical debt as you will have to, at the very minimum, restore the codebase to its original form when the part is fixed. At worst, you will have to fix the part, quick fix, and make necessary modifications to the codebase. In other words, it's all unnecessary added time.

Creating workarounds for misbehaving parts is just one form of technical debt. The remainder of this chapter is going to be dedicated to exploring other forms of technical debt and best practices that we can use to improve code. The first set of best practices we're going to explore is naming conventions.

Understanding naming conventions

Naming program attributes such as variables, functions, function blocks, methods, and more is an art. In the old days of computer programming and even in the modern automation world, it is not uncommon to find variable names such as x and y. Thirty years ago, there was logic behind these names. In the old days of programming, computing devices had limited memory and storage. This means that a name such as motorSpeed could be very taxing on a system. However, modern computing devices, such as quality PLCs, have come a long way and can easily handle a reflective name.

Before we start exploring naming conventions, it is very important to understand that these best practices should be thought of as guidelines. Though naming conventions are very important there are times when they will be ignored in the name of complexity, teaching, documentation, and so on. In other words, context will impact naming. With that, there is a lot to naming an attribute. The first set of rules we're going to look at is casing conventions.

Casing conventions

Believe it or not, one of the most important attributes of a name is the way it is spelled. Most attribute names will be a composition of multiple words. To properly format these complex names, most programmers will use one of the following casing conventions to spell the names of the program components:

- **Camel casing**: Arguably, the most common casing convention is camel casing. For this convention, the first word is spelled in all lowercase letters, and the first letter of each subsequent word is capitalized. This type of casing is commonly used for variables and methods. An example is motorSpeed.

- **Pascal casing**: In terms of Pascal casing, each word in an attribute name is spelled with an uppercase letter. This format is commonly used for things such as function blocks. An example is `MotorSpeed`.

- **Snake casing**: Snake casing is probably the least used in codebases. It is mostly used for naming directories, projects, environment variables, and so on. You will commonly see snake casing when working with operating systems such as Linux. For snake casing, each word in the name is separated by an underscore. An example is `motor_speed`.

As was seen throughout past examples in this book, we're mostly going to stick with camel casing; however, you can use what you want. The convention you use will mostly depend on your style and the style guidelines of your organization. That being said, if you opt to use a convention for an attribute, be sure to keep it consistent throughout the program, as this will make the program easier to read and understand.

Understanding how to spell an attribute's name is only half the battle. The other half of the battle is understanding how to properly name an attribute. Therefore, in the next section, we're going to explore naming variables.

Proper variable names

A variable represents an object of some type, such as a person, value, or general thing. This means that a variable's name should typically be a noun or noun-based statement, but what matters more is how descriptive the name is. For example, suppose you're working with a variable that represents the weight of a bag. A logical name for this variable might be `bagWeight`. A bad name for this variable would be something like `weight` or `wgt`; `weight` is not very descriptive, as there could be many weights in a program, and `wgt` is ambiguous at best. In other words, when naming a variable, you want to clearly identify what the variable is and what it represents. By doing this, codebase readers can quickly identify what the variables do, which will cut down on the time it takes to troubleshoot the codebase. In other words, it will lower the technical debt. A caveat to this is with teaching or documentation applications. When presenting information, it is often okay to use a short generic name to get the point across. Though a quality program starts with properly named variables, functions and methods also need proper names.

Properly naming methods and functions

Put simply, a method or function should usually be a verb name; however, this is one rule that has some wiggle room, especially in controls/embedded programming. Having a verb is considered a best practice, but as we've seen so far, what's more important is that the method or function name be descriptive. Suppose we're working on a function/method that will turn on a robot. A logical name would be startRobot(). Here, our verb is start and our statement is *start robot*. This name clearly reflects what the attribute does and what system it targets.

An example, of a non-verb name would be something like fanOn() which reads more like a noun or state description. Regardless, we have a short, intuitive name that clearly demonstrates the attribute's purpose, turning on a fan. If we were to change the name to something like turnOnFan(), we would have a longer name that would contribute little outside of making it harder to type.

Much like the other program components, function blocks also must be named well.

Naming function blocks

Function blocks represent things; this means that, like variables, they too should have a noun name. Unlike methods, this rule is a little stricter as you don't want to give a function block a verb name. It is important to remember that a function block is a blueprint for a thing and having a verb name would work against its intended nature.

A lot of organizations will either add an FB prefix or a postfix to the name. This prefix or postfix clearly signals what the attribute is. Whether you end up using a prefix or a postfix will depend on your codebase and organization. Though there are no hard and fast rules for which to use, it is usually the norm to use a postfix if at all. With that, most function block names use Pascal casing. A common name for a function block might be something like CarEngineFB.

There are other rules for other attributes such as enums, properties, and so on; however, the rules for those can be found easily on the internet. This section was just to give you a taste of what names for common attributes should be, and that proper naming lowers your codebase's overall technical debt and helps with another debt lowering technique, documentation.

Exploring code documentation

Nothing makes programmers groan louder than being told they must write documentation. Unfortunately, documentation is directly linked to the longevity of a codebase. There are many ways to document a codebase. The three most common methods are code comments, external documentation, and the code itself. The first type of code documentation we're going to look at is self-documenting code.

Utilizing self-documenting code

A well-written program will provide its own documentation. This means that if written properly, a program should be easy to follow with minimal aid. To demonstrate this, consider the following pseudocode:

```
Input temperature
overHeating = 100
targetTemp = 90
If temperature > overHeating then
    ovenFan = on
Elseif temperature <= targetTemp then
    OvenFan = off
```

This program is very clear about how it works, and the variables' meaning in the program is easily identifiable. This is essentially what self-documenting code is. Contrast this program with the following:

```
x = input
If x > 100 Then
    Fan = on
Elseif x < 90 then
    Fan = off
```

This version of the program has very little context as to what's being measured and what is being turned off or on. There is no context for what x is because the variable has an illogical name. There is also no context as to what fan is being turned on, assuming there is more than one fan in the system. Most importantly, the hardcoded values are very ambiguous. In the second program, there is no meaning behind the values. That is, what does 100 actually mean in the context of the program? If we were to look at the first program, we can easily see that 100 degrees is assigned to a variable called overHeating which tells us in the context of the program that anything above 100 degrees is considered overheating. In contrast, the targetTemp variable clearly states that 90 degrees is the ideal temperature for the device, and in the context of the program, there is no need for the fan to turn on. Overall, if you were asked to troubleshoot the second version of the PLC code in the middle of summer in a very hot shop, chances are you would get frustrated very quickly trying to figure out what the code is doing.

When writing code, you want your program to have a very refined flow, meaning you want the paths in your program to be neatly defined and easy to follow. You also want to name your attributes in such a way that you only need a quick glance to determine what the code is doing and how it works. With that, there is a trick that you can use to help make changes and add context at the same time. That trick is coding to variables!

Coding to variables

It is typically a good idea to avoid using hardcoded values in your program. Any time you need to declare a value, it is usually a good idea to create a variable and assign the value to that, especially if that value is used in more than one location. Consider the following code:

```
Input weight
If weight > 100 then
    Display too heavy
Elif weight == 90 then
    Display optimal weight
Else
    Display too light
    Wait 5 minutes
If weight > 100 Then
    Send to reject line
Else
Send to accept line
```

By just looking at this snippet, is there any context for what 100 and 90 are? Basing our assumption on the surrounding code, we can assume they are weights of some kind; however, what are they weights for? From this code, it's almost impossible to tell.

Also, we have the 100 value used in multiple locations. If you look at the codebase, we have two sets of control statements that perform a check on a value that we can assume is some type of weight. Since that value is in two different locations, if that threshold ever changes, we have to change it in multiple places. This means this program can be prone to errors because if we ever must change the value from 100 to, let's say, 1000, we will have to modify the value in multiple places. In a nutshell, we can easily miss the value in one or more locations. By missing a change for a value as critical as the one in the example, the machine will behave in erroneous and even possibly dangerous ways. A third type of error could occur from mistyping the value. When developers are in a pinch or have worked a long shift, accidentally typing 10 or 1000 instead of 100 is very common; however, it is a mistake that could have potentially life-threatening consequences.

The best way to ensure we don't have any programmatic or contextual issues is to set the value in a variable. By using a variable, we have a single point of truth for the value. This means that if we ever need to change the value, we only must do it in one place. To demonstrate this, let's look at the following pseudocode:

```
Input weight
overWeightCementBag = 100
idealWeightCementBag = 90
If weight > overWeightCementBag then
    Display too heavy
Elif weight == idealWeightCementBag then
    Display optimal weight
Elseif weight < idealWeightCementBag then
    Display too light
    Wait 5 minutes
If weight > overWeightCementBag Then
    Send to reject line
Else
Send to accept line
```

Looking at this version of the program, we can see that the 100 value represents an overweight or overfilled bag of cement, while 90 represents an ideally filled bag. If any of these values ever need to be changed, we only need to change one line of code! Therefore, by coding to a variable, we can quickly and accurately change values with minimal effort and cut down on the codebase's overall technical debt.

Self-documenting code is just one way of documenting a codebase. Though it is true that a well-written codebase should be able to guide a developer, sometimes we need a little more. In the next section, we're going to look at writing proper code comments.

Code commenting

Another form of documentation that will help your codebase last the test of time is commenting. There are many PLC and traditional codebases that have no comments. These programs will usually take extra time for developers new to the codebase to understand, and it can ultimately mean that the codebase will become unmaintainable after a while.

Logically named program components can only go so far. Many times, extra context must be given to explain the underlying logic for the code. This is where comments come into play. Comments are very important for software development; however, commenting is a bit of an art that must be practiced. To start, let's look at what a good comment is!

Good comments

Comments are notes in the source code for yourself and other programmers. Comments should be short, simple, to the point, and, above all else, provide context. A comment's job is to provide the reader with a quick summary of what a block of code does. For example:

```
//this code turns the conveyor motor on and off
IF speed >= motorSpeedCutOff THEN
    motorOff := TRUE;
ELSIF speed < motorSpeedCutOff THEN
    motorOff := FALSE;
END_IF
```

The comment in this code block is one line and directly conveys what the code does. It can be argued that using more descriptive names, such as `conveyorMotorSpeedCutOff`, would serve the same purpose, and it could. Regardless, the comment provides a clear and concise description of the purpose of the code, and it removes any mental gymnastics that a reader may have to perform to understand its purpose. Though code comments can reduce technical debt and add context to code, they can be somewhat detrimental if not implemented properly. In the next section, we're going to explore what a bad comment looks like.

Bad comments

Just because your code has a lot of comments doesn't mean that it is well documented. Too many comments can be as detrimental to a codebase as too few. Too many comments can clutter up the source code and overwhelm the reader. Here is an example of a poorly documented program:

```
//this code turns the conveyor motor on and off
IF speed >= motorSpeedCutOff THEN
    //when the motor speed is greater than the cut off speed
    turn the motor off
    motorOff := TRUE;
ELSIF speed < motorSpeedCutOff THEN
    //when the motor speed is less than the motor cut off
    speed turn the motor on
    motorOff := FALSE;
END_IF
```

The two comments in the IF statements are completely unnecessary. The self-documenting nature of the variable names provides enough information to the code reader as to what the IF statements do. The explanatory comments do little more than bloat the code file and possibly confuse the code reader. In a situation like this, it is best to remove them. As can be seen, there is a bit of a balancing act between having just enough code comments to be productive while not overloading the code reader. Proper code commenting takes practice, and many organizations will have guidelines on code commenting.

> **Note**
>
> One exception to this rule is education. Many times, code will be heavily commented for technology documentation purposes or classroom demonstrations.

Code documentation is one crucial way of lowering technical debt and creating quality codebases. Regardless, one codebase killer that often pops up in automation programs is dead code!

Understanding and eliminating dead code

Dead code is, without a doubt, the cancer of codebases. It adds nothing to the program but, like a vampiric leech, consumes vital resources from the PLC. So, what is dead code? Dead code can best be summarized as code that adds nothing to the program. This could be just adding two numbers together for no reason or having a function that runs a bunch of lines of code but contributes nothing to the overall success of a program. An example of dead code can be viewed in the following pseudocode:

```
Input age
Input temperature
age = age
    If temperature > 100 Then
        Turn on fan
Endif
```

The first line takes an input for age, and later, the value of age is reassigned to age. As can be seen, the reassignment does nothing for the program; the operation is essentially useless. If we removed the variable, the program would in no way be impacted.

If you have code that runs but contributes nothing, you can run into the following issues:

- Raise overall complexity of the program (increase technical debt)
- Introduce security risks
- Use valuable resources such as storage, memory, execution time, and so on unnecessarily
- Make it difficult for new developers to understand the code

There is another concept called **unreachable code** that is sometimes confused with dead code. Unreachable code is code that exists in a codebase but is never executed. This could be functions that are implemented but never called, branches in a control statement that can never be reached, or any other type of code that will never run. Much like dead code, unreachable code is a cancer that, over time, will degrade the quality of the codebase. Many development systems will typically alert you to the presence of both unreachable code and dead code. If you see a warning about either of these during the compilation process, it is important that you fix these issues as quickly as possible.

Dead code and unreachable code often contribute to what is called **code rot**. Code rot is essentially the decline in quality of a codebase in relation to complexity, support, maintenance, and so on. This degradation is caused by several issues, such as the aforementioned issues, outdated dependencies, poor maintenance, or a general lack of documentation. One way to help fight code rot is to keep it simple!

Keeping it simple

In software engineering and engineering in general, there is a principle that should always be followed. This principle is **Keep It Simple, Stupid (KISS)**. As the name suggests, the KISS methodology encourages engineers to avoid over-complexity. This is often a concept that is counterintuitive for many new programmers. As stated earlier, it is a common trait for many new team members to want to show off their chops and make a good impression, especially when their team leadership is metric-oriented and sees complexity as something that should be admired. However, this can drastically overcomplicate a PLC program and send it to the cyber trash heap well before its due time.

An example of an anti-KISS methodology is the following:

```
userAge : INT
age : INT
Input age
userAge = age
```

```
If userAge <= 21 Then
    User not allowed
Else
    Allow user
```

In this case, the userAge = age line is completely unnecessary. This line does contribute to the overall operation of the program and does run, so it's not dead or unreachable code, but it is overall needless in terms of engineering. It would make much more sense to just use the age variable instead of making userAge.

This is just one very simple example of where the KISS methodology mindset should be applied. In real life, KISS violations may not be that obvious. The biggest violator of KISS usually comes in the form of over-engineering. Common things to look for that can increase the complexity of code are as follows:

- Using unnecessary libraries
- Bloating the codebase
- Adding unnecessary parts

These are just a few points that can increase complexity; however, there are many more. As a good developer, you should be able to spot these and eliminate them in the almighty code review.

What to look for in a code review

Most quality organizations will have their employees conduct what's called a code review. A **code review** is like a peer review for code. The goal of the exercise is to ensure that your peer's code makes sense, works, and has no major flaws that could impact the overall quality of the codebase. It is typically a good idea for any organization to have its developers implement code reviews. Code reviews will help developers grow and ensure that best practices are followed and the overall technical debt is at a minimum.

When you're performing a code review, there are a few things you want to keep an eye out for:

- **Poor naming conventions**: The first thing that I like to look at are names. When conducting a code review, I want to ensure that attributes are named properly and follow the correct casing and naming schemas. For instances where the program uses camel casing for variable names, you want to ensure that the programmer didn't use snake casing for any of them. Outside of that, you want to ensure that the programmer is using the proper pre- and post-suffixes to denote attributes. If the program is using FB at the end of each function block name, forgetting to put the suffix at the end of a function block can confuse future programmers.

- **Logical flow and structure**: Ideally, you want to check the flow of a program to ensure that it makes sense. Evaluating your own logic is a lot like proofreading your own papers, and it helps to have a second or third pair of eyes on it. When you are reviewing someone else's code, you want to ensure that their program is structured well, has no random jumps, flows neatly from top to bottom, and makes sense overall. If something doesn't sit right with you, it probably won't sit right with others in the future.

- **Dead and unreachable code**: You want to keep a keen eye out for dead and unreachable code. As stated before, dead and unreachable code can and usually will degrade a code project. If the compiler does not pick up dead or unreachable code, it's your job as a reviewer to do so! You want to pay particular attention to project files that are not used, such as function blocks, methods, GVLs, structs, enums, and functions.

> **Note**
>
> Some standards require default branches even if they don't always run or run very seldom. This is not true dead or unreachable code because it can still run under certain conditions.

- **Poor comments/documentation**: Comments can bloat a codebase very easily. You want to ensure that everything is logically documented. That is, what needs to be commented is, and unnecessary comments are either shortened or removed. When it comes to code comments, it's better to be over-documented than under-documented, but in all, it is a balancing act. You also want to pay attention to the self-documentation aspects of the codebase. You want to ensure that the program itself reads like a storybook. That is, you can look at the code and get a good idea of what the code is doing just by looking at it.

- **Over-engineering**: Over-engineering can kill a project just as easily as under-engineering. When reviewing code, keep KISS in mind. You want to ensure the program does its job, but you don't want to go overboard. You'll want to keep an eye out for things such as variables that are slated for future use, unnecessary attributes, and so on. Another key aspect to look out for is attributes such as function blocks, methods, and general functions that are a little too specific.

Learning how to conduct a good code review takes time and practice. There are many strategies that can be used to ensure a clean code review. For example, some teams will create checklists while others will use automation tools to assist in the process. Before we get into our final project and conduct a code review, we're going to look at some things that we should never do.

Things to avoid in software engineering

Just as there are best practices, there are also bad practices. In this section, we're going to explore a few bad practices that are common in the software/automation industry.

Fitting a problem into a solution

Not too long ago, it was not uncommon to see a lot of fly-by-night companies that claimed they were going to change the world with their product. This was mostly in the traditional side of software engineering, but the automation industry was equally guilty of this. At a high level, it was not uncommon for a company to go to a customer and try to tailor a customer's problem to fit their product. This is a terrible practice, as the company's product often got a reputation for not working, and the customer's problem was not solved, or at least not adequately. This issue is not unique to corporate madness; engineers often do this too at the lower levels of development. It is not uncommon for developers to want to use what I like to call *resume technology* for a project. In these cases, developers will pick whatever the hottest technology is and, regardless of whether it's truly a good fit for the project or not, use it for the project. The result is usually a hacked-together mess that will have a limited lifespan. To avoid this, only use a specific technology if it makes sense to do so and not because it's the new, cool kid on the block. Using the wrong technology will create a lot of technical debt and, in the worst-case scenario, sink the project altogether. Along these lines is another common problem: using software as a means to fix hardware.

Fixing hardware with software

Automation engineers typically have a hardware-first approach to machine building. As stated before, software is typically treated as a second-class citizen and as a means to drive hardware. This is the worst philosophy a PLC programmer can have. Software needs to be thought of as the brain of a machine. It's easy to tout that your machine uses the latest gearboxes, best encoders, and fastest PLC that money can buy, but without software, all that cool hardware is nothing more than a paper weight. Software drives the hardware. Without software, you have nothing but expensive circuit boards. Under no circumstances should software ever be used to compensate for malfunctioning hardware. It is not uncommon for code to be changed to accommodate bad encoders, malfunctioning power supplies, and many other things. This is the definition of technical debt. This practice will kill your codebase by making it unmaintainable. If you have broken hardware, fix the actual problem; don't compensate for it at the expense of the software. Finally, the last bad practice that we're going to explore is the code review mentality.

Having only one code reviewer

Code reviews are a way to ensure code quality, help teammates grow, and promote general collaboration. However, they can also get very political. If there is only a single coder reviewer and they like another teammate, chances are they will have a higher probability of approving their code. On the other hand, if a code reviewer does not like another teammate, they will probably be more critical of that person's code and reject it more often. Either way is counterproductive for the organization.

Typically, it is best to have the whole group review the code, or at the very least, use a minimum number of reviewers. Having more than one pair of eyes on a codebase will help pinpoint more issues, provide checks and balances, and allow others to learn from the mistakes and successes of others. On top of that, code reviews can get bottlenecked really easily if only one person is performing them. In all, having only one reviewer can stifle a team due to politics, limit the growth of others, and cause delays.

By this point, you should have a decent overview of some best and bad practices. This means that we can move on to our final project and perform a simulated code review.

Final project: Performing a simulated code review

For the final project, we're going to use what we learned in this chapter and apply it to a simulated code review. The code we're going to explore is as follows:

```
1 Function Block motorControllerFB
2     XAxis : int
3     YAxis : int
4     Method motor_on()
5         Move to XAxis
6         Move to YAxis

7 Function Block power_supply
8     ps : bool
9     Method turnMotorOn()
10        ps = true  //this turns the power supply on
11    Method turnMotorOff()
12        ps = false //this turns the power supply off
```

```
13 PLC PRG
14    If btn1 = true then
15        Turn on power led
16        PowerSupply.TurnMotorOn()
17        1 + 3
18    If btn1 = false then
19        Turn off power led
20        Powers_supply.TurnMotorOff
```

Based on what we have explored thus far, there are a number of issues here. In short, we have quite a bit of technical debt. First and foremost, there are casing issues. Some attributes use camel casing, some use Pascal casing, while others use snake casing. This is a major issue as it raises technical debt. Another quick look at the function block names shows that some (*line 1*) are using the postfix FB, while another function block (*line 7*) is not. This will cause inconsistencies and possible confusion for other developers.

The first function block is passable in terms of the code. The code is clear, and the operations are logical. However, with the power supply function block, there are a few things that we need to note. The variable names are not reflective. The name ps (*line 8*) is not clear. For this codebase, a better name would be powerSupply or something akin to that. If we changed the name of that one variable, we could eliminate the comments on *lines 10* and *12*, which, as of now, aren't totally necessary because we can still kind of infer what the lines of code do.

The PLC_PRG routine is the worst of all three attributes. First, having two If statements (*lines 14* and *18*) is overkill and can lead to issues down the road. A better option would be to convert the second If statement to an elif command. This will ensure that only one of the control statements can be on at a given time. As of now, a weird edge case could potentially trip up the program. Second, there are some issues with the method calls. In each If statement, they are spelled differently. This means that either the developer did not test the code, or something was cached on their system. Finally, there is a line of dead code in the first If block (*line 17*).

Summary

This chapter has been about best practices and lowering technical debt. To lower technical debt, we explored best practices such as conducting code reviews, naming conventions, commenting, casing conventions, dead code, and more. There is a lot involved in developing quality code, and what was explored here were just some common code issues that could arise and sink a codebase. These best practices are very important to understand because, in the next chapter, we're going to explore creating libraries.

Questions

1. What is the KISS methodology?

2. What is dead code?

3. What is unreachable code?

4. If you needed a variable for the elbow joint of a robot, what would be a good name?

5. What does a good comment look like?

6. What does a bad comment look like?

7. When should you use a variable?

8. What is camel casing?

9. What is Pascal casing?

10. What is snake casing?

Join our community on Discord

Join our community's Discord space for discussions with the authors and other readers: `https://packt.link/embeddedsystems`

7

Libraries: Write Once, Use Anywhere

Developing an application will typically involve integrating many different functionalities into one project. For example, if you consider a modern industrial application, it will often be a composite of many different components from many different vendors. Writing custom code to properly interact with all these third-party devices will, at the very least, be a daunting and time-consuming task. At worst, it could be impossible to accurately figure out how third-party devices operate to programmatically interact with them. To remedy this problem, the device manufacturer or other parties will often create what are called **libraries** to integrate modules into a project.

Almost all modern software projects, whether traditional or industrial automation applications such as PLCs or HMIs, use libraries. In fact, it is nearly impossible to build a modern application of any kind without the assistance of libraries. Libraries are very powerful programming tools that can drastically reduce the complexity of a project, as well as the overall development time needed to pull the project off. To master libraries, we will explore the following:

- What is a library?
- Libraries versus frameworks
- Installing a library
- Using a library in Ladder Logic
- Distributing a library
- Guiding principles for developing a library

To round out the chapter, we're going to create a simple library for a robot.

Technical requirements

As per all previous chapters, this chapter will require nothing special other than a copy of CODESYS. If you have skipped ahead and have not read the past chapters, you will need to download and install a copy of CODESYS.

The code for this project and all other examples can be found at the following URL: `https://github.com/PacktPublishing/Mastering-PLC-Programming-Second-Edition/tree/main/Chapter%207`.

Investigating libraries

A library is a piece of software that is designed to provide functionality for another project. In common programming lingo, a library is often called a dependency because the main program is dependent on the code that resides in the software module. This may sound complex, but under the hood, a library is a collection of function blocks and other attributes that are designed to cut down on new code.

In a very basic sense, a library is a code module that can be imported into a project. These code modules are used to help developers do many things, such as cut down on redundant code that is used across different projects, reduce bugs, integrate niche functionality into a project, and much more. Essentially, when you're working on a project, you want to focus on accomplishing the project. In other words, if you're working on an industrial 3D printer that needs to read **Comma Separated Values** (**CSV**) files, you want to focus on building the 3D printer, not on a program that can read CSV files. This is where libraries come into play. Libraries allow you to add certain functionality to your code without losing focus on the main project.

Common uses for libraries include the following:

- To cut development time by using prebuilt and tested code
- To interface with custom or proprietary components
- To augment existing code with niche functionality by using third-party libraries
- To easily distribute code to other developers

As stated before, libraries are the backbone of any modern programming language. For many programming languages, such as C++, libraries are required for any new application. For example, for most C++ programs, the **Standard Template Library (STL)** must be imported to use many of the basic features of the language. Libraries in CODESYS and other similar systems are a bit more optional compared to languages such as C++; however, they are no less important. Libraries can be responsible for several things, and some common examples include the following:

- Communicating with IoT devices
- Communicating with hardware interfaces
- Integrating machine learning/artificial intelligence
- Using communication protocols and conversions
- Interfacing with cloud service providers

There are many more uses for libraries. To get the most from libraries, there is a key distinction that needs to be made between a library and what's called a framework.

Libraries versus frameworks

In everyday programming lingo, the terms library and framework are used interchangeably; however, the two terms refer to two different things. To be a quality programmer of any kind, it is very important to understand the difference between the two types of modules. To begin this exploration, we're going to explore what a library is at a fundamental level.

Understanding libraries

As was alluded to earlier, a library is a code module that augments your code. In other words, a library is a collection of prebuilt function blocks and attributes that your program can call. If you were to view a library through the eyes of biology, a library can be thought of as a performance-enhancing drug for athletes. A library does not require any specific structure and can be used in any type of application. This is in contrast to the other type of code module, which is called a framework.

Understanding frameworks

A framework is like the skeleton of a person or animal. The purpose of a framework is to provide the necessary structure to ensure your program does a certain thing. For example, a common application for a framework in a traditional app would be the Django framework for Python. This framework essentially turns the Python language into a web server programming language. A framework will provide all the necessary scaffolding for your program to do a certain task. Compared to a library, a framework can restrict what you can and cannot do. Your custom code will serve as little more than the necessary straps to ensure the framework's tasks are carried out. A common theme that you will hear when discussing frameworks with other developers is that the framework will morph the language into what can be thought of as a derivative or flavor of the primary language. Overall, where your code calls a library, a framework calls your code.

In terms of automation, you are much, much more likely to encounter a library than a framework. Though you can, as we will see later on in the chapter, build your own custom library, you are much more likely to import a library from a source. So, in the next section, we're going to practice importing a library into a program.

Importing a library

We have been using the term *third-party library* quite a bit. Generally, the context for the term, at least for this book, is a library that is distributed by a person, organization, or so on. These libraries can usually be downloaded from sources such as GitHub, vendor websites, or any other download source.

In terms of PLC programming, many libraries come from a vendor. However, you can still get libraries from sources such as GitHub. If you opt to use a library that you get from a source such as GitHub, you must import it. To demonstrate this, we are going to use a custom library that has one function block that consists of one method, which adds two numbers. This is a custom library that was developed for this book. The library can be downloaded at the following link: `https://github.com/PacktPublishing/Mastering-PLC-Programming-Second-Edition`.

The library is named `adderLib`. For this example, you will need to pull down the library to your development machine.

Installing a library

The following are the steps necessary to install a library. The adderLib library is written using CODESYS and will only work with that system. Also, the following steps are for CODESYS; if you find yourself using a different development system, the steps may differ, but the spirit of the operation will likely be similar. It is best to consult the documentation for the system of your choice.

> **Note**
>
> The library for this tutorial will be included in the GitHub repo; however, as new versions of the system are released, it may be incompatible with the version you're using. If you find yourself with an incompatible version of CODESYS, skip to the *Using a library in Ladder Logic* section of this chapter to build a library you can import.

The steps for importing a library in CODESYS are as follows:

1. Create a new project, click on the **Tools** menu, and select **Library Repository...**, as shown in *Figure 7.1*:

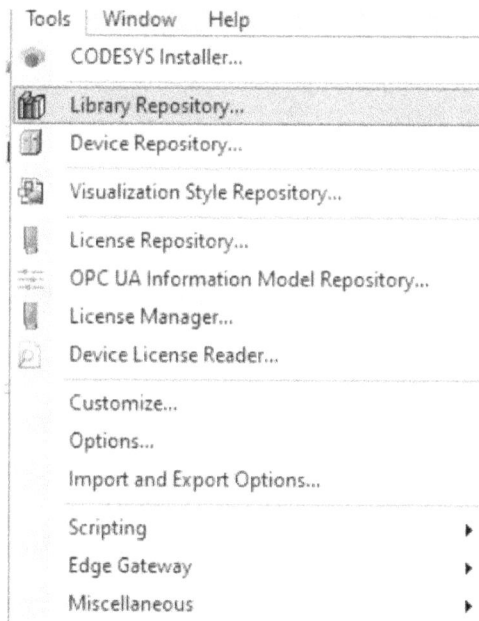

Figure 7.1: Tools drop-down icon

When you click the icon from *Figure 7.1*, you will be met with a screen similar to this:

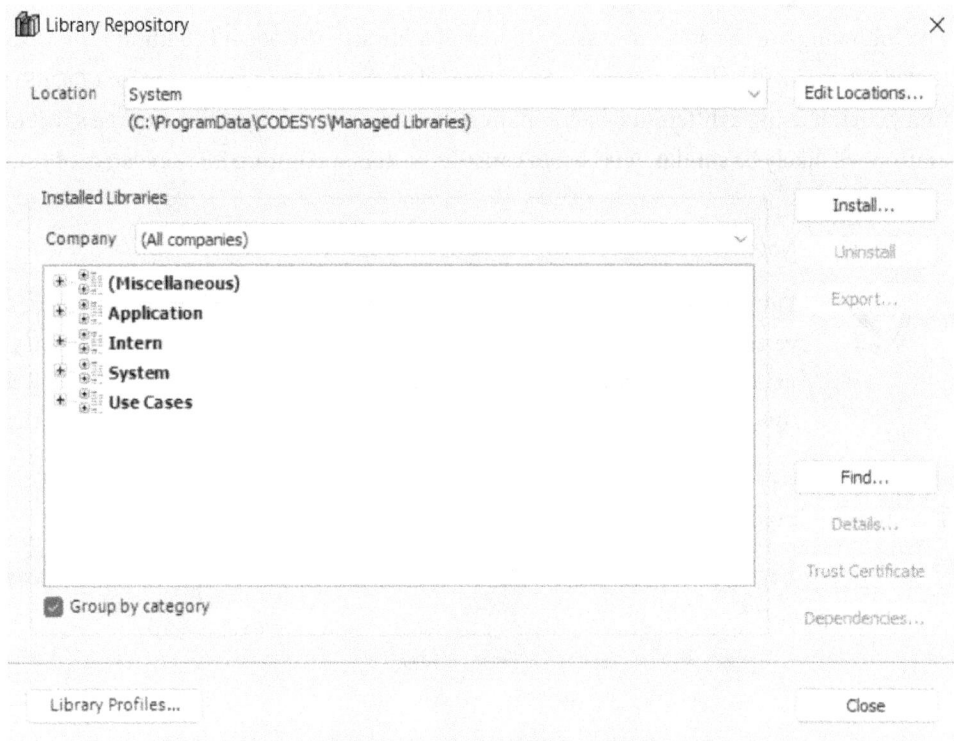

Figure 7.2: Library repository screen

2. Once you see this screen, click on the **Install** button. This will open a normal **File Explorer** window.

3. Navigate to where you downloaded the library and select it.

4. Once you select the library, you must import it into the project. To import the library into your project, click on **Library Manager** in the project tree.

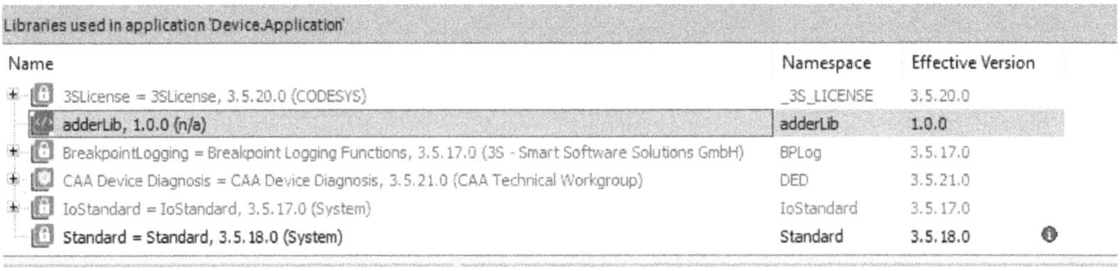

Figure 7.3: Library Manager

5.　Once you see this screen, click on the **Add Library** button, as in *Figure 7.4*:

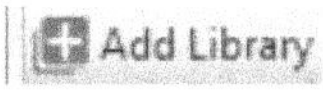

Figure 7.4: Add Library button

6.　Expand the **(Miscellaneous)** dropdown and select the entry called `adderLib`:

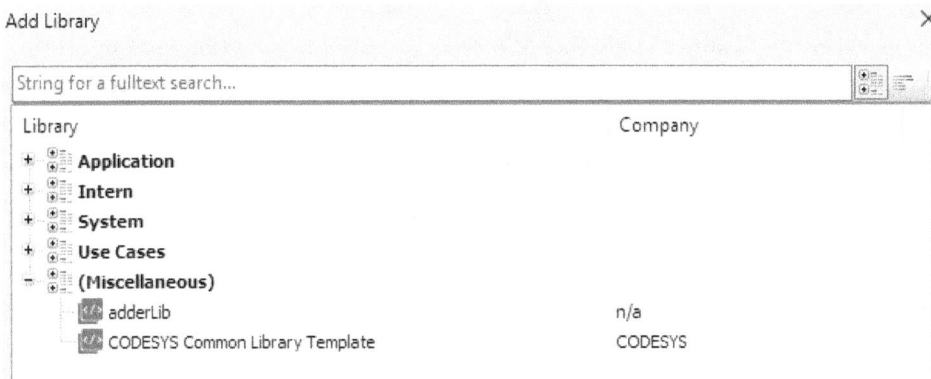

Figure 7.5: Library selection

7.　Once you select the library and click the **OK** button, you should be all set up.

8.　To use the library, navigate to the PLC_PRG file and create the following variables:

```
PROGRAM PLC_PRG
VAR
    Addition : additionFB;
    Result : INT;
END_VAR
```

You will also require the following logic:

```
result := addition.sum(33,33);
```

When you run the application, you should see an output similar to the following:

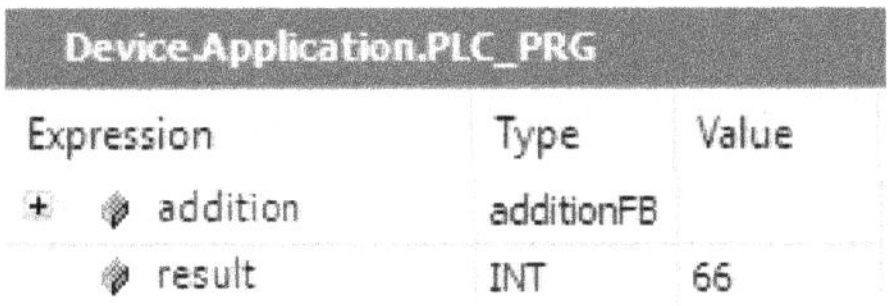

Figure 7.6: Library output

This is a general way to install a third-party library. This library was written and demonstrated in Structured Text. Just because a library is written in one programming interface doesn't mean you're married to that language. In the next section we're going to use a Structured Text library in Ladder Logic

Using a library in Ladder Logic

It's no secret that **Ladder Logic (LL)** still rules the automation programming world. Though **Structured Text (ST)** is a vital language and is slowly taking over the automation programming landscape, we still need to be able to integrate the two. For this example, we're going to take a library written in ST and use it in an LL program. To follow along, create a new Ladder project and download and install the LadderAdderLib library that is included in the GitHub repo. You will need to import it with the same steps that we used before.

The code for the library is simple. The project is simply a function block that adds two numbers and has a method named diff that subtracts two numbers. The only attributes for the main function block (AdderFB) are two input variables called a and b, and a third called sum. The code for AderFB is as follows:

```
FUNCTION_BLOCK AddFb
VAR_INPUT
    a : INT;
    b : INT;
END_VAR
VAR_OUTPUT
    sum : INT;
END_VAR
VAR
END_VAR
```

While the logic is simply this:

```
sum := a + b;
```

For the diff method, the variables will be as follows:

```
METHOD PUBLIC diff : INT
VAR_INPUT
    a : INT;
```

```
    b : INT;
END_VAR
```

The subtraction logic for this method will be this:

```
diff := a - b;
```

To use the library in the main or consumer project after you install and import it, simply use the following variables:

```
PROGRAM PLC_PRG
VAR
    adder   : AdderFB;
    diff    : INT;
    a       : INT := 500;
    b       : INT := 100;
END_VAR
```

In the **Ladder Logic** section of the PLC_PRG file, navigate to **ToolBox**, add two **Box** components, and configure them to match *Figure 7.7*:

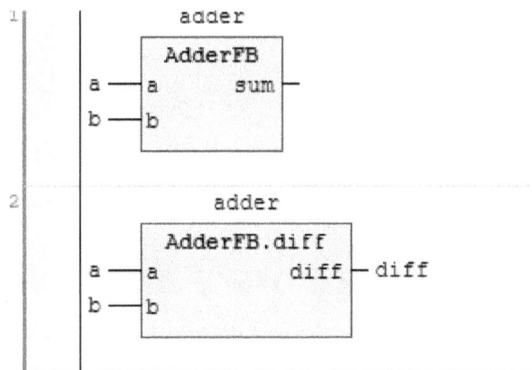

Figure 7.7: Configured library function block

Once you have all that configured, run the project, and you should be met with the output in *Figure 7.8*:

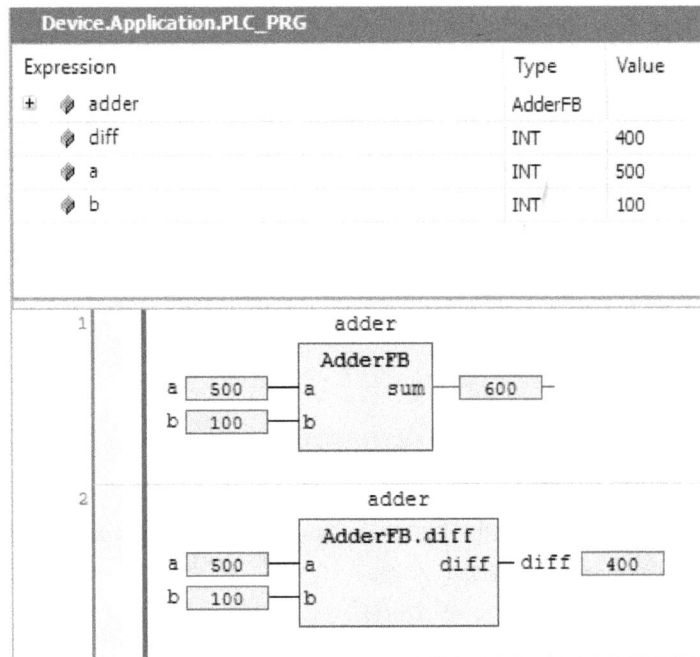

Figure 7.8: Library output

Now that we have a basic understanding of how to implement a library, we can create one ourselves. To begin, let's investigate the architectural principles of creating a library!

Guiding principles for library development

Developing an effective library can be tricky. Where you'll have a clear-cut application in mind when developing a PLC program for a machine, you'll have to make certain assumptions when developing a library. You will not know ahead of time who will use the library, how they will use it, or what they will use it for. Hence, creating a good library can be a very tricky and daunting task. There are no clear-cut ways to create a perfect library, but there are a few rules that I came across that have helped me develop some decent ones in the past.

Rule 1: Remember KISS

The first rule of any software project, especially a library, is to keep it as simple as possible. When creating a library, you want to ensure that you're following the KISS methodology. A complex library can become impossible to use; therefore, it is imperative to ensure that the library is as simple as possible. As we will see with *Rule 3*, a simple way to reduce the complexity of a library is to employ the Façade pattern to hide complexity.

In terms of the KISS methodology for libraries, the following are a few guardrails to help keep it simple!

- **Single responsibility**: Much like a function block, a library should do one thing and one thing only. If you pack too much functionality into a library, that can make it very hard to use. This means if you're making a library for a motor drive, the library should only have exposed functionality that supports operations to control the motor. Adding functionality to support things such as temperature sensing or something else that isn't motor control into the library can complicate it beyond use.

- **Remove useless functionality**: A good cliche to remember is *if you don't use it, lose it*. A killer of any library is junk functions or function blocks that serve no real purpose. This is a common problem for many libraries. Having useless or redundant attributes can kill the usability of a library, as it can cause confusion about how to use it.

- **Use reflective naming**: Do not get clever with the names of attributes. Ensure that each public attribute's name clearly reflects what its purpose is. Going off the rails with a name can cause a lot of confusion and overcomplicate the usage of the library.

- **Limit dependencies**: A library will often depend on other libraries to operate. These dependent libraries are often referred to as transient or downstream dependencies. These dependencies can pose many problems because the end user will need to have them for the library to operate. This means that if the end user does not have access to the correct transient libraries you used to create the module, it is essentially broken.

With this rule in place, we can move on to abstraction and encapsulation!

Rule 2: Abstraction and encapsulation

Going with the theme of removing the possibility for your end user to shoot themselves in the foot, all of the function block attributes should be well encapsulated with a decent level of abstraction. In short, when developing a library, it is very important to show the consumer the absolute minimum they need to use the module. My general rule of thumb for all attributes, especially ones that are in a library, is if I'm not planning on calling it from outside the function block, it gets an access specifier of PRIVATE.

Due to the nature of the library and the wide variety of applications, it is important to hide as much of the inner workings as possible. Generally, I like to teach my students to write a program for the most inexperienced person in the room. This principle is even truer in library development. Expanding on my general rule, I usually tend to create what I like to think of as an **entry point** for the methods. This entry point will have all the necessary arguments and return types; however, if there is dependent logic that breaks my one-sentence rule, I will break that out into other PRIVATE methods and use the entry point to orchestrate them. This will create a system where an outside consumer will only have to make one call to accomplish a task. Sometimes this will be possible, but other times it won't. Regardless, in my experience, it is best to have a single method call that can accomplish the task than to burden the end user with multiple other method calls to accomplish the same task. This principle kind of leads to the concept of design patterns.

> Note
>
> This is a major reason why it's important to think of abstraction and encapsulation in terms of hiding function block components! It helps clear clutter.

Rule 3: Use the Façade pattern liberally

When it comes to library design, the most useful pattern, in my opinion, is the Façade pattern. This pattern can greatly reduce the complexity of a library when it comes to its overall usage. For example, if you're working on a library for a robot, it may require a complex operation to start, turn off, or operate the machine. Since the operations are complex, it can be very hard for the programmer to remember the correct sequence to carry out a particular operation. This is where the Façade pattern comes into play.

Consider the following pseudocode:

```
Function Block Starter
    Method battery()
    Method ledOn()
Function Block Twist
    Method motorOn()
    Method rotateMotor()
```

For a program like this, all the methods will need to be called at some point to move the robot. As can be deduced, this could be hard to pull off, especially when the operations are more intricate and complex. The Façade pattern can be used to help alleviate some of the complexity. In short, a Façade pattern could be employed, like the following:

```
Function facade:
    Method turnOnRobot():
        Starter.Battery()
        Starter.ledOn()
    Method moveRobot():
        Twist.motorOn()
        Twist.rotateMotor()
```

With this setup, if the programmer needs to move the robot, all they have to do is call the turnOnRobot method and then the moveRobot method. In other words, we went from needing to call four methods to only two, which are easier to remember and use!

Rule 4: Documentation

You could develop the greatest library in the world; however, if it is not documented, it will be about as good as useless. It is important to remember that a library is typically a compiled project, and in many (especially older or simpler) systems, ordinary comments will not be visible to the consumer. You must use other means to communicate to other developers how to properly use the library. There are many ways to document the proper usage of a library, including custom documentation such as PDFs, websites, GitHub pages, and so on. You can also provide documentation in CODESYS itself.

There are a few things that must always be documented in a library, which are as follows:

- **Library information:** It is necessary to provide information on what the library is designed to accomplish.
- **Function blocks:** You will want to provide a simple synopsis of the function block.
- **Methods:** You will want to provide a synopsis of what the method does and provide information such as return types and arguments.
- **Variables:** You will also need to document the exposed function block and method-level variables. You will want to provide information on what the variables are meant for.

All these attributes can be easily documented in CODESYS with minimal effort. In terms of providing code documentation, there are many ways to document things; however, the syntax that I usually gravitate towards is the following:

- **Declaration header**: Denoted with `///`. The triple slash is typically the safest to use for declaration headers because these comments will always show up in the documentation, and by default, at a minimum, you want this feature notated. I also recommend keeping these features to a one-sentence summary.

- **Member/attribute header**: Denoted with `(*<comment>*)`, this is a way of creating a multi-line comment, which is often useful if an in-depth explanation of the attribute is needed. You can also use the standard `//`; however, to use either syntax, the system must be configured to read them.

> **Note**
>
> I've never been a fan of using `//` in my documentation. I typically advise using multiline comments for attribute documentation because 1) it's more flexible, and 2) it makes it easier to write more detailed descriptions of the attributes.

To demonstrate this, let's modify our example library.

Open the `LadderAdderLib` library project and modify the code to match the following:

```
///This function block adds numbers
FUNCTION_BLOCK AdderFB
VAR_INPUT
    a : INT; (*Input 1 INT*)
    b : INT; (*Input 2 INT*)
END_VAR
VAR_OUTPUT
    sum : INT;
END_VAR
VAR
END_VAR
```

Once you are done with that, import the library into an example project. When you do this, double-click on the library in **Library Manager**, and you should be met with something similar to the following:

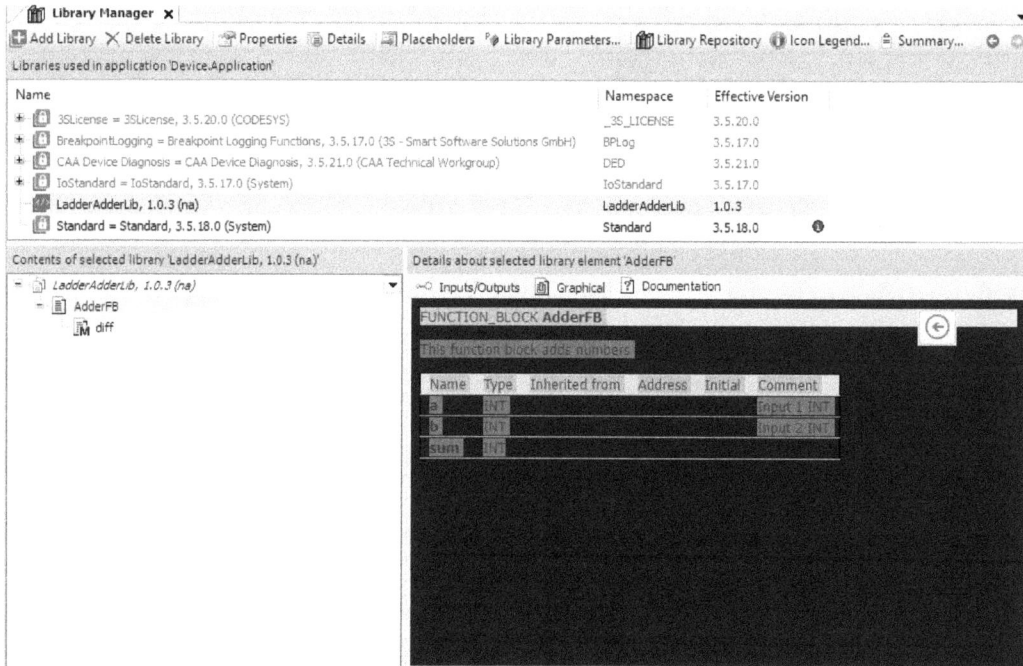

Figure 7.9: Method documentation example

The takeaway from the screenshots is that the triple slash (///) will generate a general message at the top. In other words, the triple slash is more of a general attribute description, while the parentheses are used more for the general description of variables.

There are other ways to add documentation, such as putting logically related function blocks into folders and documenting what the module(s) are meant to do. This is accomplished by right-clicking the folder and then clicking on **Properties**, then finally, selecting **Documentation**, which will render the following:

Properties - addition   ✕

Common   Documentation   Build   Access Control

This function block has one method called sum that will add two numbers and return the
sum of those numbers.

OK   Cancel   Apply

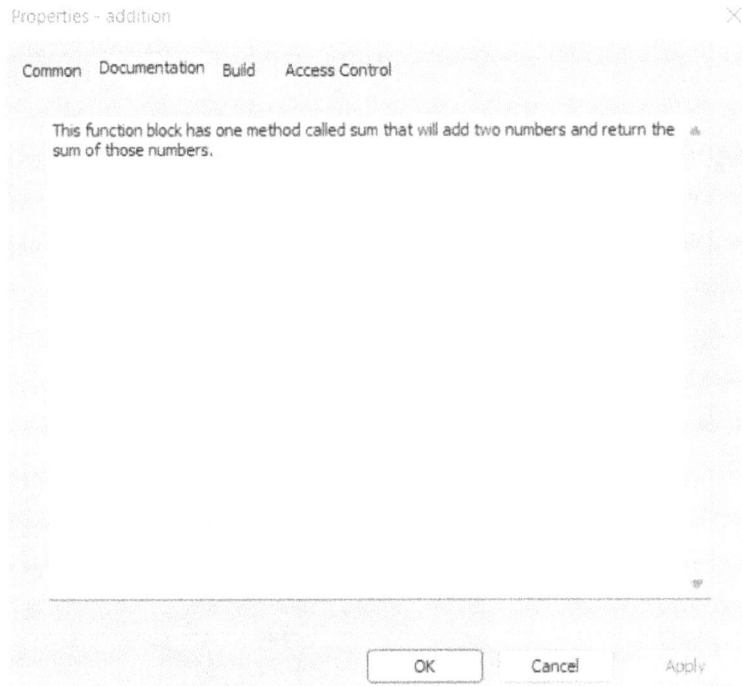

Figure 7.10: Folder documentation window

When you click the **OK** button, the documentation will be generated. As with the other methods, you will be able to view the documentation after the library has been imported. To view the documentation, click the **Library Manager** icon again and click on the folder. You should see something similar to the following screenshot:

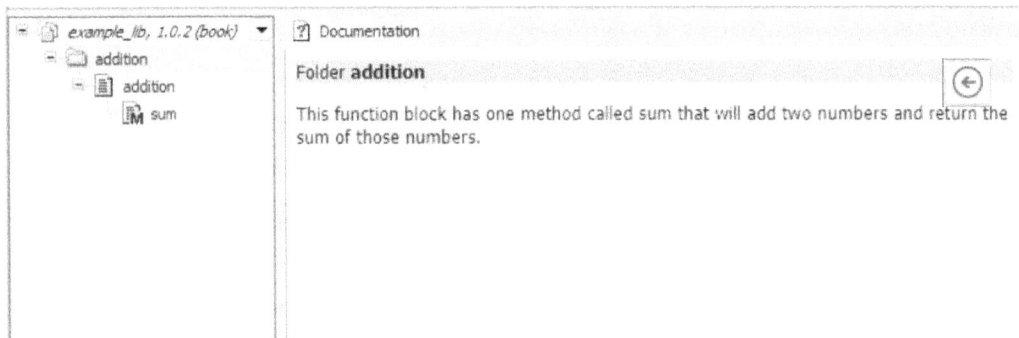

example_lib, 1.0.2 (book)   ▼   ? Documentation
    addition   **Folder addition**   ⬅
        addition
        M sum   This function block has one method called sum that will add two numbers and return the
                                          sum of those numbers.

Figure 7.11: Folder documentation

The final aspect that needs to be documented is the general information about the library, or, as is commonly known, metadata. To do this, you will click the **Project Information** section in the library project area, which will generate a popup like the following:

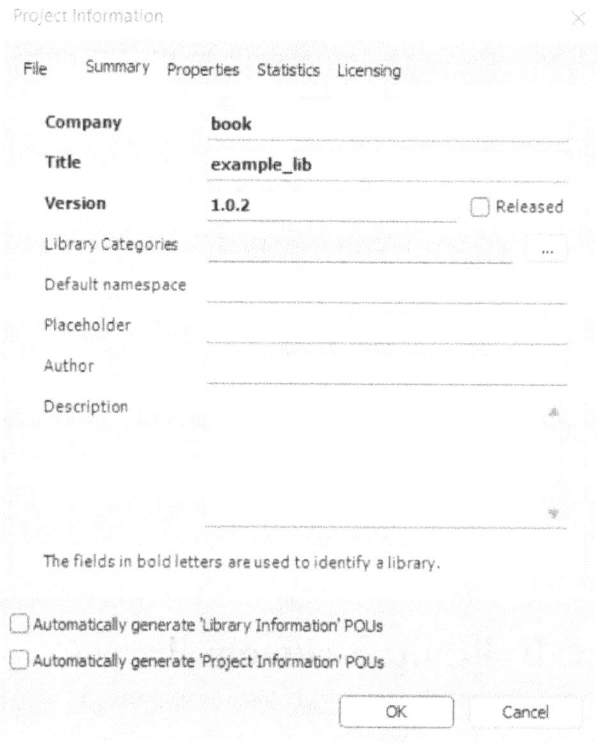

Figure 7.12: Library documentation

Arguably, the two most important fields are the name and the version. The name is important for obvious reasons; the end user needs to know what they're working with. The other important piece of information is the version number. Quality version numbers use a concept called semantic versioning, which signals to the user if the version of the library they are using is compatible with their system. Therefore, the next section is going to look at how to properly version your library.

Semantic versioning

One of the most important things you need to consider when developing a library is the version number. Put simply, the version number should follow a `<major.minor.patch>` scheme. This scheme is called semantic versioning. The meaning of each part of the scheme is summarized in *Table 7.1*:

Version components	Meaning
Major	Changes will break backward compatibility
Minor	Changes will add backward-compatible features
Patch	Changes are backward-compatible bug fixes only

Table 7.1: Semantic versioning scheme

Before we move on, it is important that you understand the principles that were explored in this section. Developing a library for deployment is not like developing a normal program. Things must be named, documented, and architected well. Above all else, the library must be easy to use. Once you understand these principles and have a grasp on them, we can attempt to create a simple library in our final project.

Final project: Building a custom library

For our final project, we're going to build a library for a robot. This library is going to be simple and easy to implement; however, the final aspect of the project will be for you to clean it up and make it easy to use! Therefore, to begin, let's explore the needed requirements.

Requirements

For this project, we are going to need a simple library that can perform the following functions:

- Home the motor
- Turn the motor on
- Turn the motor off
- Stop the motor
- Position the motor

This will be a very simple library and will not require complex architecture. For a library as simple as this, we don't have to worry about complexities such as design patterns; however, the Façade pattern can make the library easier to use. So, let's break down the methods we will need:

- Zero: Will zero out the motor
- Home: Will return the motor to its home position
- Turn on the motor: Will zero out the motor and put the motor in a standby state
- Turn off the motor: Will zero the motor and turn the motor off
- Stop the motor: Will halt the motor without zeroing it out
- Position the motor: Will move the motor to a position

Implementation

The first thing we need to do is to create a new project; however, unlike creating a normal project, we are going to do the following:

1. Select **Libraries** and **Empty library**, as in the following screenshot:

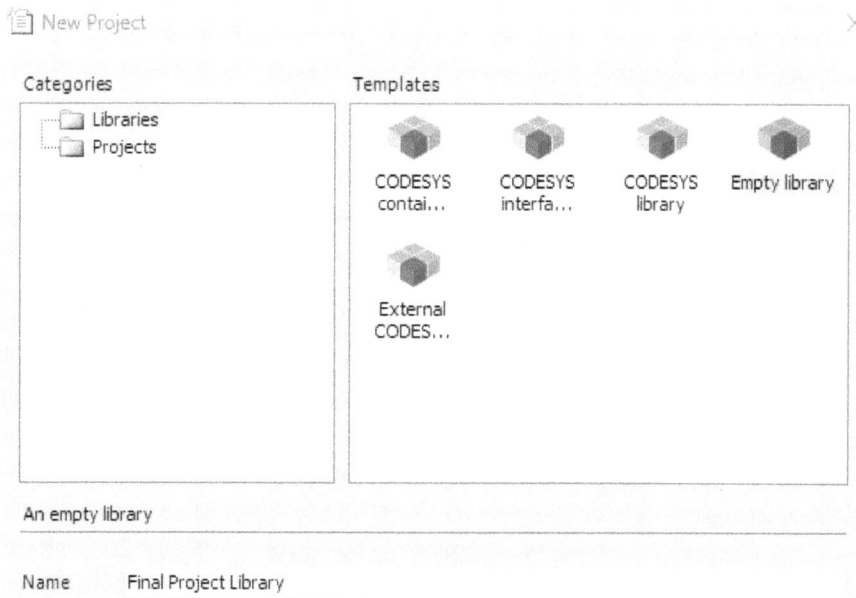

Figure 7.13: Library creation

There are other ways to create a library with a full structure, such as by selecting **CODE-SYS library**. However, this option will give you a full project with potentially unnecessary files and structure. You can opt to use this if you would like, but for now, use an empty library to remove bloat.

2. After you complete that step, you should see a project tree like the following:

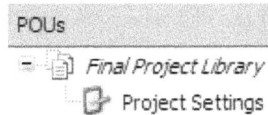

Figure 7.14: Library project tree

3. Next, right-click **Final Project Library** and add a function block named MotorControl with the methods in *Figure 7.16*. Set the return type of all the methods to BOOL and the access specifier to PUBLIC.

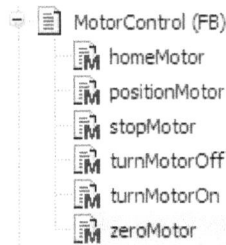

Figure 7.15: Library project tree

4. Now that we have all the methods set up, we can start implementing the code. The first thing that we need to do is declare a variable that holds the motor's positions. For this, we will need to go into the MotorControl function block and set a variable, as in the following code:

```
FUNCTION_BLOCK MotorControl
VAR_INPUT
END_VAR
VAR_OUTPUT
END_VAR
VAR
    motorPosition : INT;
    motorOn : BOOL;
END_VAR
```

5. Next, we will implement the zero method with the following code:

```
motorPosition := 0;
```

6. This will be all that is required for this method, as no internal variables will be set.

7. Next, we will implement the turnMotorOn method with the following:

```
ZeroMotor();
motorOn := TRUE;
```

8. After you have set up the turnMotorOn function, we will now implement the turnMotorOff method with the following:

```
ZeroMotor();
motorOn := FALSE;
```

9. stopMotor will consist of only the following:

```
motorOn := FALSE;
```

10. The homeMotor method is also quite simple. This method will turn the motor on if it is off and then call the zeroMotor method. Once these operations are complete, the motor will be shut down. To do this, implement the following code:

```
IF motorON = FALSE THEN
    motorON := TRUE;
END_IF
ZeroMotor();
motorON := FALSE;
```

11. The next method to tackle is the positionMotor method, which will take in an argument and set the motorPosition function block variable to it, as in the following code:

```
METHOD PUBLIC positionMotor : BOOL
VAR_INPUT
    motorPos : INT;
END_VAR
```

The logic for this method is as follows:

```
motorPosition := motorPos;
```

12. Now that all the variables are set up, we will need to save the project as a compiled library.

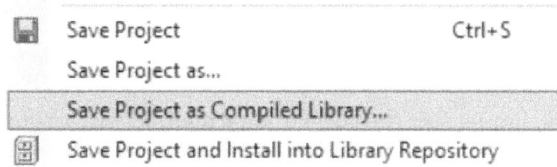

💾	Save Project	Ctrl+S
	Save Project as...	
	Save Project as Compiled Library...	
📇	Save Project and Install into Library Repository	

Figure 7.16: Saving the library

13. When you save the library, you will be met with a **Project Information** screen, as in *Figure 7.17*. Input the following information from *Figure 7.17* to save the library:

Project Information   ✕

| File | Summary | Properties | Statistics | Licensing |

Company	**learning**
Title	**motorContorl**
Version	**1.0.0**	☐ Released

Figure 7.17: Information fields for the library

This will create the `Project Information` file. This file will hold the metadata for the library. You can change the version number, name, or anything else by double-clicking the file once it is created.

14. At this point, your library is now saved and ready to be imported, similar to how we did in the *Third-party libraries* section. Once you import the library, you can modify the `PLC_PRG` file with the following to consume the code:

```
PROGRAM PLC_PRG
VAR
    motor : MotorControl;
END_VAR
```

At this point, we should be able to access all the methods in the function block. For example, we now have access to the following:

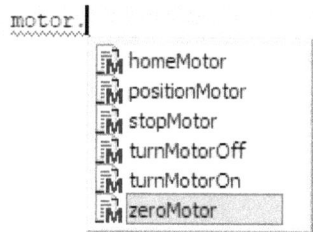

Figure 7.18: Library methods

To give our project a test drive, implement the following reference variable if you have not already done so:

```
PROGRAM PLC_PRG
VAR
    motor : MotorControl;
END_VAR
```

The motor variable is a simple object reference variable for the MotorControl function block in the library. To test a few methods, we're going to rig the PLC_PRG POU file with the following:

```
motor.turnMotorOn();
motor.positionMotor(motorPos := 120);
```

When the code is run, you should get the following:

Figure 7.19: Outputs

In this example, we only called two of the library methods. We only called two to see it in action and prove that it works.

Project improvements

The final project demonstrates that the code works; however, we did not document the library members. Also, if a developer were to perform a task such as positioning the motor, they would need to call multiple methods. For example, to position the motor correctly, they would likely need to do the following:

- Call `stopMotor` if the motor is running
- Call `homeMotor`
- Call `positionMotor` to reposition it
- With the current setup, the library user would have to remember a bunch of steps each time they need to position the motor. For this cleanup, do two things. First, document what the methods do. Second, rework the library to use a Façade method to carry out the described positioning operation.
- This exercise is meant to simulate real-world library development. To carry out this clean-up, you need to assume that 1) you do not know who is going to use your library and 2) you do not know how the library is going to be used. This means you must really generalize your solution. When it comes to a task like this, there is no right or wrong solution. Essentially, you can only do your best to make it as usable as you can for everyone.

Documentation hints

As part of your documentation challenge, ensure you're following *Rule 4* and have the following covered:

- Ensure the purpose of your variables is clearly defined. For example, if they are inputs for some numerical calculation, stipulate it.
- Define any return types and argument types. CODESYS does this automatically, but you need to ensure you get in the habit of notating it in case you ever find yourself working with a system that doesn't.
- Describe your methods with a clear sentence and apply the *one-sentence rule*.
- Clearly denote what the function block does and how it should operate. Though it is not totally necessary, some developers will list out the methods as well.

Now that you've made and cleaned up your revolutionary library, it's time to share it with the world. In the next section, we're going to explore some common considerations that need to be addressed before you ship your code!

Distribution

If you were to ship a library like the one we just built, you would need to determine how you're going to share it. Typically, many developers will opt to deploy their code to a platform such as GitHub. This will allow others to pull the project down and use it. This route assumes you're making your project free to use. Depending on how you set the project up, this route will also allow others to contribute and work on the project with you.

Sometimes you can even sell your project. You can do this by creating a website and charging users to download it. In some PLC ecosystems, you can even opt to deploy your code via official channels. In many cases, you can distribute your project for free or for a charge. However, one important aspect to consider when deploying your code is licensing.

Licensing

Another consideration that you must consider is licensing, as it will dictate how users will be allowed to utilize the library. Licensing is very important whether you're using or creating a library. There are many different licensing types to choose from. Common licenses are MIT, BSD, and Apache licenses.

If you opt to use a third-party library, you need to pay attention to the licensing agreement. Many libraries and plugins are free to use; however, this may not mean that you can freely distribute them. In other words, just because you don't have to pay for a library, it doesn't mean you can use it in a product that you're going to ship. This is an issue that is more related to traditional programming languages, but can still bite you if you're not careful.

As with many traditional programming languages, you can download third-party libraries from vendor websites, GitHub, or anywhere else. From many downloadable sources, the plugin is free; however, it will come with a license agreement that will tell you how you can use the library and distribute projects that utilize it. For some licenses, you can do whatever you want to with the library; for others, there are restrictions on modifications, while others are much stricter on what you can and cannot do. It is wise to remember that there are many different interpretations of *free*, and you would be well advised to understand the types of licenses the software you're employing has.

Summary

In this chapter, we explored libraries. We learned what they are, how to use them, what third parties are, basic development principles, and so on. You should now be able to use libraries from external sources or create your own. What you will find is that by using libraries, you can truly port code to different projects that use a compatible system and cut down on your overall development time and effort.

There are a lot more to libraries, such as namespaces and so on, that were not explored in this chapter. A whole book could be dedicated to this subject. It is recommended that you explore libraries more on your own. This chapter was just a crash course to get you familiar with the concept and consumption of libraries.

The key to maintaining your library is version control. Whether you're deploying through GitHub or selling your project, you will need to keep it organized and versioned. In the next section, we're going to explore how to do this with Git!

Questions

1. What is a library?

2. Why is documentation important?

3. What are three common types of software licenses?

4. What are some good design patterns to use in a library?

Part 2

Software Engineering
for Automation

In this section, you'll broaden your skills beyond programming and dive into the essential tools, practices, and architectural techniques that support professional software development. You'll learn how to manage code effectively with Git, navigate the software development lifecycle, and use UML to design clear, scalable architectures. You'll also explore methods for testing, debugging, and leveraging AI to troubleshoot issues, along with applying SOLID principles to produce high-quality, maintainable automation software. By the end of this part, you'll have the practical knowledge needed to collaborate confidently, architect solutions, and ensure your code is robust and resilient.

This part of the book includes the following chapters:

- *Chapter 8, Getting Started with Git*
- *Chapter 9, SDLC: Navigating the SDLC to Create Great Code*
- *Chapter 10, Architecting Code with UML*
- *Chapter 11, Testing and Troubleshooting*
- *Chapter 12, Advanced Coding: Using SOLID to Make Solid Code*

8

Getting Started with Git

Many small and mid-sized automation companies, for whatever reason, will not invest time or money in version control. Unfortunately for them, this is a recipe for disaster. Source control is as much a part of modern software as gasoline is to car racing. In short, an organization will not be able to scale without adequate version control software.

There are many different types of version control platforms available. Many of these platforms are geared toward traditional software development; however, most of these can still be used to great success in the automation world. It is true that without proper plugins, some features of the version control software may be unavailable; regardless, it can still be leveraged to great success within an organization.

This chapter is going to be dedicated to understanding one of the most, if not *the* most, popular version control platforms on the market: Git. To do this, we're going to take a hands-on approach to using the Git **command-line interface (CLI)** and learn to use the software via a terminal. If you don't have any experience working with a terminal, don't fret; it may seem scary, but it is quite simple to use. The key to working with any terminal program is simply learning how to talk to it. To do this, we're going to look at the following:

- Understanding what version control is
- Understanding what version control is not
- Understanding Git
- Understanding GitLab
- Using the Git CLI
- Understanding branches
- Exploring PLCopen XML

Finally, to round out the chapter, we're going to create a project for a simulated car wash to experiment with creating branches and pushing changes.

Technical requirements

To follow along with this chapter, you will need a few things. First, you will need Git installed on your machine in some way, shape, or form. On the off chance you have access to a Linux computer, you can install or use the prepackaged version of Git for Linux. However, in the more likely scenario of you using Windows, you can use a program called Git Bash, which can be downloaded at the following link: `https://git-scm.com/downloads`.

Git Bash is the CLI that will allow you to interface with a repository management system. With that, you will also need a Git repository manager. For this project, you can set up a free account with GitLab and use it for the following tutorials. However, you are not limited to GitLab; you can use pretty much any system, such as GitHub, Bitbucket, or anything else that also supports Git, so the only real consideration there is when picking a repository manager for this book is to ensure that it is compatible with Git. You can sign up for a free account with GitLab at the following URL: `https://about.gitlab.com/pricing/`.

What is version control?

When I first started out as an automation engineer, I would often be required to modify software, install software, or do any number of things to a machine's codebase. However, I came to realize that this was often quite difficult as the correct codebase was usually stored on someone else's computer or lost in the vacuum of cyberspace. This meant that I would have to spend needlessly long periods of time at a customer site doing tasks that were already done. In other words, I would have to rework jobs, and assignments that would otherwise take a few minutes would turn into weeklong affairs. As I grew and expanded my horizons as an automation engineer at various companies, I found this behavior to be common. For anyone else who has ever experienced this, all hope is not lost. In fact, there is a relatively simple solution to this problem. That solution is called **version control**.

Version control is a way of storing software projects as well as other documentation in repositories that share a centralized URL and can keep track of all changes that have been made to the files. What we're going to consider to be version control or version control systems will consist of a true version control software like Git and a repository manager like GitLab. In day-to-day speech, when someone refers to version control, they are normally referring to a setup like this.

The terms *source control* and *version control* are often used interchangeably, but there is technically a difference between the two. However, it can be argued that the differences are mostly splitting hairs, so we're going to stick with the normal vernacular and use the terms interchangeably.

Version control is especially useful when there is more than one person working on a project or if someone other than the initial developer needs the source code. The key here is that version control makes the code be in one centralized point, which means there is one source of truth for the codebase. This means that if there is a permanent change to the codebase, it can easily be made available to everyone working on it. In turn, this means that accidentally losing changes to a codebase, confusing different codebase versions, and requiring extensive rework becomes a non-issue.

The true power of source control stems from its versioning nature. As the name suggests, source control software will version the files in a codebase. This means that if you or someone else makes a modification to the source code and the changes are committed, the original code is not lost. If needed, the code can be easily rolled back to its original state before the changes were introduced. This feature is especially handy when it comes to automation. Often, a customer will need to temporarily change out a machine component, which means the source code will have to be altered to accommodate it. As every automation engineer knows, the customer will typically want to roll the changes back and put the original part back in. For an organization that doesn't utilize source control, that could easily equate to an extensive rewrite to restore the code, in other words, unnecessary work. However, for an organization that does use source control, rolling back the changes can be as simple as pressing a few buttons or copying and pasting some changes.

Traditionally, PLC code is not supported well in standard repository management systems. This means that certain features available for common languages, such as Java or C#, are not available for PLC projects. This is at least true for repository management systems such as GitLab or GitHub. As the line between traditional computer programming and automation programming blurs, newer automation focused version control systems are emerging, but haven't made a huge impact in the industry yet. Nonetheless, a version control program can greatly increase the collaboration and productivity of a development team.

Note

Automation software is generally not supported in the same way that traditional programming languages like C++ or Java are by systems like GitLab or GitHub. There are some features such as viewing source code in the browser that are typically not supported.

Now that we've explored what source or version control is, we can move on and explore what it is not.

What version control is not

There are a lot of myths and misunderstandings about what the role of source control is. A lot of these misguided truths stem from older engineers, managers, and business owners who do not have a solid background in the software development process. Regardless, many myths and partial understandings have hamstrung a lot of small to mid-sized organizations. In this section, we're going to look at some of these to learn what source control is not.

Source control is only for large teams

This is arguably the most misguided myth of them all. Source control can and will increase productivity regardless of whether the development team has 100 engineers on it or just one. The primary purpose of source control is to version files and prevent them from being locked away on an employee's development computer. Regardless of the number of engineers on a team, version control will allow an organization to have a single point of truth for the software, allow for the housing of the software in a centralized location, and above all else, keep a record of all the changes in the code since it was originally pushed.

Source control is a security risk

This is one of my favorite misconceptions about source control, mostly because of how silly it is. Many older automation engineers and managers who have no experience with source control or cloud storage often misunderstand what cloud storage is and how version control works. This myth is multifaceted, with the first misconception stemming from what cloud storage security is. Many older engineers believe that cloud storage is unsafe because the company using the service has no control over security. This is a noble but very misguided assumption. First, it is important to understand that though no organization can guarantee perfect security, the security that many of these cloud storage solutions offer will be significantly better than anything a small to mid-size automation company will realistically be able to pull off. It is important to remember that a respected source control host houses many projects from many companies and individuals, and though they can never guarantee foolproof security, it is realistically safer to store your code online than on an in-house system that many small to mid-size automation companies have to offer.

Another facet of this myth is that source control must be stored in the cloud. Again, this is unequivocally untrue. Most respected source control vendors will offer the ability to deploy an instance of the software on a local server. Typically, a modern way that vendors will distribute the version control software is with what's known as a container or a standard download. Containers will utilize a system such as Docker or Podman to run the application on a local server. The nature of Docker and how to run a Docker instance go well beyond the scope of this book. Regardless, it is possible to deploy and run your own version control instance on your server. Utilizing a container or download deployment is not recommended for smaller companies, as containerization and general management of a source control system will require in-depth IT experience and resources. Housing your own version control instance will require a hefty server computer to store the code, an admin who can maintain the version control instances (especially if it is a container deployment), and, ideally, a backup server somewhere that can act as a **disaster recovery** (**DR**) system. Many larger companies that have more resources and budget often opt to run their own deployments, but for mid-size and, especially, small companies, it's usually cheaper and safer to use a version control service in the cloud.

Finally, the last facet of this myth is that anybody can see and pull the code. Again, this is naively untrue. All quality version control systems will allow users to cherry-pick who they want to view and access their source code. This means that if person X is no longer employed by the company or no longer needs access to the codebase, an admin on the system can simply revoke their permissions and prevent them from seeing or accessing the code.

Version control is the same thing as a shared file system

This is, unfortunately, the philosophy of many smaller organizations that do not have a solid foundation in software development principles. A filesystem on a server is not, and cannot replace, a version control system. A shared filesystem can, at best, house different versions of a program in different folders. This, unfortunately, is only as good as the least organized developer. If someone uploads their code to the wrong directory, that is a directory that houses an older version of the codebase; the old codebase is lost for good.

Overall, a traditional version control system is not going to give the same benefits for a PLC project that it would for a general-purpose programming language, but it can greatly increase productivity and promote collaboration across a development team. Version control will create a much more consistent and cohesive environment to work in and prevent half-baked changes from finding their way into the codebase. So, how can we start using a version control system?

Understanding Git

There are many ways to interact with a source control system. Most systems will support some type of UI or have a plugin for an IDE such as Visual Studio or CODESYS. These plugins can get you through the day; for example, you will be able to easily upload and download changes as well as create branches. However, to effectively use and understand a source control system, it is important to understand how to use it via a CLI. With that, we're going to look at installing Git.

Installing Git on Linux

> **Note**
>
> If you are using Windows, you can skip ahead to the *Installing Git on Windows* section.

Okay—very few people are going to develop PLC code on a Linux system; however, automation is much more than PLC software engineering. There are niche systems, especially advanced robotics systems, that often use or require a Linux distro. Regardless of why you're using Linux, you are in luck when it comes to Git. Git often comes packaged in most Linux distro repositories; therefore, it is very easy to get Git up and running.

As any Linux user knows, each distro has its own package manager. The two most common Linux distros are of the Debian variety, such as Ubuntu, and the Fedora variety, such as Rocky, **Red Hat Enterprise Linux** (**RHEL**), or CentOS. For the most part, the following commands can be used for these distros. Keep in mind that each Linux distro is different and could use a different package manager, so if these commands don't work for you, a simple internet search will reveal the correct command.

Fedora installation

To install Git on a Fedora-based system, you will usually need sudo privileges. Once you secure those privileges, you can use either of the following commands:

```
sudo yum install git
```

You can also opt to use the dnf package manager if it is available on your system, with the following:

```
sudo dnf install git
```

Debian installation

If you are using a Debian distro such as Ubuntu, you can use the following command:

```
sudo apt install git
```

Now, if you're using Windows, you will want to use Git Bash. For this book, I'm going to use a Linux machine; however, the commands will be exactly the same for Windows.

Installing Git on Windows

More likely than not, you're going to be developing on a Windows machine. This is especially true if you are using a development system such as CODESYS or TwinCAT.

Git Bash installation

Regardless of what development system you are using, you will need to download a program called Git Bash. The link to download this program is in the *Technical requirements* section of this chapter; to follow along, please follow the link and download the latest available version. If you follow the instructions properly, you should get a program called Git Bash that will be a simple terminal program.

The terminal program is an interface for interacting with repositories. After installing Git Bash, you should be able to use git commands either via a PowerShell terminal or the terminal program that was installed. If you opt to use the Git Bash terminal, it's important to note that it will use Linux commands. If you're more comfortable with PowerShell, you can apply the same git commands that we're going to explore, but you will have to use PowerShell commands to navigate around.

WSL installation

Another way you can install Git on your machine is to use a WSL instance. A WSL instance is a way to run a Linux instance, such as Ubuntu, on your Windows PC. Essentially, with WSL, you get a full version of a Linux distribution such as Ubuntu on your computer. WSL is becoming a very popular choice with all types of developers because you can get the best of both worlds!

Regardless of which option you use to set up your Git instance, the commands will be the same. You can pick a method to practice with now and easily transfer those skills to any of the other media we explored here. With that, once you've set up your CLI, we can move on to setting up a remote repository.

Understanding GitLab

You may be somewhat confused as to what the role of Git is compared to that of a repository. Many inexperienced engineers confuse the Git system with actual repositories; however, they are different. The differences between the two are summarized next:

- **Repository**: A repository is like a house where the files that make up the codebase live. For a system such as CODESYS, the program will be contained in a singular unit. In other words, unlike with a traditional programming language, you will not be able to see the individual files by default. A repository is also version-controlled when used with a system like Git. This means that there is a clear record of the changes that were made. In other words, if you need to revert to an older version of the codebase, you can.

- **Git**: Git is a program that allows you to interface with the repository. Git allows you to push files to the repository, pull changes, merge, and more. There are other alternatives to Git, such as SVN, but Git is the most popular, and most repository systems support it by default.

As stated before, when someone refers to version control in the wild, they are usually referring to the repository and Git/SVN. This book will use GitLab as a repository manager; however, the concepts explored here can be used across any Git-based system, such as GitHub, Bitbucket, or the like. The procedure for setting up the repository will assume that you're using GitLab, but setting up a repository in any system is easy.

To begin with, use the link in the *Technical requirements* section to navigate to GitLab and set up a free account. Free accounts are more restrictive with things such as the number of users, project sizes, and so on; however, for practice, a free account should suffice.

Once you create an account, you can create a new project by selecting **Create Blank Project**. This will take you to a form that will prompt you to enter information such as the project name. As with many things in the IT space, this form has a good chance of changing, so it is best not to memorize the form but rather understand what the required fields are!

- First, the project name is the name of the project. This field should mirror the name of the software project. For example, if you're working on a welder project for Company X, you could use a name such as Welder_companyx. A name such as this reflects what the software is for.

- Once you input a name, the system will automatically generate a project URL. The project URL will be used later to access the code. You will have some freedom when it comes to this URL, but it is recommended to use the generated link.

You can leave all the other fields blank and click the **Create Project** button, and your repo should be set up. Most version control systems will mirror this setup. Setting up a repo is straightforward and follows the same basic pattern.

Once you have the repo set, you need to choose one of the Git interface methods and install Git. Again, it is highly recommended that you install Git Bash via the link provided in the *Technical requirements* section if you're using Windows. Installing Git Bash is very straightforward and is just like installing a typical Windows program. Once you have Git Bash installed, you can opt to use either the PowerShell CLI or the terminal Git Bash comes with. Either will work for the remainder of this tutorial.

If you opt to use PowerShell, run the following command:

```
git -v
```

This command should provide the version info, as in *Figure 8.1*.

```
git version 2.44.0.windows.1
```

Figure 8.1: Git version info

Once you see version info, as in *Figure 8.1*, you should be good to use Git. Now, it is important to know that the version you see may differ; that is normal.

Once you have the repo and Git Bash set up, you can start exploring version control!

Using the Git CLI

Before we get started, it is important to note that systems such as CODESYS do have tools that can integrate with GitLab and other systems. These tools are GUI-based, integrate directly with the platform, and are easy to use. However, they are often specific to a singular system and vary in their functionality and operations. Some of these plugins are also proprietary, which can further hamper their adaptation. So, if you're using one of these plugins, especially in a production environment, be careful you are not violating any terms of usage.

For this chapter, we're going to stick with the classic Git command line because it's vendor-neutral as far as programming languages go and can be used with pretty much any file type. Git has a lot of functionality to offer, and learning the CLI can give us great control over what we're doing. This section is going to cover the basic operations that Git offers. To begin getting our feet wet, we're going to explore cloning a repo.

Cloning a repo

Arguably, the most common Git operation is cloning a repo. Cloning a repo is a fancy way of saying that we're going to pull the project from our remote or online repository. The core of cloning the project is the project's URL that was created when you set up the repo. To view this URL, navigate to your project and click the **Code** button. This button is typically to the right of the screen.

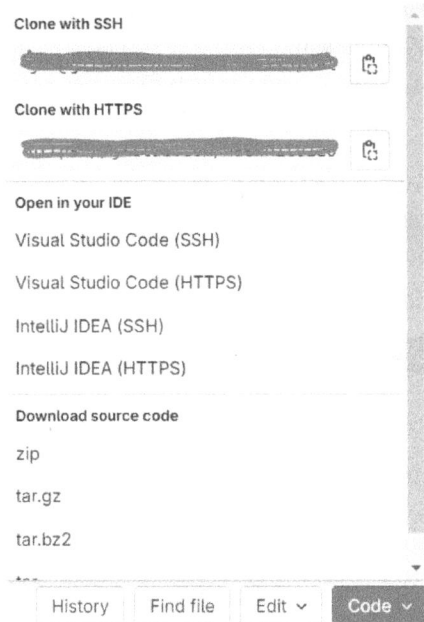

Figure 8.2: Clone menu

From here, you will have two options. You can clone the project with SSH or HTTPS. SSH typically requires a little extra configuration, but once configured, it will not prompt you for login credentials. On the other hand, HTTPS will require login credentials but requires no further setup; therefore, in this chapter, we will use HTTPS.

To clone the project, copy the HTTPS URL by clicking the button with the clipboard. Then, run the following command:

```
git clone <url>
```

Add the URL you copied in <url>.

If you type in pwd, you should see the file path where your project was copied to. To open the project, navigate to that directory. Depending on the system you're using, you may get a message saying that you have cloned an empty repo. This is fine because this is a fresh repo with nothing in it. GitLab typically preloads a README.md file into the repo, so if you're using GitLab, you may not get this particular warning.

Once you have the repo cloned, create a new PLC project on your system of choice and add the following code:

```
PROGRAM PLC_PRG
VAR
        x : INT;
        y : INT;
        sum : INT;
END_VAR
```

The main logic

```
x := 3;
y := 2;

sum := x + y;
```

The code for this particular tutorial doesn't matter much, so don't worry if it isn't perfect.

Once you have the code in place, save the project and copy it to the directory that was cloned down. Essentially, you should have something akin to *Figure 8.3*.

Figure 8.3: Git project structure

When the project is uploaded, run the following commands in your terminal one by one.

This command will add all new files and projects to Git:

```
git add .
```

The following command will commit the project to Git's history:

```
git commit -m "Initial Commit"
```

This command will take a message. When you commit, you want to have a descriptive message about what you're committing. For example, if you added the logic for a motor drive to the project, you could have a message such as *Motor drive code added for welder*. In this case, we're working with an initial commit, so just using that phrase as a commit message is fine for our purposes.

Finally, run the following command:

```
git push origin main
```

This command will push the code to the remote repository; that is, it will push it to GitLab. More specifically, the code will be pushed to what is called the main branch. We're going to cover this more when we explore branches later in this chapter.

Once you run the final command, navigate to your project in GitLab. You should be met with something like the following:

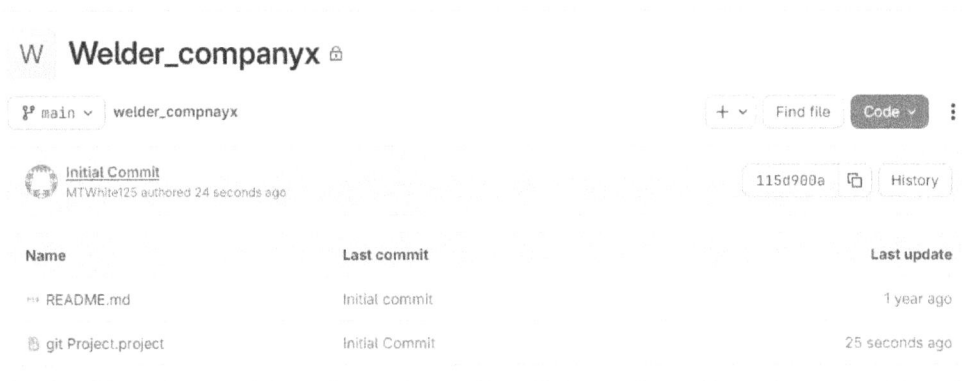

Figure 8.4: Project repo

The project can now be cloned by anyone who has access to the repository. The ability for anyone to access the project is both a pro and a con. On the one hand, anyone who needs access to the project has it, but on the other hand, anyone can modify the project. So, what can you do to ensure that your stable project hasn't changed? You can use branches!

Implementing branches

When done right, a project in a source control system should conceptually look like a tree. That is, you'll have the main trunk of the tree – in this case, the main branch that we just pushed to – and you'll have other branches. These other branches are logically isolated subprojects. That is, a branch is the same source code that lives in the main branch, but it is meant to be altered in some way. Typically, any time you need to address a bug or add a new feature to your codebase, you'll first create a branch in your repo.

To create a branch, all you need to do is clone your project and run one of the following commands:

```
git checkout -b "<name of branch>"
git switch -c "<name of branch>"
```

The lateral command is newer but serves the same purpose, so you can use either. For this book, to keep things consistent, we're going to use the checkout version.

To test this out, we're going to create a new branch for our welding project called function_block. We can accomplish this with the following command:

```
git checkout -b "function_block"
```

To test whether our branch was created successfully, you can run the following command:

```
git branch
```

This command will show all our local branches, that is, branches that exist on our computer. If the branch was successfully created, you should see something like *Figure 8.5*:

```
* function_block
  main
```

Figure 8.5: Branch

Now that we have a branch created, we're going to open our project and add a function block named mulFB with the following variables in it:

```
FUNCTION_BLOCK mulFB
VAR_INPUT
    a : INT;
    b : INT;
END_VAR

VAR_OUTPUT
    product : INT;
END_VAR

VAR
END_VAR
```

Once the variables are in place, be sure to add the following logic:

```
product := a * b;
```

Modify the `PLC_PRG` file to include the following variables:

```
multiply : mulFB;
prod : INT;
```

Also, add the following logic:

```
multiply(a:=3, b:=3);
prod := multiply.product;
```

Save the file and run the following Git commands in the terminal of your choice:

```
git add .
git commit -am "Added mulFB"
git push origin function_block
```

Once you run those commands, you can navigate over to GitLab or whichever system you opted to use and look for your new branch. In GitLab, you will want to navigate to your project and click the drop-down menu that says `main`. When you click the drop-down menu, you will see the branch you created, as in *Figure 8.6*:

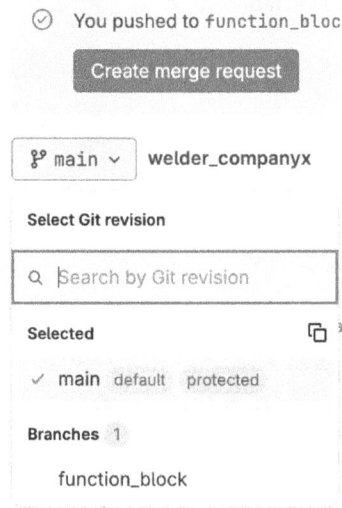

Figure 8.6: Branch drop-down menu

Once you click the branch name, you should see files similar to the ones shown in *Figure 8.7*:

Name	Last commit	Last update
README.md	Initial commit	5 days ago
git Project.Device.Sim.Device.Application...	Added mulFB	5 minutes ago
git Project.Device.Sim.Device.Application...	Added mulFB	5 minutes ago
git Project.Device.Sim.Device.Application...	Added mulFB	5 minutes ago
git Project.project	initial commit	4 days ago
git Project.~u	Added mulFB	5 minutes ago

Figure 8.7: Project files in Git

> **Note**
>
> Notice that a lot of files were committed. These are cache files that are generated by CODESYS. You can exclude these files by creating a `.gitignore` file and listing them.

When you perform a Git clone, you will, by default, pull the main branch. If you're working on a specific branch, you will need to check out that branch.

Checking out a branch

To access code in a branch, you need to check out or switch to that branch. This is a simple operation, and to do this, the first thing we're going to do is delete the project from our directory. After you delete the project, run the following commands in your terminal sequentially:

```
git clone <url>
cd welder_companyx
git checkout function_block
```

The `cd` command will simply change the directory to the repo you just cloned. If you changed to a different directory before you ran the second command, you will need to navigate back to the directory you cloned the repo in. Once you complete these operations, open the project, and you should see all the code we just added. In other words, you pulled down your branch!

Once you complete your code changes, the branch needs to be merged into the main branch. Typically, this is done after a code review. The main drawback to using a raw system such as GitLab with a PLC programming system is that you can't view file changes natively like you can with most traditional programming languages, such as C++ or C#. PLC projects that use a raw system such as GitLab without additional Git plugins will require your peer reviewers to pull your branch for inspection. Some repository plugins will allow users to view code in remote repos; however, as was stated before, support and functionality for these tools will vary. This technique can be thought of as a quick and cheap methodology when plugins aren't available. Regardless, to signal your code is ready to be merged into the main branch, create what's called a **merge request**.

Merging code changes

A **merge** request is a request to consolidate code changes in your branch with the main branch. To do this in GitLab, navigate to your branch and click the **Create merge request** button. This button is usually located at the top of the branch page but could be subject to change. At the time of writing this book, the button looks like the one shown in *Figure 8.8*.

Figure 8.8: Create merge request button

This will navigate you to a form-like page. Typically, you will assign your reviewers here and add information in the description box, such as what the changes were for. For now, just click the **Assign to me** link, scroll toward the bottom, and click the **Create merge request** button.

Next, click the **Merge** button shown in *Figure 8.9*.

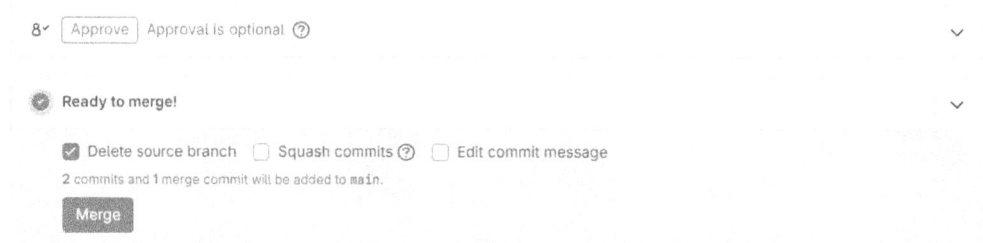

Figure 8.9: Merge button

Once you complete these steps, you can test your merge by deleting your project again and cloning down your code. Once you have cloned your code, open your project and you should see all the new code you created!

This was just a quick crash course on how to use the basics of Git. When it comes to a production environment, you need to consider how your group implements branches and reviews, so we'll look at this next.

Understanding branches

Branches are very important when it comes to source control. A poorly implemented branch strategy can lead to a corrupted main branch. Therefore, here are some basic tips to help you create great branches and keep your repo nice and clean:

- **Naming**: A branch's name should be reflective of what it's meant to do. For example, if the goal of your branch is to fix a specific bug, name your branch something along the lines of `motor_bug_fix`. If your branch is supposed to integrate a new feature, name it something such as `UDP_Support`. If your group uses a ticketing system such as Jira, it is usually a good idea to put the ticket number on the branch as well. It's an even better idea to link the branch to the ticket if possible. It is also important to ensure that your branch name does not contain any spaces. For example, `function block` is not a good name and could cause issues with the version control system you're working with. It's a good idea to ensure that words in a branch name are separated by either a dash or an underscore, such as `feature-branch` or `feature_branch`.

- **Check your branch**: It is important to know which branch you're currently on. Working with different branches can get confusing quickly, and it can be easy to make mistakes. Typically, your terminal program will put * by the branch you're on and change its color. It is important to understand that when you first create a branch, it is what's called a local branch which means it lives on your computer. When you push the branch, it becomes a remote branch which means it lives on whatever server your source control instance is on. To see all your local branches, you can use `git branch`; to see all the remote branches you have access to, you can use `git branch -r`; and finally, to see all the branches you have access to, whether they be remote or local, you can simply use `git branch -a`.

- **Reviewers**: It is usually a standard practice to have at least one other person review your code before it is merged. Systems such as GitLab allow you to set a minimum number of reviewers. Essentially, this feature will block any merger attempts made until the minimum number of approvals is met.

- **Intermediate branches**: Another good strategy is to have an intermediate branch to merge with. This methodology is commonly called release branching or staging branching. Many organizations, especially Agile ones, like to have releases on a specific schedule. So, it is not uncommon for them to have a branch that everyone merges their code into for a sprint or a given interval of time. Typically, engineers will merge into this branch, then that intermediary branch will be merged into main. This strategy is more for projects that are on a timed schedule and have a lot of people working on them. For a small, one-off project, this may be more trouble than it's worth.

- **Clean up after yourself:** Not every branch will get merged. Sometimes a new feature will be forsaken after the branch has been made; other times a branch may have been created to test a concept that was never meant to be a part of the project. Branches will also quite often need to be deleted after they have been merged. In any case, these branches need to be removed. A simple way to manually delete a branch is with the following command:

```
git push origin --delete <branch_name>
```

Oftentimes, you will either be prompted to delete a branch when it's merged or the branch will automatically be deleted. In either of these cases, you will not need to manually delete it with the preceding command. If, for any reason, the branch is still lingering after you merge your it, you can still resort to using the command just shown.

> Note
>
> It is very important to remember that once a branch has been deleted, it and any code/data that lived in the branch will also be permanently removed.

- **Pull often:** Sometimes, you may find yourself working with other engineers in the same branch. Though it can be argued that this isn't the greatest practice in the world, it does happen. If they make a change, you need to ensure your codebase reflects it. To do this, you can use `git pull`. This command will fetch all the content from the remote branch (branch in GitLab or whichever system you're using) and put it in your project. Now, this can be easier said than done. If that person has made changes to their code, such as removing something, it could cause a conflict. There are ways to get around this; however, when it comes to PLC code, it is usually easier to simply delete your project, re-clone the code, and check out the branch again.

Recently, a new technology has been introduced to help provide some cross-compatibility to PLC projects. Essentially, for compatible systems, this technology generates a lightweight file to store in a system such as Git.

Exploring PLCopen XML

Most PLC systems are not compatible, meaning you typically can't port a project from one PLC system to another; however, a standard exists that allows developers to export their project in a specialized **eXtensible Markup Language** (**XML**) format. XML is a data exchange language that used to be popular in the early days of the internet. Though it lost its popularity to more modern data exchange languages such as JSON, it is still used quite a bit as a generic means of formatting data that can be used for parsing at another time or by another system. Some of the more advanced PLC systems are now offering a feature that allows you to export your program and certain other metadata in an XML format. This means if your programming system is compatible with what's called PLCopen XML, you can, in theory, port your project from one system to another.

Note

This is a fairly new technology in terms of practical application, and you could encounter bugs that stem from XML files generated in different systems. For example, a program could use a function block or syntax that is not universal or supported by the standard.

When it comes to source control, you can store your whole project as we just did; however, it would be better to export your project when necessary and upload the PLCopen XML file. This will keep your repos lighter, and if your group is familiar with DevOps tools such as Coverity, SonarQube, or the like, you can create whole CI/CD pipelines to scan your PLC code, albeit you will most likely need to create your own custom plugins. New research is currently underway on how to leverage PLCopen XML and CI/CD pipelines, but at the very least, it is a reliable means of storing your PLC project in Git and possibly creating CI/CD pipelines around it.

To export a program in CODESYS, you can navigate to the option highlighted in *Figure 8.10*:

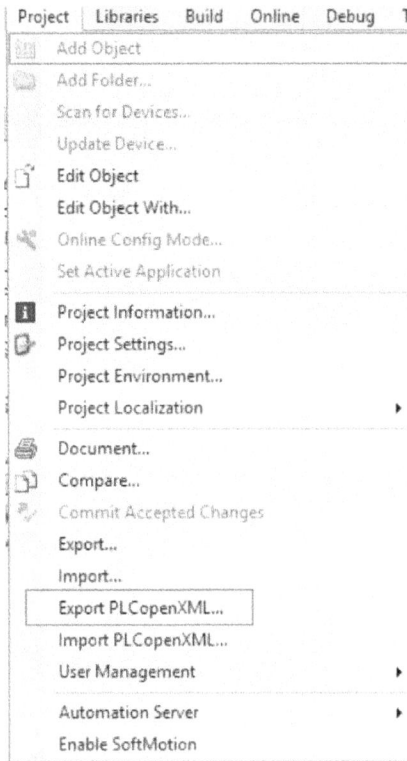

Figure 8.10: PLCopen XML export

This will export the program in XML. You will have the option to customize your export; however, you can export everything on the device by selecting **Device**, shown in *Figure 8.11*:

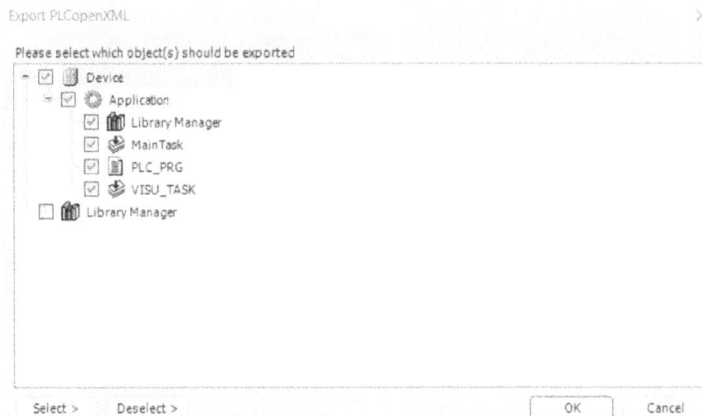

Figure 8.11 -Export selection

Importing one of these files is as simple as selecting **Import...** from the menu and selecting the file.

Now that we've explored the basics of Git and PLCopen XML, we can move on to our final project!

Final project: Modifying a project

For our final project, we're going to modify a project. We're going to make a branch and push a small modification to the branch and finally merge the change. To begin, let's create a new project called CarWash in GitLab:

1. Navigate to the root page and click the + button in *Figure 8.12*:

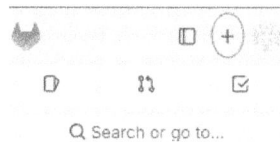

Figure 8.12: Creating a new project

2. Now, create a blank project, fill out the project name, and click **Create**.

3. Next, add the following PLC code to a new project:

```
PROGRAM PLC_PRG
VAR
    msg : WSTRING;
END_VAR
```

4. For the main logic, add the following:

```
msg := "at the car wash yeah!";
```

5. Clone the project down, add the PLC program we just created to it, and push the project.

With that, the project is now set up to begin the final simulation.

In this simulation, we will add another message to the code that says, "oh boy". Follow these steps:

1. Pull down the code.

2. Create a branch. In this case, name it new-message.

3. Make your code change.

4. Push the code into your branch.

5. Merge it.

Before you read on, try to do this on your own!

Final project: Solution

Were you able to get it right? If not, here's the solution:

1. To pull down the code, you should have utilized the following command:

    ```
    git clone <url>
    ```

2. The following command should create a new branch named new-message:

    ```
    git checkout -b new-message
    ```

3. Next, you should have modified your PLC program by adding the following:

    ```
    PROGRAM PLC_PRG
    VAR
        msg : WSTRING;
        msg2 : WSTRING;
    END_VAR
    ```

    Of course, you should have also added the following logic:

    ```
    msg2 := "oh boy";
    ```

4. You should have then saved the project and executed the following commands:

    ```
    git commit -am "oh boy code added"
    git push origin new-message
    ```

5. From here, you should have created a merge request. At this point, the code should be in the main branch!

Summary

This chapter was a short tutorial on how to carry out basic operations in Git. Git is a very powerful source control tool, and it can mean the difference between haphazard projects strewn across people's computers or everything in one central location. Though many GUIs can be used to perform Git operations, any engineer worth their salt should have a basic understanding of how to use the Git CLI. In all honesty, though some people will use a Git plugin to perform these operations, it is often seen as a crutch. Therefore, you should become very familiar with at least the basics of Git.

Source control, and, by extension, Git, play a pivotal role in the life cycle of software. Software, as you have probably deduced at this point, is way more than just writing awesome code. In the next chapter, we're going to explore key concepts to ensure that your code is designed well as we explore the software development life cycle!

Questions

1. What is a branch?

2. What is Git?

3. What is the difference between Git and a system such as GitLab?

4. How do you clone a project using Git?

5. What is the difference between a local and remote repository?

6. What does `git pull` do?

7. What is the difference between `git branch -a` and `git branch -r`?

8. How do you create a branch with the Git CLI?

9. What does `git add .` do?

Further reading

- *XML Exchange*: `https://www.plcopen.org/standards/xml-echange/`

- *Git cheat sheet*: `https://git-scm.com/cheat-sheet`

Join our community on Discord

Join our community's Discord space for discussions with the authors and other readers: `https://packt.link/embeddedsystems`

9

SDLC: Navigating the SDLC to Create Great Code

The greatest challenge that any inexperienced programmer will have to overcome is the urge to create. This may sound oxymoronic because that's the purpose of engineering: to create something new. Couple that mentality with a hardware-first attitude, and it's not hard to see why so many entry-level programmers muck up projects. Software development is much, much more than writing code. Whether you're a traditional programmer working on a social media app or a PLC programmer programming a smart factory, coding is arguably of medium importance.

Saying that coding isn't the most important part of software development may sound like utter blasphemy or ignorance to those looking to create. However, software development is actually a series of steps, and coding is towards the middle. This process is called the **Software Development Lifecycle (SDLC)**. The SDLC is the flow that needs to be followed to ensure that your project will not only work but survive the test of time.

If your goal is to grow as a programmer of any kind, you must understand the SDLC and how it works. Many developers fall into the trap of thinking that if they know all the latest PLC and automation technologies that are on the market, they are "good." However, knowing how to use those technologies versus when to use those technologies is two radically different things. Being able to understand and implement this concept is what separates an engineer from a coder, and it all starts with the SDLC.

This chapter is going to be dedicated to understanding the SDLC and how to leverage it to produce quality code. To do this, we're going to explore the following concepts:

- Understanding the concept of the SDLC
- The general steps of the SDLC
- Understanding how to implement the SDLC

We will end the chapter with the deployment of a working temperature conversion program similar to the ones we have developed in the past. However, this time, we will build one properly using the steps of the SDLC.

Technical requirements

The source code for this chapter can be found in the GitHub repo for this book. You can use the following URL: https://github.com/PacktPublishing/Mastering-PLC-Programming-Second-Edition/tree/main/Chapter%209.

Understanding the SDLC

The **SDLC** comprises steps in the software development process. Much like any other engineering process, the SDLC is the process that should be followed in some way to ensure you are correctly building the correct program. Depending on who you ask or what you read, the number of steps in the SDLC can vary; however, the SDLC is usually broken down into the following:

1. Gathering requirements
2. Designing the software
3. Building the software
4. Testing the software
5. Deploying the software
6. Maintaining the software

Some models will only use five steps, and some will use more; however, no matter the model, the steps are the same, just broken down differently. For this book, we will stick with the six steps outlined. Graphically, the SDLC can be pictured as in *Figure 9.1*:

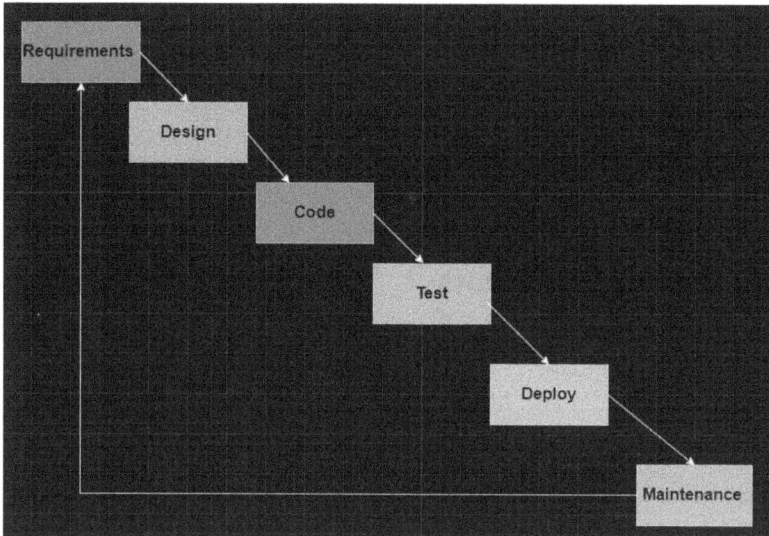

Figure 9.1: SDLC

To many PLC programmers, the SDLC is as exotic a concept as alien life is to zoologists. Sadly, this comes from the mentality that software is an unimportant component of automation. However, to properly implement the concepts that we have covered thus far and to take our PLC software to the next level, we need a clear understanding of the SDLC.

In the introduction, we kind of touched on why the SDLC is important. In this section, we're going to go a little further. Many PLC programmers are usually non-software engineers. Chances are, if you're reading this book and are not a student, you're probably something akin to a mechanical engineer, electrical engineer, electrician, technician, or so on. Chances are also in favor of you having written PLC code in the past. If you are not an experienced software engineer, chances are that code is just a complement to your hardware. As we have touched on in this book, this is a very poor ideology to have. Software must be treated like any other engineering project.

As with any other well-engineered project, you will need to know what the product is meant to do and whether it has a competent design, construction, and test routine before it is deployed. After all that is done, you will need to make modifications and repairs to the product according to the customer's wishes. In traditional software development, it is common for developers to try to shape a problem to fit a solution or produce a solution that is so poorly designed that it cannot be adapted to meet new challenges that the end users will encounter. As with any other process, there are proper ways to implement the SDLC.

The general steps of the SDLC

There are many frameworks to implement the SDLC, such as Agile, Waterfall, and so on, that are very important to understand. Before we can dive into those concepts, we first need to understand what each step of the SDLC is responsible for. Though we touched on this a bit, this section will be dedicated to exploring the steps in the SDLC so we can implement them.

Gathering the requirements

If you ask a person off the street or an inexperienced software developer what the most important aspect of developing software is, chances are, they will answer *coding*. It makes sense since software development is about developing software. However, the following two steps must be completed before you or another developer even thinks about touching a keyboard: *requirements and planning*. Some break this phase into two distinct phases, while others simply call this an analysis phase. Regardless of whether you consider this a phase or phases, it is the backbone of the project. In short, without this step, you simply do not have a project. If the SDLC is a building, the requirements/planning phase is the very foundation, and like any other building, if your foundation is poorly planned and implemented, your building will eventually crumble.

Requirements can be thought of as a punch list of functionalities that are required by the program. In other words, the requirements are the functionalities that are necessary for the program to solve the problem. Without properly gathering the correct requirements, there is little chance that the program will solve the problem at hand, and you'll end up having to fit the problem into the solution, which is quite literally the worst thing you can do.

During this phase, you will plan out the rest of the SDLC, choose the technologies you are going to use, and so on. This phase is the most pivotal step in the development of your software. If this phase is not completed properly, regardless of what methodology you're using, your project is as good as sunk.

During this phase, you will want to think about the following deliverables:

- Technology stack
- A plan of how the SDLC will be executed
- Who will work on what components
- The requirements of the project
- User acceptance criteria

In traditional software engineering, some groups will pick the technology stack during the design phase. In automation, customers and projects can be pickier about the tech stack. For example, they may only use Allen-Bradley or Beckhoff PLCs and may only want specific software packages for HMI or SCADA systems. These requirements need to be established at the beginning of the project, which is why the technology stack needs to be identified as soon as possible. Waiting until a later stage of the SDLC can run the risk of losing thousands of dollars, if not more, in components and software, as well as running the risk of significant redesign.

As we established, software development is so much more than coding. Fast, efficient code is not the key to a quality project; having the code meet the customer's needs is. It is all too common for developers to try to fit a problem to a solution. Young, edgy programmers are usually more concerned with showing off their chops than getting the job done. A good manager and programmer can put that aside and develop a strict set of requirements to meet the customer's goal.

The following are some tips for collecting requirements:

- **Know your end users**: During the requirements/planning phase of the development life cycle, it is important to develop a good picture of what you're trying to accomplish and who your end user or users will be. If you can, it is a good idea to try to communicate with your end users during this stage. For example, I would generally like to speak with the operators in a one-on-one setting just to get a feel for who they are and what they know. Writing a program is a lot like writing a paper – you want to gear the program toward your audience. Some developers like to target leadership for this conversation. Though it is important to have their input, I found it much more productive to interview the people who actually work the machine.

- **Create user stories**: In the Agile methodology, there is a concept that is known as the **user story**. Essentially, the user story is a single sentence that describes the role of the user, the action, and the added value of the action. A general user story is as follows:

```
As a <role> I want to perform <action> to get <value>.
```

For example, we can write a user story such as the following:

```
As a technician, I want to reset the sensitivity of the sensor to
get better readings.
```

- **Write down your user stories**: Writing user stories is an Agile technique that is often not used in a PLC programming environment; however, in my opinion, this is one of the most powerful requirement-gathering techniques, regardless of which methodology you choose to develop your software with. Writing down user stories is an excellent way of conceptualizing functionality, especially in an environment that may consist of user-restricted operations. They also help prioritize functionality while allowing progress metrics to be used with proper Agile practices. Just from a pure organizational point of view, I would strongly recommend trying user stories in at least one project.

There is a lot to the art of gathering requirements. Learning how to gather requirements is a skill that takes practice. Put simply, to gather requirements, you have to learn what questions to ask, how to make customers feel comfortable, and understand your end user. There is a lot to the art—so much so that whole books have been written about this. Now that we have a general idea about the requirements, it is time to look at turning these requirements into a coherent design that will allow for expandability.

Designing the software

It is not uncommon for hiring managers to ask junior-level engineers in interviews how many phases of the SDLC they've worked across. Most junior engineers will typically spout off how much they code and love coding, but very few will dip into their design experience. This is mostly because, to many junior developers, coding is designing. Unfortunately, this is another attitude that can sink a project. Put simply, without a quality software design in place, you can utilize the best hardware and write the fastest code, but in the end, your project is doomed.

Outside of collecting requirements, the design phase is by far the most important.

I generally like to think of a program design as three parts:

1. **Overall architecture**: This is a high-level design of how the system will behave as a whole. This level will determine the necessary subsystems and their responsibilities, how data will flow between the subsystems, and the more granular technologies that will be used at both the hardware and software levels.

2. **Component architecture**: This level is concerned with how a subsystem will look and operate. This could be fleshing out how an HMI should work for a subsystem, getting database schemas together, implementing high-level software and hardware designs for the subsystem, choosing third-party libraries, and coming up with a general action plan to implement the system.

3. **Implementation of components**: This level is dedicated to designing the guts of a sub-system. In contrast to the component architecture phase, this phase is concerned with the details, such as producing pseudocode for a program, fleshing out an HMI wireframe, producing basic proof of concepts for how parts should work, and so on.

There are many ways to set up designs for a system. Common design techniques are as follows:

- **Flowchart designs**: A graphical representation that shows the flow of the program
- **Pseudocode designs**: A word-based representation of code that is used for drafting out a program without needing to worry about syntax
- **Wire frames**: A rendering of how the UI/HMI should look
- **Block diagrams**: How components interact with each other in a system
- **Electrical and wiring diagrams**: Electrical schematics
- **Mechanical diagrams**: Mechanical schematics

> OOP design
>
> When it comes to object-oriented programming, there is another design tool that can be used to flesh out a program's overall object-oriented design. This design method is called **Unified Modeling Language (UML)**, and it will be explored in *Chapter 10*.

This phase is very document-driven. Though it is not uncommon to produce some working proof of concept products, this phase will be primarily organizing thoughts and designs.

For now, it's important to understand that there is a major difference between the building phase (coding phase) and the designing phase. Your design is the foundation for both the hardware and the software of your project. If this phase is not fleshed out well, the rest of your project will have a very shaky foundation. However, when you do flesh out your design, you will be able to move on to every developer's favorite phase, the building/coding phase!

Building the software

After a quality design is produced, you can move on to what most developers live for: *coding*. Since everyone knows what programming is, and due to the amount of time we've already spent on programming in this book, this section will be relatively short. Coding is not the most important aspect of developing an excellent program. The heart of a program resides in the requirements and design. When those two steps are done correctly, the code should, for the most part, write itself. Now, this does not mean that writing code should be considered unimportant because, obviously, it is. No code ultimately means no product. The point of this section and chapter, in general, is to merely hammer home the point that code is not everything when it comes to software development. Coding is very important, but it is not the most important aspect of the development process.

With that in mind, how do we know whether our well-designed, well-crafted program meets the requirements and is working the way it should? The answer to that is testing. In the next section, we are going to cover what testing is and some types of testing.

Testing the software

Testing is never at the top of a developer's wish-to-do list; however, it is one of the most important steps in the SDLC. You may have just written the most sophisticated and fastest code ever, but unless you test it, it's trash. For many, testing is often confused with debugging, and many new programmers will often assume that if they debug the code enough to make it run, it's been tested thoroughly. I have personally seen seasoned PLC programmers with the same attitude. Regardless, this is a major fallacy that has killed many great projects before their time.

Software testing is not debugging. As we will explore in *Chapter 11*, debugging is the act of finding bugs, whereas testing is the act of ensuring the quality, reliability, and functionality of a program. As we will explore in *Chapter 11*, there are many ways to test a program, with some common, high-level test types being the following:

- Functional testing
- Non-functional testing
- Regression testing
- Acceptance testing

There are many more types of testing, and each of these has many of its own sub-categories. For now, it is enough to be aware of these.

Testing is a vital phase in the SDLC as this stage will ensure that your product, both hardware and software, will not only last the test of time but meet the customer's needs. A lack of under-standing of this phase is a deadly trap, as inadequate testing can not only ruin your organization's reputation but can lead to legal and, in extreme cases, criminal liability.

Once you have all the testing out of the way, you can move on to what many would consider the final phase of the SDLC, deploying your product.

Deploying the software

Depending on who you talk to, the deployment phase may be considered the final step in the SDLC. This is the phase where the magic happens. By that, I mean this is where you allow the intended user access to the product.

When it comes to PLC programming, this phase can be one or two things. When it comes to automation programming, this phase can be either installing your machine at its intended location or simply uploading a new program or patch to the PLC. Regardless, this is where your customer, whoever they might be, will get to use your product.

For a PLC integrator or machine builder, this phase will typically involve shipping and installing a machine at a customer site or a production environment of some kind. Ideally, in-depth testing would have been conducted in front of your customer or operator before you reached this phase, but if that did not happen, this phase could be easily marred with micro-modifications to the software and even the hardware at times. If you're deploying a whole machine, you can expect at the very least a few days of on-site support and training. Typically, when working out the details, it's a good idea to budget for at least one engineer or technician, who can plan to be on-site for at least a few days, if not a few weeks, depending on the machine.

In terms of deploying only software, this phase may be a little different. Deploying PLC software can be somewhat less invasive. In today's day and age, it is not uncommon for engineers to be able to remotely upload software to a PLC. If your organization offers a software product for a specific machine, the deployment phase may simply be uploading a PLC program from a desktop halfway around the world. On the other hand, if you're working on something such as a tool or a library, deploying it might be as simple as pushing your working project to a GitHub repo or some type of app store where it can be downloaded. Regardless, once you deploy your software, it's time to start maintaining it!

Maintaining the software

The final phase of the SDLC can vary based on who you talk to. For example, some consider the deployment phase to be the final phase, while others like to add a decommissioning phase to remove old systems. Regardless, the maintenance phase is often considered the final stage in the classic SDLC implementation. In this stage, you're going to be concerned with monitoring, fixing, and generally improving your product. In terms of a physical system, this could be fixing burned-out components or upgrading components such as PLCs, motor drives, sensors, and so on. In terms of software, this phase usually involves making changes to the codebase, such as bug fixes, upgrades, logic to support new hardware, and the like.

With the path that industrial technology is taking, this phase is radically changing for users who are using new and advanced computer-based PLCs and technologies in general. Until a few years ago, it was not unusual to have to send a programmer out to the machine to upgrade the software. A task such as upgrading the software for every machine in a factory or for every customer that owns a model of the machine could easily take weeks or months to complete, especially for machines in geographically dispersed regions. Nowadays, astute organizations can leverage automation tools and DevOps practices to instantly upgrade and patch machines around the world with the click of a mouse.

Many version control systems, such as GitLab, will come packaged with or support a third-party orchestrator that will allow you to create what's called a CI/CD pipeline. These pipelines will allow you to create automation scripts that will do various tasks to your code, such as performing basic tests and so on, and deploy it to an end user. There are also automation tools, such as Ansible or Puppet, that will allow you to configure devices remotely. These technologies are very advanced and are not designed with PLCs in mind, especially PLCs that are not PC-based, and their usage goes beyond the scope of this book. However, with the proper know-how and investment, these technologies can be leveraged to drastically cut down on your overall maintenance and even deployment efforts.

In all, for today's day and age, the organization that can learn to master these new technologies and DevOps will shape the future of the automation industry. More importantly, the engineers who can leverage these technologies and integrate them will be worth their weight in gold, especially for engineers who are working for organizations that are applying Industry 4.0 technologies.

Understanding the steps of the SDLC is only half the game. Implementing the SDLC is a challenge in itself; however, there have been a number of methodologies that have sprung up over the years that have made this task much more manageable. In the next section, we're going to explore three commonly used methodologies.

Understanding how to implement the SDLC

Knowing how to implement the SDLC is as important as knowing how to code. In modern-day software development, there are two core methodologies that are often used to implement a project. One is called the Waterfall method, and the other is called Agile. To begin our exploration into implementing the SDLC, we're going to first explore the older Waterfall method.

The Waterfall method

The **Waterfall method**, or **Waterfall model**, is one of the earliest documented SDLC implementations. It is a very rigid and error-prone methodology that, if not implemented correctly, can, and usually will, cost significant time in rework. The Waterfall methodology can best be summarized with *Figure 9.1*.

The Waterfall method uses the same core phases as the SDLC and is often depicted as a staircase. To get to the bottom, you must first descend down each previous step to get to the next. If you miss one step, you can easily tumble down to the bottom and get hurt. If there are any mishaps in any of the steps, the project is at risk of failure. To make matters worse, if there are any mishaps, development will have to backtrack to the phase that caused the mishap, and all subsequent steps must be redone.

This development methodology sounds like it could be a nightmare, and in many cases, it can be, especially with large and complex projects. However, this rigidness is often considered a benefit of the model. The Waterfall model works very well for mission-critical applications, such as things like space vehicles, medical devices, financial systems, and so on. This methodology is often considered to be very detail-oriented and document-driven, which makes it great for those applications and their respective industries. A derivative of this methodology is sometimes used is called the V-model. In the following section, we're going to take a quick look at the V-model.

The V-model

A derivative of the Waterfall method is the V-model. In the V-model, sometimes referred to as the Verification and Validation model, testing is done in parallel with implementation. This allows defects and general flaws to be uncovered faster and more effectively without the overhead of the Waterfall method. The flow for the V-model can be viewed in *Figure 9.2*.

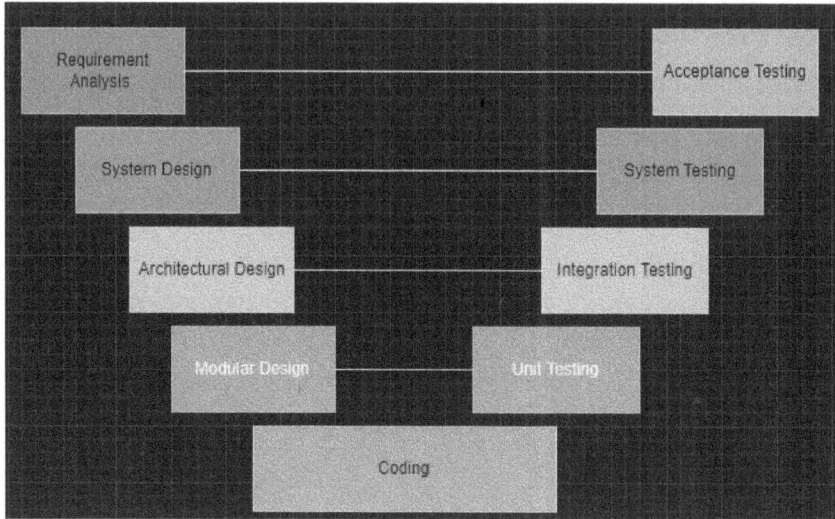

Figure 9.2: V-model

In this model, a development task on the left side of the V is associated with a corresponding testing task on the right.

The V-model is seen as an improvement over the Waterfall methodology, but it's not the dominant SDLC framework. Recently, there has been a relatively new model on the block that is becoming the new norm in implementing the SDLC.

The Agile framework

Agile has, without a doubt, become a major buzzword in the IT world. In fact, Agile has become so popular that many non-IT industries have adopted it. For example, it has become a staple in the legal and publishing industries. In its simplest form, Agile is a mindset or framework for working iteratively and collaboratively to deliver value and quickly adapt to change. In other words, Agile is like treating a project as a series of puzzle pieces.

In an Agile environment, you will break a project down into meaningful parts that can be completed in a certain amount of time, called a sprint, which is usually two to four weeks. As new pieces are delivered, they are assembled until the project is complete.

The key to Agile is collaboration and cross-functionality. For example, one group may be responsible for implementing welding robots for one assembly line, while another team might be responsible for the sanding robots of another. Both teams have the skill to do the other team's job, but they each have their own responsibility. In essence, these teams are not siloed; rather, they are feature focused.

To facilitate collaboration and cross functionality, Agile will typically use a series of ceremonies that attempt to accomplish this. A ceremony is usually a time-boxed meeting with a specific goal. Common Agile ceremonies are as follows:

- **Sprint planning**: This phase is designed to pull in the work the team hopes to accomplish in a sprint. Typically, there are features in what's called a backlog that need to be implemented. During the sprint planning phase, the team will decide on the most appropriate features to complete during the sprint and commit to finishing them during that allocated time slot.

- **Daily standup**: The daily standup or daily scrum is a short meeting that is held each day. The goal of the meeting is to allow engineers to give a status report, highlight any blockers that are causing them problems, and generally stay aligned. Most Agile purists will timebox this meeting to about 15 minutes; however, depending on the size of the team, it can easily creep into the 30-minute to one-hour range in extreme cases. In my opinion, this is arguably the most productive meeting because it provides a time when engineers can seek guidance while providing managers and team leads with a clear status.

- **Sprint review**: This is a meeting that is typically held at the end of a sprint. The goal of this sprint is to showcase the work completed and allow stakeholders to provide feedback. Potential adjustments are discussed here in relation to the stakeholders' feedback. Depending on the length of the sprint, this meeting is usually one to two hours, though it can go longer, depending on the workload and discussions.

- **Retrospective**: This is another meeting that is held at the end of the sprint. The goal of this meeting is to determine what went well and what could be improved for the next sprint. This meeting should be time-boxed to about one to two hours.

- **Backlog refinement**: This is an ongoing activity that prepares the backlog (an ordered list of work items—such as features, bug fixes, and improvements—to be implemented) for the next sprint by prioritizing items. This is usually a one-to-two-hour meeting.

Depending on the flavor of Agile that you use, there could be more ceremonies as well. It is also important to note that you don't have to follow all these ceremonies to be Agile. Agile is a framework, which means you can pick and choose what works best for your team. Many organizations feel that Agile has too many meetings, while others feel it does not have enough. In short, it can be customized to fit your needs.

Agile is a cultural thing. This means that it will only work as well as the people who follow it. If you opt to adopt an Agile environment, you want to roll it out slowly. Rolling the full Agile process out too fast will cause resistance from employees who are not used to it and can lead to the utter failure of its implementation. Remember to take your time with implementing the process and listen to any feedback the team(s) have.

In all fairness, Agile is not for every organization. Most Agile resources will say that Agile can be for small to large groups, which is very true. However, for very small teams, Agile ceremonies such as the daily standup can be just as much of a hindrance as a way to increase productivity. Due to the nature of many automation shops, Agile may not work well, especially if your projects are relatively short and you have a small overall engineering team (usually less than five engineers). If the engineering staff for a project is over five people and the project is large enough to be meaningfully broken down, Agile might be something to explore. In all, whether Agile is right for your organization is up to you. It has many great features, such as the daily standup and backlog refinement, but ultimately, the choice to implement Agile is up to your team.

There are many other SLDC implementations, such as the Spiral model, the Iterative model, and more. For now, we're going to switch gears and get some practice implementing the SDLC. For our final project, we're going to use the SDLC to deploy a simple program!

Final project: Creating a simple temperature converter

Now that we have explored the SDLC, we are going to apply what we learned and build a full project with those principles. The following section will be dedicated to building a temperature conversion program.

Gathering requirements for the program

As we have discussed in this chapter, the first thing we need to do is determine the requirements for the project. Our goal is to create a temperature converter like the one we built before. The program will need to be able to convert between all temperature units. We can say our requirements are the following:

- Should convert Fahrenheit to Celsius and Celsius to Fahrenheit
- Should convert Celsius to Kelvin and Kelvin to Celsius
- Should convert Fahrenheit to Kelvin and Kelvin to Fahrenheit

With these requirements, we can move on to the design phase.

Designing the program

This temperature converter program shouldn't be that complicated to build. For the most part, all we need are a few methods in a function block to pull this off. Therefore, due to the simplicity of this project, a pseudocode program to flesh out our ideas will suffice. For this project, we can use a simple function block that we'll call TempConverterFB, which will house all the necessary methods to convert temperatures.

For this project, we could use the following:

```
TempConverterFB Function Block
CtoF method:
    CtoF = ((temp * 9) / 5) + 32;
CtoK method:
    CtoK = temp + 273.15;
FtoC method:
    FtoC = ((temp - 32) * 5) / 9;
FtoK method:
    FtoK = (((temp - 32) * 5) / 9) + 273.15;
KtoC method:
    KtoC = temp - 273.15;
KtoF method:
    KtoF = (((temp - 273.15) * 9) / 5) + 32;
```

As can be seen in the pseudocode, each method is simply going to be a single line that will perform a math calculation. With this simple design in place, we can move on to coding it up!

Building the project

Now that we have finalized the requirements and design, we can move on to doing what developers love to do: writing code. Since we have a decent design in place, we can now easily start implementing the code. Based on the pseudocode, our program structure will look akin to the following:

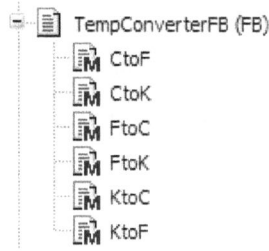

Figure 9.3: TempConversion project structure

Each of these methods will have a single input argument that we'll call temp. The temp variables will be of the type REAL; hence, be sure to include that in each of the methods' variable blocks. Also, each method will have a return type of REAL and an access specifier of PUBLIC. To implement these methods, we are going to use the following code:

- CtoF method:

  ```
  CtoF := ((temp * 9) /5) + 32;
  ```

- CtoK method:

  ```
  CtoK := temp + 273.15;
  ```

- FtoC method:

  ```
  FtoC := ((temp - 32) * 5) / 9;
  ```

- FtoK method:

  ```
  FtoK := (((temp - 32) * 5 ) / 9) + 273.15;
  ```

- KtoC method:

  ```
  KtoC := temp - 273.15;
  ```

- KtoF method:

  ```
  KtoF := (((temp - 273.15) * 9)/5) + 32;
  ```

This will be all the code needed for our project. The next thing we need to do is test the code.

Testing the program

This program is incredibly simple. For a program of this size and complexity, unless specifically told otherwise, it is probably easier to unit test the code manually, that is, input values ourselves as we run the code. To do this, we need to create a series of simple method calls, record the output of the program, and compare it to the expected values.

The steps for testing our program are as follows:

1. Add the following variables:

    ```
    PROGRAM PLC_PRG
    VAR
        tempConverter : TempConverterFB;
        unit1 : REAL;
        unit2 : REAL;
        unit3 : REAL;
    END_VAR
    ```

2. Next, modify the PLC_PRG file of the program to match the following:

    ```
    unit1 := tempConverter.FToC(33);
    unit2 := tempConverter.FToC(-100);
    unit3 := tempConverter.FToC(500);
    ```

When the code is run, you should get what's shown in *Figure 9.4*:

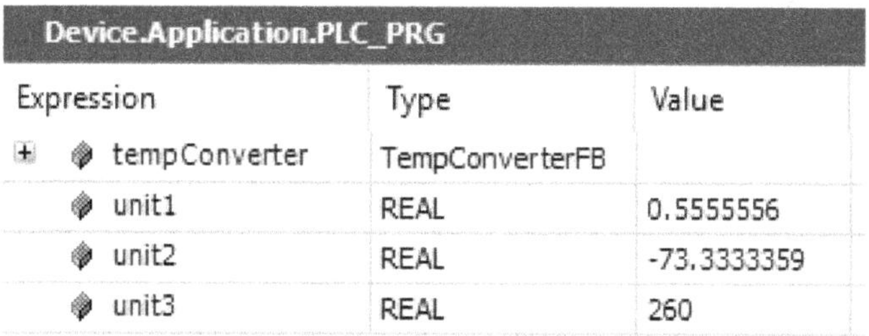

Device.Application.PLC_PRG		
Expression	Type	Value
⊞ ◈ tempConverter	TempConverterFB	
◈ unit1	REAL	0.5555556
◈ unit2	REAL	-73.3333359
◈ unit3	REAL	260

Figure 9.4: Unit test output

To verify the output, crunch the numbers manually. That is, create a spreadsheet with the equations and inputs, or simply run the numbers by hand. What you'll notice is that the values you calculate should match the program's output. If all goes well, the program will work, and we'll be ready for deployment!

Deploying the project

At this phase, we have tested the program, and we are ready to send it out into the world. For automation engineers, this usually means shipping your machine or installing your patches. Depending on what you're working on, this could also mean deploying your code to a repository such as GitHub or some other public venue for others to download.

> Do it yourself
>
> For this book, we are only going to make the software available on GitLab. As an exercise, create a new repo, as we did in *Chapter 8*, and push the project to it. This will simulate deploying a project to a public repo, as engineers typically do with GitHub.

Maintaining the program

Now that you have a working product that you have deployed in some way, you can look at how you want to expand it or fix any bugs that may arise in the software. This is a prime opportunity for you to take the program we built and expand on it while keeping true to the SDLC. Ideas you can use to expand the program would be to include some of the following:

- Temperature input limits
- Add other unit conversion function blocks
- Improve the general usability of the software

At this point, you should try to think of something to do to modify the software. Whatever you do, start at the first phase and work your way down to re-deployment. I would recommend trying to add a few different features using this methodology just to get the hang of everything.

Summary

It is pivotal for any software developer to understand the full gamut of the SDLC, as the SDLC is a guide to properly flesh out software. No matter what you're doing, you should always follow the SDLC as closely as you can so that your software will be easy to build, fix, and expand upon in the future.

This chapter was a crash course in the SDLC, the methodologies that govern it, and the steps that it encompasses. Of all the chapters, I would argue that this is one of the most important. Too often, developers get caught up in what I like to call the *code culture* of just blindly building things with no roadmap of where they are, where they've been, or where they're going. Being able to navigate the SDLC will set you apart from those developers, as in-depth knowledge of the SDLC is what separates an engineer from a programmer. With these principles under your belt, you can step out and build software that will be extraordinary.

When it comes to OOP and design, pseudocode alone can be insufficient to highlight the relationships in a program. Understanding how your function blocks will interact with each other is very important as your project ages and scales. To highlight these relationships, there is a design technique called UML that we touched on previously. This technique can be used to show how your function blocks relate to one another. In the next chapter, we're going to leverage UML and use it to design a program!

Questions

1. What is Agile?
2. What are the main ceremonies in Agile?
3. Define the SDLC.
4. How many steps are there in the SDLC?
5. What is the Waterfall method?
6. What is the most important phase of the SDLC?
7. Name all the phases in the SDLC.
8. What is the build phase of the SDLC?

Further reading

- *Waterfall model*: https://en.wikipedia.org/wiki/Waterfall_model
- *Agile Alliance, The 12 Principles Behind the Agile Manifesto*: https://www.agilealliance.org/agile101/12-principles-behind-the-agile-manifesto/

Get This Book's PDF Version and Exclusive Extras

UNLOCK NOW

Scan the QR code (or go to packtpub.com/unlock). Search for this book by name, confirm the edition, and then follow the steps on the page.

Note: Keep your invoice handy. Purchases made directly from Packt don't require an invoice.

10

Architecting Code with UML

In the previous chapter, we explored the SDLC in all its glory. As we learned, designing a system, whether it's electromechanical or software, is vital to the success of a project. Throughout this book, we've also used pseudocode as a means of designing our software. Pseudocode is an excellent design tool; however, the key drawback is that its main purpose is to help us flesh out an algorithm. It's true that it can be used for object-oriented designs, but it's not ideal for that task. The most optimal way to flesh out an object-oriented design is to use a design system called **Unified Modeling Language (UML)**.

UML is a design tool that can be used to show the relationships between function blocks. In other words, if each function block in an object-oriented PLC program is a puzzle piece, the UML diagram is the assembled picture. UML is an excellent tool to help us understand the design, understand what components rely on what, and avoid costly mistakes in the long run.

To learn about UML, we're going to explore the following topics:

- Understanding what UML is
- Understanding what UML is used for
- Understanding why UML is important
- The basics of drawing a UML diagram

To wrap things up, we're going to use UML to design an object-oriented program that will model a car.

Technical requirements

UML diagrams can be drawn with almost any medium. You can use programs such as Microsoft Visio, PlantUML, or even a pen and paper. In many brainstorming situations, engineers often use a whiteboard to flesh out ideas using UML. What method you use for this book is up to you. However, the diagrams in this book have been rendered with a website called **draw.io**: `https://app.diagrams.net/`.

If you want to follow along and use it, you are free to, but if you feel more comfortable with a different medium, you can use that. In the world of UML, it really doesn't matter what you use, so long as your diagrams convey your program well!

Understanding UML

In its most basic sense, UML is a method of designing object-oriented programs on paper. Moreover, it's a design tool that allows engineers to map out how function blocks or classes will interact with each other. UML diagrams also help engineers plan out what methods each function block will have, as well as their access specifiers. In a well-crafted UML diagram, a method's return type and arguments will also be stipulated.

An example of a very simple UML diagram can be viewed in *Figure 10.1*:

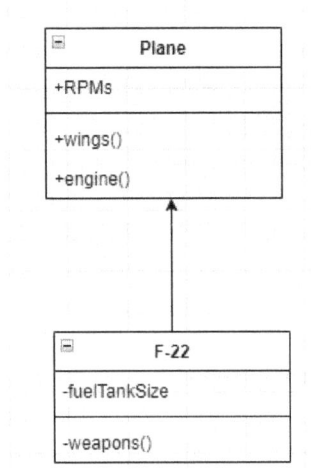

Figure 10.1: A simple UML diagram

A real-world diagram would be more complex; however, what *Figure 10.1* demonstrates is the basic concept of a UML diagram. To really appreciate the concepts surrounding UML, we need to explore why they are important.

What is UML used for?

An electrical or mechanical engineer would never begin working on a project without an electrical schematic or blueprint of some kind. A big oversight that many automation programmers, and even traditional software engineers, make is that they do not invest the time or effort into properly designing their software before they start writing it. This mentality harks back to software being treated as a second-class citizen in automation, something that has been explored throughout this book! Essentially, not using UML to map out your software project is like trying to build an electrical circuit without drawing an electrical schematic. Can a program be built without a UML diagram? Sure; however, doing so will drastically increase the probability of flaws occurring in a system and will likely reduce the overall lifespan of the program.

There are a few types of UML diagrams that can be used for different things. Arguably, the most common is the **class diagram**, an example of which is shown in *Figure 10.1*. As has been alluded to, this type of diagram is used to demonstrate the relationships between classes or function blocks.

The following are other common types of UML diagrams:

- Sequence diagrams
- Object diagrams
- Component diagrams
- Activity diagrams
- Timing diagrams
- Communication diagrams
- Package diagrams
- Profile diagrams
- Use case diagrams
- State machine diagrams

For this book, we're going to focus on using class diagrams; however, sometimes, we will use a stripped-down variation of class diagrams that only reference the function block's name and not necessarily the function block components, such as the methods and variables. These stripped-down UML diagrams are very common in high-level designs and design discussions.

Why is UML important?

Object-oriented projects can quickly devolve into sheer chaos if they are not properly planned. When not planned out, function blocks can easily turn into dumping grounds for methods. Relationships between function blocks can easily become ways of side-stepping OOP rules and make no logical sense while the overall codebase can become nearly impossible to troubleshoot and maintain. When the SDLC is followed properly and UML diagrams are produced, these difficulties can be alleviated. By taking a little time upfront and mapping out the relationships and internals of a function block, copious amounts of time can be saved in troubleshooting and scaling.

At the end of the day, UML is just a tool. Just because you UML out a program doesn't mean that your project is going to automatically become a quality codebase. Nonetheless, UML diagrams can help ensure quality by exposing potential trouble spots and bad relationships between function blocks, relaying design information to other engineers, and more. Therefore, the next step in fleshing out UML diagrams is learning how to draw them.

The basics of drawing a UML diagram

As we saw in the previous sections, UML drawings are essentially composed of a series of boxes and arrows. Based on that information, what can be thought of as a box is representative of a function block for PLC programs or a class in a traditional OOP language. As explored previously, a quality UML diagram will contain the methods and variables that are contained in the function block. To begin our exploration of UML, we're going to explore how to represent a function block in UML.

> **Note**
>
> Drawing a UML diagram is like drawing an electrical schematic. Everyone will usually have their own style, and the only real requirement is to ensure the reader can follow the diagram and implement what is being conveyed.

Representing function blocks in UML

Even though function blocks or classes are very easy to represent, a few details need to be followed to ensure the function block is represented accurately. To begin our exploration of UML function blocks, let's examine *Figure 10.2*.

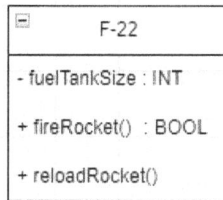

Figure 10.2: F-22 function block

As can be seen, there is some information here that we need to unpack, starting with the function block's name.

UML name

The first key piece of information that is conveyed in a UML diagram is the function block's name. Typically, the function block's name is at the very top of the box. In this case, it's F-22. This name should stand out in the diagram. There will usually be many function blocks in a given diagram, and it can sometimes be hard to pinpoint a specific function block under the best of circumstances. To help make the function block easier to find, you will typically want the name to stick out. As such, it is common to have the name in bold or a slightly larger font, centered at the top of the UML representation to help make the function block easier to pinpoint.

Names are just one piece of information a UML representation of a function block is meant to convey. Two other vital components that UML helps convey are methods and variables.

Representing methods and variables

The key components in any modern function block are its methods and variables. Therefore, it is very important to accurately represent these components in the UML diagram. To do this correctly, there are two general rules you must follow:

1. Variables go below the function block's name but above the methods
2. Methods go below the variables and are usually the last entities to be defined in the UML representation of the function block

Another key piece of information that we need to convey in a UML diagram is the access specifier of a component. In the next section, we're going to explore how to represent these in UML.

Representing access specifiers

With one of the core aspects of OOP being hiding data/complexity and having the ability to expose certain function block entities, we need to clearly depict what is exposed and what is hidden in the diagram. To do this, drafters will usually place a symbol next to the function block component to represent its access specifier. Typically, the following symbols are used to indicate whether a function block is private, public, or protected:

- +: Public
- -: Private
- #: Protected

Our function block, shown in *Figure 10.2*, contains a variable and two methods. In this case, the `fuelTankSize` variable has a minus sign by it. This means that the variable is meant to be private. On the other hand, the `fireRocket` and `reloadRocket` methods have plus signs by them, which means they are meant to be public.

UML data types and arguments

A well-drafted UML diagram will usually have a variable's data type next to it, and a method will usually have a return type and arguments. For example, the UML block we have for F-22 has `INT` by `fuelTankSize` and `BOOL` by the `fireRocket` method. In this case, the unique method is `reloadRocket`, which has nothing by it. In our scenario, this is meant to signal that the method returns nothing.

Arguments and data types are details that are often overlooked by many engineers as they are typically not of vital importance to the overarching design and will often change during implementation. Sometimes, leaving the arguments, variable types, and return types out of the diagram can make the diagram a little more flexible, as it is often hard to nail down that level of detail in the design phase of the software. A technique that I like to use involves making a small chart off to the side that contains the data and return types for function block components. This keeps the diagram's layout more generic and will allow flexibility for these changes. It also helps keep the diagram cleaner as it prevents clutter when there are methods with many arguments.

Throughout this book, we will typically take a shorthand approach to drawing UML diagrams. For example, to demonstrate things at a high level, we may only use the name of the function block. As we've explored, this shorthand approach is commonly used for brainstorming sessions and rough outlines. Regardless, one key piece of information that always needs to be conveyed in the diagram is the function block relationships.

Understanding UML relationship lines

When it comes to drafting UML diagrams, the symbols your rendering system uses may vary. UML programs are like electrical CAD programs; each program may have slightly different representations for the same concept. So, be aware that the design of the symbols that we're about to explore may vary a bit from program to program!

The basic UML relationship symbols

As an automation engineer, you will arguably use inheritance and composition the most. Therefore, we're going to focus on those two relationships. To start, we're going to explore how to represent inheritance in UML!

UML for inheritance

To represent inheritance, we typically use either of symbols shown in *Figure 10.3*:

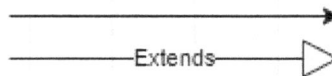

Figure 10.3: The inheritance symbols

This arrow should point at the parent function block. For example, suppose we're working on a bank program. Our bank program may need to support both a savings account and a checking account. In this case, we can implement the core functionality for both accounts in one function block and the details of the other types in two other specific function blocks. In this case, the specific function blocks will need to use the base functionality of the parent. A rough UML diagram would look like this:

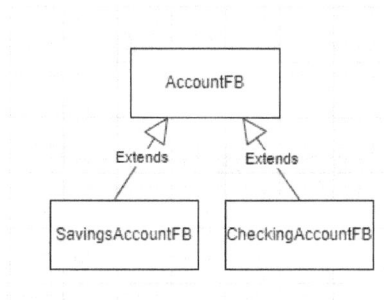

Figure 10.4: A simplified UML diagram that shows inheritance

Generally, composition relationships will look similar.

UML for composition

The composition relationship has a similar arrow operator. To denote composition, we typically use the symbol shown in *Figure 10.5*:

Figure 10.5: Composition arrow

Much like the inheritance symbol, the diamond part of the arrow will point to the composite function block. Suppose we have a UML diagram for a car. Since a car has an engine, brakes, and a transmission, the UML diagram would look like what's shown in *Figure 10.6*:

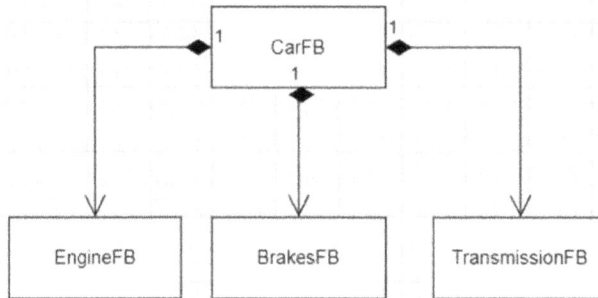

Figure 10.6: A UML diagram that shows composition

The goal of *Figure 10.6* is to construct a car function block. Since the car is composed of a transmission, brakes, and engine, the diamonds from those blocks point to that function block.

Drawing a diagram is only one phase of the UML process. To use a UML diagram, we need to be able to convert one into code!

Converting UML diagrams into code

To learn how to convert a UML diagram into code, we're going to take a UML diagram for a vehicle and turn it into pseudocode that can later be converted into a PLC program.

Figure 10.7 shows a vehicle UML diagram with six function blocks:

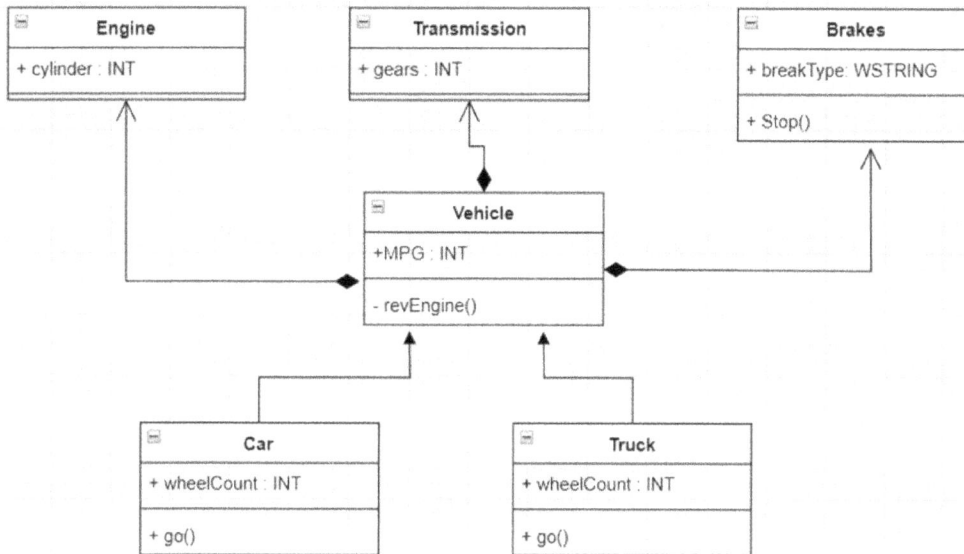

Figure 10.7: A vehicle UML diagram

The heart of the program is the Vehicle function block. This function block is composed of an engine, a transmission, and brakes. Two vehicle types are derived from the Vehicle function block: Car and Truck.

When it comes to converting UML diagrams into actual code, you'll want to implement a strategy to keep yourself organized. Everyone follows a different approach, but in my opinion, the easiest thing to do is to convert each block one at a time. Therefore, we're going to start by turning the three function blocks at the top of *Figure 10.7* into pseudocode.

Engine function block

This function block only contains a single public variable:

```
Function Block Engine
Vars:
    Public:
        cylinder : INT
```

Transmission function block

Similar to the engine function block, this function will also only contain a single public variable:

```
Function Block Transmission
Vars:
    Public:
        gears : INT
```

Brakes function block

This function block is a little more complex; the UML diagram shows that this function block will have a stop method and a brakeType variable. The variable is pretty standard; however, notice the method. As we can see, the diagram did not show any logic. This is often considered a drawback of UML since diagrams of this type will usually only show the overall structure of the program, not the granular details of what the methods do and how they work. For this example, we're going to assume that this function is only going to stop the vehicle:

```
Function Block Brakes
Vars:
    Public:
        brakeType : WSTRING
Methods:
    stop():
        stop vehicle
```

Vehicle function block

This function block is structured similarly to the Brakes function block in that it has both variables and methods. A big difference though, is this function block will contain references to the other function blocks. Notice that there were no reference variables in the UML diagram; this is because those references can be inferred from the arrows. Some drafters will insist on including reference variables; however, if you follow proper naming conventions and understand the symbols, you should be able to easily infer the references:

```
Function Block Vehicle
Vars:
    Public:
        MPG         : INT
        transmission: TransmissionFB
```

```
        brakes      : BrakesFB
        engine      : EngineFB
Methods:
    Private:
        revEngine():
            rev the engine
            use other function block code
```

Car function block

This function block is a derived function block. To showcase this in terms of pseudocode, we can use the EXTENDS keyword to signal inheritance. Therefore, we can draft out the pseudocode with the following:

```
Function Block Car EXTENDS Vehicle
Vars:
    Public:
        wheelCount : INT

Methods:
    Public:
        go():
            Make car go
```

Truck function block

This function block is almost a copy of the Car block. It has the same methods and variables, with the only differences being the function block's name and the logic in the go method. This function block is also derived from Vehicle, and we can represent it with:

```
Function Block Truck EXTENDS Vehicle
Vars:
    Public:
        wheelCount : INT

Methods:
    Public:
        go():
            Make truck go
```

Chapter challenge

Practice makes perfect. To practice what you've learned so far, convert the following code into a UML diagram:

```
Function Block Engine
Vars:
    Fuel : WSTRING
Methods:
    Public:
        consumeFuel()
            consume fuel


Function Block Plane
Vars:
    Public:
        engine    : Engine
        planeType : WSTRING
Methods:
    Private:
        liftLandingGears()
            tuck landing gears
    Public:
        turnPlaneOn()
            start plane


Function Block F4Phantom EXTENDS Plan
Vars:
    Public:
        Weapons : WSTRING
Methods:
    Public:
        fireRockets()
            Fire rockets
```

Once you've finished this challenge, you can move on to the final project!

Final project: Modeling a program representing multiple cars

For our final project, we're going to rework the car example. To do so, we're going to model a program that represents multiple cars. This program will model an electric car and an internal combustion engine-based car.

Getting started

Let's take a look at the basic requirements for each type of car. For this example, we will assume that each type of car (each solid bullet point) is a function block.

All cars have the following:

- A steering wheel
- Tires
- Brakes

All electric cars have the following:

- A battery
- A charger

All internal combustion engines have the following:

- A gas tank
- A piston
- An oil pan

With the requirements established, the next thing we need to do is analyze them to determine the relationships between the function blocks.

Relationship analysis

Just from the general outline, we can see that an electric car and an internal combustion engine car are both cars. This means that these function blocks have an "*is-a*" relationship with a Car function block. Therefore, for these two functions blocks we will utilize inheritance with a more abstract Car function block.

Analyzing the requirements further, we can see that all cars have a steering wheel, tires, and brakes. Notice the "*has-a*" relationship. This means that for these function blocks, we will use composition. If you use this same logic for the other bullets, you'll see that they all share the same relationship.

Relationship summary

The relationships for the function blocks can be summarized as follows:

- **Car**: This will be composed of the brakes, tires, and steering wheel blocks. This function block will serve as a parent for both the electric and the internal combustion engine car.

- **Electric car**: This will be composed of the battery and charger function blocks. This function block will also inherit from the Car function block.

- **Internal combustion engine car**: This function block will be composed of the gas tank, piston, and oil pan function blocks. Like the electric car, this function block will also inherit from the Car function block.

Now that all the relationships have been fleshed out, we can create the UML diagram.

Basic UML diagram

Figure 10.8 is a general representation of the program.

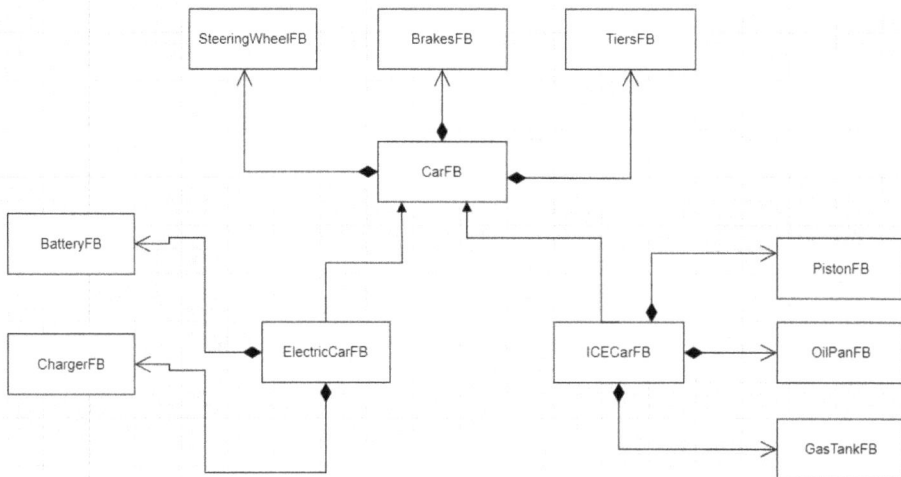

Figure 10.8: Skeleton diagram

As can be seen, all we have implemented are general function blocks and their relationships. As can be deduced, this diagram does not contain any methods or variables. This mostly stems from no methods or variables being stipulated.

Chapter challenge

For the final challenge, redraw the diagram so that it includes the following function block attributes:

- Steering wheel:
    - Methods:
        - Public: `directionToGo(WSTRING)`
        - Private: `honkHorn(WSTRING)`
- Brakes:
    - Methods:
        - Public: `stop()`
- Tires:
    - Variables:
        - Public: `make : WSTRING`
- Car:
    - Methods:
        - Private: `rev()`
        - Public: `shift()`
- Electric car:
    - Variables:
        - Private: `type : WSTRING`
- Internal combustion engine car:
    - Variables:
        - Private: `gasType : WSTRING`

For the charger and battery function blocks, add two private variables of the `WSTRING` type. For the combustion engine block, add variables for `oilPan` and `piston` that are of the `WSTRING` type. Finally, add a private variable called `tankSize` and a getter and a setter method, respectively.

Summary

This chapter has been a crash course on creating UML class diagrams. In this chapter, we explored what UML diagrams are, what they are used for, how to draw them, and more. UML is a great design tool that can help flesh out things such as function block names, method names, and variables. It is also a very handy tool for fleshing out function block relationships.

Many engineers skip designing their programs. For a large project such as one that would represent a piece of industrial equipment, this is a recipe for disaster. It is very important to flesh out function blocks and their relationships. Not doing so can mean introducing function blocks with illogical relationships, names, and more, which can utterly kill the long-term survivability of a program.

Regardless of whether you use UML, no program is ever perfect. There will always be misunderstandings, code that does not work, and so on. So, in the next chapter, we're going to explore testing and troubleshooting code!

Questions

1. What does UML stand for?
2. Name three types of UML diagrams.
3. What order is information defined in a UML function block?
4. What symbol is used to represent inheritance?
5. What symbol is used to represent composition?
6. Why is UML important?
7. What is class diagram UML for?
8. What symbol is used to denote a private function block?
9. What symbol is used to denote a public function block?
10. What symbol is used to denote a protected function block?

Join our community on Discord

Join our community's Discord space for discussions with the authors and other readers: `https://packt.link/embeddedsystems`

11

Testing and Troubleshooting

No code will ever work right on the first go. No matter how good a programmer someone is, there will always be errors in a program. This is where testing and debugging come into play. Testing and debugging are vital concepts in software engineering and, by extension, automation programming.

The terms testing and debugging are often used interchangeably, but they refer to two radically different techniques that are designed to tackle two different problems. This chapter will explore and contrast the two concepts.

The goal of both testing and debugging is to produce high-quality code that works. To understand this, we're going to explore the following concepts:

- The difference between debugging and testing
- Verification and validation
- Various types of testing
- Debugging tools and techniques
- How to use systems such as ChatGPT to debug code

Finally, to close out the chapter, we're going to test and debug a program!

Technical requirements

To follow along with this chapter, you're going to need the code, which can be downloaded from this URL: https://github.com/PacktPublishing/Mastering-PLC-Programming-Second-Edition/tree/main/Chapter%2011.

Also, to follow along with the AI aspect of this chapter, you're going to want to create a ChatGPT account if you do not already have one. You can do that here: `https://openai.com/chatgpt/overview/`.

If you have experience with another AI platform, such as Copilot, you can use that as well.

Difference between debugging and testing

Testing and debugging are two terms that are often used interchangeably. It's not uncommon to hear an engineer say that they need to test their code when they mean they need to debug it, and vice versa. The first step in truly understanding the differences between the two concepts is to explore what debugging is.

What is debugging?

Debugging is the art of finding and removing bugs from your code. Every developer will debug many times throughout the life cycle of a program. It is very rare for a program to compile and run correctly the first time, especially after a long day. It is not uncommon for a developer to accidentally miss a period, comma, or misspell a word when writing their code. On top of that, even if the code does compile, there is no guarantee it will run as intended.

Common types of bugs are as follows:

- **Syntax errors**: This type of bug is triggered by invalid syntax. Syntax errors will usually prevent your program from compiling and running. Normally, CODESYS, as well as other **integrated development environments (IDEs)**, will produce a red squiggle line under the offending syntax.

- **Logic errors**: Similar to syntax errors, logic errors are very common. These are issues in your logic that will cause wrong outputs, wrong computations, and so on. These bugs can be a little more dangerous than syntax errors because you won't know about them until the program runs, and since these are caused by compliant syntax, you usually won't get a red squiggle line. These defects can cause infinite loops, crashes, and software failures, which, in terms of automation, will cause machine failures and other potentially dangerous situations. These are the bugs that are best found via a debugger tool.

- **Functional errors**: Functional errors are bugs in the program that prevent the software from working as expected. For example, a functional error may come in the form of a button that does not turn on the correct assembly line when pressed. These errors are not the result of bad syntax, nor are they the result of logical errors. Depending on the severity of the bug, the issue can range from a minor inconvenience to dangerous machine malfunctions. Normally, these bugs are discovered during the functional testing phase of the machine's development. In other words, these bugs are usually discovered when you're testing the machine to ensure it works as expected. Debugging tools can also help find and fix these defects.

Debugging is concerned with finding these errors and removing them. This is in contrast to testing.

What is testing?

While debugging is the art of finding defects in a program, testing is the art of ensuring that a program works the way it is intended to. This may seem similar to debugging, but it isn't. An example of debugging would be to figure out why a program isn't compiling. In contrast, testing is the art of ensuring the system can return a valid value. In other words, suppose you're working on a packaging robot; testing would be concerned with ensuring the robot packages sacks with the correct weight of material and pushes them down the correct line. There are various types of testing, and each type has its own goals to ensure the software works. In the following section, we're going to explore these types.

Verification and validation

Testing generally falls into two categories. One is known as verification testing and the other is validation testing.

What is verification testing?

Verification ensures that the software is high quality and works. When you employ some form of validation testing, you're ensuring that the program is as bug-free as possible and works as expected with no issues. Verification boils down to answering the question, are we building the system correctly?

What is validation testing?

Validation is the process of ensuring that the program solves the original problem. In short, validation is the process of making sure that the program meets the user's needs. In contrast to verification, validation boils down to answering the question, are we building the right system?

Engineers, including me, will often try to box different types of testing into verification and validation testing. However, whether a particular type of test is a validation test or not is mostly theoretical, and there are gray areas. Some types of testing, such as regression testing, can often fit in both categories depending on the context they are being used in. In the next section, we're going to explore some of the basic types of testing and their use cases.

Various types of testing

Some types of testing are more common than others. However, each type of test has its use and will attempt to determine whether an aspect of the system meets its requirements.

Exploring test cases

The backbone of any type of testing is the **test case.** A test case is essentially a procedure for testing a unit of the system. A test case can be thought of as a record of sorts that details the following information:

- The test date
- Who performed the test
- The code blocks that were tested (if possible—not necessary)
- The input values
- The expected output
- The actual, recorded output
- Whether the test passed or failed

You technically don't have to keep a physical copy of the test results, at least for many industries. However, some industries will require you to keep records, especially when it comes to mission-critical applications such as medical devices.

In terms of producing a test case, you don't need anything fancy unless it's specifically requested. For many automation organizations, simply using a spreadsheet that contains the aforementioned information will suffice. However, some industries, such as the medical and banking industries, may require you to use specialized software. Remember, though, that in terms of automation, this is not typically the norm.

To fully understand why we need a test case, we need to understand what a test is. To do this, we're going to explore what is arguably the most common type of testing: unit testing.

Unit testing

One of the most common types of testing is **unit testing**. Unit testing is testing code blocks that provide meaningful value. This can be testing functions, methods, or whole function blocks. Generally, you are testing the smallest block(s) of code that can return a meaningful result.

When possible, I like to test at the method or function level, especially when these blocks are public. If possible, you will want to find and use a framework that will allow you to write automated unit tests; however, as mentioned earlier, in automation programming, it is not always possible to do this due to the nature of the industry. Therefore, you will often need to test manually. Usually, what I like to do—and this could be argued to be a bad practice—is to modify my code with a series of output statements or return values to track the code flow and output, respectively. For example, if I'm working on a math library that consists of add, subtract, multiply, and divide methods, I would normally feed in simple test values and watch the outputs in the variable window as we have done so many times in this book.

> Note
>
> When unit testing, it is a good idea to either find a unit test form online or create a spreadsheet that will keep track of the information that was noted in the *Exploring test cases* section.

When it comes to unit testing, you want to shoot for at least 80% coverage. This means that, at a minimum, you want to ensure that at least 80% of the code is tested with your unit tests. This does not mean that one test has to cover 80% of your code. What it means is that you have a series of tests that, when combined, have run at least 80% of the code in the codebase. To calculate your code coverage, you would use the following equation:

*Code Coverage = (Number of lines executed by unit tests/Total number of lines in program) * 100*

Depending on what you're working on, 80% may not be enough. For example, if you're working on a medical device or a device for the military, the code coverage may need to be increased depending on industry standards. You will also want to run the same unit test multiple times with different values, including values that may accidentally be entered.

> Note
>
> Remember that 80% code coverage is usually the minimum amount of code coverage for a project. While some systems may require more than an 80% code coverage rating, it is usually rare to see anything above 90%.

Dead code and unreachable code can act as a gotcha for this metric. Essentially, dead code and unreachable code cannot run but are still in the codebase. This means that those types of code will be factored into the calculation, especially if static code analysis tools are used, which in turn can throw off your coverage metrics. As a result, you could have a much higher code coverage percentage, but your metrics might appear to be much lower. This is a prime example of why it is important to remember that, in terms of code, if you don't use it, lose it! Where unit testing is largely about testing the actual code, functional testing is more about testing the system as a whole.

Functional testing

Put simply, **functional testing** validates your software—in the case of automation, hardware too, against the requirements that were set out during the requirements phase. For this type of testing, you want to use someone who might be familiar with the overall gist of the machine but probably not someone who spent a lot of time on the code. This isn't always possible, but it is ideal if you can manage it. Generally, functional testing is considered a form of **black-box testing**. During this phase of testing, you are not concerned with the code anymore; you are concerned with the behavior of the code and how it relates to the overall objectives of the project. This means you are testing the machine in the manner that the operator would use it.

For this type of testing, you will also need test cases, but these test cases will be overarching and system-related. For example, instead of worrying about how fast, secure, and so on your code is, you are going to be concerned with the outcome of running what would be considered a real-world process. Suppose you're programming a packaging machine and your machine does the following:

1. Opens empty bags
2. Fills the bags with cement
3. Seals the bags
4. Weighs the bags
5. Sends bags with the correct weight to a holding area
6. Sends bags of the wrong weight to a recycling bin

To initiate this process, you will need to input the following information into the HMI:

- Input the number of bags into the HMI
- Select cement products to bag
- Input the weight

The actual test would consist of ensuring that the correct number of bags were produced, the proper cement was bagged, and the bags were of the correct weight.

Now, all of these testing methods that we have explored thus far are used when the product is being developed—in other words, pre-deployment. However, what about when an application has been changed after deployment? In that case, you need to consider regression testing.

Regression testing

Another kind of testing that we need to explore is **regression testing**. Regression testing is, for the most part, testing your system after the program has been changed in some way. This change may be an upgrade to the software, a bug fix, or any other time a developer changes a line of code in the system. This is very important in the automation world as software usually changes as the customer's processes change. Usually, if a plant or other machine owner decides they are going to change their process, they will update the PLC software to accommodate the new process. Requested changes will be either a bug fix or a software modification, such as changing a feature, adding a feature, or making any change to the codebase. In short, the moment you change or add any line of code, you will need to retest your program.

Luckily, when it comes to regression testing, you already have most of the test cases, especially if you're just fixing bugs that are found in the program. The goal of regression testing is simply to ensure that the modified code does not negatively impact the code that was not changed. If you're just fixing bugs, you can use the same test cases you used to validate the system before you deployed it, and if you are adding new functionality, all you have to do is create new test cases for the newly added functionality to use in conjunction with the original test cases.

> Note
>
> Regression testing blurs the lines between validation and verification testing. Some sources cite it as a form of verification testing, while others cite it as a form of validation testing. It will boil down to the context for which you are using it.

The next form of test to explore is integration testing!

Integration testing

Integration testing is pretty straightforward in concept. Typically, your machine will have at least two software components. You will more than likely have an HMI and the PLC code. In some cases, you may have other software components, such as databases, logging software, or any number of other software components. In any case, you will need to make sure the components are seamlessly working together. This is where integration testing comes in. Much like unit testing, there exist many frameworks for integration testing; however, much like unit testing, in my experience, integration testing is often done by hand in automation.

Integration testing is testing modules. Though integration testing is traditionally concerned with software components, if your software is controlling hardware or integrating with the hardware, you must factor that in. This can add an extra layer of complexity to integration testing as a failed test case may be caused by either a faulty software component or a faulty hardware component. You will need to have an understanding of both the overall software components and hardware components to be able to troubleshoot problems that integration testing uncovers.

To conduct integration testing, it is common to write test cases that collect similar information that is collected in the unit test cases. However, you want to make sure that your tests will encompass all the modules you're trying to test. In other words, if you're testing the integration between your HMI and PLC code, you will need to write test cases that will encompass the HMI clicks and the PLC output. Consider the following steps:

1. Navigate to the **Homing** screen.
2. Click the **Home All** button.
3. Verify that the motor positions in the PLC variables area are all 0.

These steps are for a homing test. All we are doing is testing the integration of the HMI's homing feature and the PLC code. In this case, we are just testing that the HMI and PLC are interacting properly.

Testing can be carried out either manually or via some type of automated means. In the next section, we're going to explore the differences between the two.

Automated versus manual testing

There is a difference between automated and manual testing. Manual testing is essentially carrying out the test cases by hand. That is, you're following the steps manually by pressing buttons on an HMI, inputting data manually, or even manipulating the hardware by hand in the case of automation.

Manual testing can be error-prone because it requires a human to carry out the tasks. This means that the person can make mistakes, input wrong data, or even execute the steps out of order. These hiccups can result in erroneous results, which could result in false positives in terms of defects, or worse, false negatives.

When it comes to testing, it is usually considered safer to use what is known as automation testing, that is, creating programs that can carry out the test cases and collect the data automatically. Typically, these tools will format the results of the test in a cohesive report.

Testing tools are becoming common in automation programming. Though they are becoming common, they are not as common as they are in the traditional programming world. Often, these tools are more limited to their traditional counterparts; however, certain aspects of a system, such as an HMI, that are derived from traditional technologies such as C/C++, Java, or Python, can utilize common testing tools.

There are also tools that can directly interact with user interfaces such as an HMI. These tools can be very useful in the automation realm because they can carry out a plethora of test cases by interacting directly with the digital controls. The PLC code itself can be tested indirectly with these tools as well. You can create a simple user interface and have the testing tool digitally interact with the controls to test the PLC program.

Finally, if you are skilled with traditional programming languages, you can create your own scripts using languages such as Python. How these tools would work is up to your imagination, but in theory, you could produce very powerful and very dedicated testing systems. Doing so would require a hefty investment both financially and in terms of effort.

When it comes to testing, it is typically better to use automated tools when possible. Automated tools will typically produce more accurate results, will not make mistakes in terms of order, and, of course, will produce a report. In all, testing tools are just now becoming viable in the automation world, especially for PLC programs, so they are limited and often expensive.

One automated tool that CODESYS offers for automated testing is called Test Manager. Test Manager can be used to create automated tests for CODESYS applications and libraries. Though a very powerful tool, it is a proprietary plugin but comes packaged with the tool bundle for the CODESYS Professional Developer Edition.

Testing is just one facet of finding and eliminating defects. As stated before, testing is more for finding defects in the behavior of the system. However, when it comes to development, we also need to find defects in our code; that is, we need to explore debugging!

Debugging tools and techniques

Debugging is the art of finding defects in our code. Debugging is mostly focused on finding and fixing issues while the code is under development. Debugging is as much an art as it is a science. Effectively debugging a program takes a lot of practice and a deep understanding of certain techniques. With that, the first technique we're going to explore is the debugging process!

Breaking down the debugging process

Debugging a program is a process. A good developer will never jump in and start modifying code without a clear understanding of the bug and a roadmap to a solution. Depending on which articles you read, the number of steps may vary, but the general gist of the steps does not.

The general process for debugging a system is as follows:

1. **Reproduce the problem**: The first step in troubleshooting a bug is reliably reproducing it. Before you can start troubleshooting the problem, you need to be able to trigger the defect on command. As a PLC programmer, this can often be difficult due to the hardware components. Often, a full machine setup is necessary to fully reproduce the bug. All things considered, it is of vital importance that you can reproduce the bug at will before proceeding. If you cannot reproduce the problem, the ultimate patch that you will create may not work as intended.

2. **Isolate the problem**: Assuming that the defect stems from the software, the next step is to isolate the problem. Isolating the problem in this context means figuring out the code that is causing the issue. This can be done in a variety of ways. One of the most common is a technique called print debugging, and another uses a tool called a debugger. Depending on your experience with the codebase and the possible defect, this can be a daunting task. This phase can easily take the most time in the debugging process as you may have to sort through the whole codebase to find the offending code. To make matters worse, it is not unusual for the defect to stem from logic in multiple files, or multiple files will need to be modified to fix the issues.

3. **Analyze the problem**: Once you find the problem, the next step is to understand it. Much like the isolation step, this can also be a very time-consuming task, especially if you're working on a patch for an existing system. Often, you will be bouncing between all the past steps to fully analyze and understand the issue. Generally, this is what I refer to as playtime. During playtime, a developer will have to play with the code by passing different values, triggering different conditions, and so on, to fully understand the behavior of the offending code.

4. **Fix the issue**: Once you understand the problem and what is causing the issue, you can proceed to fix the defect. Developing a patch for a system requires an in-depth understanding of the system and how it is intended to work. In short, this is where you're going to implement your solution. If the problem is hardware-related, it is very tempting to try to compensate for the issue with a software patch. This is one of the worst things you can do as a PLC programmer. The program should assume it is controlling properly functioning hardware. If you modify the source code to compensate for malfunctioning hardware, your program is now only compatible with that particular component. This means that once the faulty hardware is replaced, the source code will have to be restored to its original state, which can be challenging depending on how much code has been changed. Malfunctioning parts can be left in machines for many years, and it's easy to forget about the patches that were used as workarounds. When parts are replaced, bugs will be reintroduced into the system. In short, this type of modification should be avoided at all costs; it is never okay to compensate for malfunctioning hardware, such as broken encoders, poor motor or motor drives, and sensors, by changing the source code.

5. **Validate your solution**: The final phase in debugging your program is testing your fix. To ensure your solution fixed the problem, you need to verify that the system works as intended. This phase will require a working knowledge of the way the machine is intended to work, as well as the issue you were trying to fix and how to trigger the issue. If your patch does not work as intended, you should start over from *step 1*. If you have pre-written test cases, you should use those to test the patch. However, depending on where and what you're working on, you may not have that luxury; therefore, you should test it the best you can. When testing your patch, it is important to ensure that the patch didn't accidentally break something else. This means it is important to thoroughly test not only the feature you're patching but also the whole system. If the machine is deployed or is functioning, it is a good idea to run a full cycle on the system. This means you should run at least one test production run on the system.

These steps will only provide a roadmap for troubleshooting a problem. The true trick in debugging is to find the bug. Depending on your experience with the codebase and the complexity of the codebase, this can be challenging.

Understanding the hardware pitfall

A major pitfall you will often see as a PLC programmer is faulty hardware mimicking software defects. The term that I like to use for these is *broken software*. Essentially, the "broken software" cliche stems from faulty hardware, such as a bad sensor or faulty circuit breaker, sending erroneous signals to the PLC. This can usually cause the system to behave in unexpected ways. Common issues that can cause these problems are the following:

- Poor power supplies
- Faulty sensors
- Short circuits from debris
- Bad and oxidized contacts
- Extreme temperatures
- Dust
- Component drift, such as capacitors decreasing due to drying out

For any system, faulty hardware can often appear to be defects in the software. Therefore, if the system has been deployed for a while and there have been no issues or software updates, faulty hardware can oftentimes be confused with undiscovered bugs. It is very important to remember this when you're going through the debugging process because in systems that are composed of both hardware and software, the hardware is sending the states to the software!

Once you have figured out that the problem is actually software-related, you need to employ strategies to find and fix the problem. The easiest of which is a technique called print debugging.

Practicing print debugging

The go-to technique for many developers to find and eliminate bugs is using what's called **print debugging**. This diagnosis technique is essentially putting messages in your code that give you clues to what's going on during the codebase's execution. This could involve placing markers in your code to see exactly where a program is failing.

In traditional programming, print debugging usually involves using that language's `print` statements to display information either to a log, the terminal, or some type of output that the developer can view. When it comes to PLC programming, where there sometimes isn't a typical terminal to display output to, you can create a `STRING` or `WSTRING` variable to hold messages that will act as a roadmap to determine where the program is at.

To demonstrate this concept, we need to see an example. Suppose that we have a simple division program. The program itself has four variables called division, dividend, divisor, and notLessThan1. The variable section for this program will look like the following:

```
PROGRAM PLC_PRG
VAR
    division        : REAL;
    dividend        : REAL := 4;
    divisor         : REAL := 2;
    notLessThan1    : BOOL;
END_VAR
```

The main logic will consist of two variables being divided and the quotient being assigned to the third. If the third is less than 2, it will mutate notLessThan1 to TRUE. The main logic for this operation will be as follows:

```
division := dividend / divisor;

IF division < 1 THEN
    notLessThan1 := TRUE;
ELSE
    notLessThan1 := FALSE;
END_IF
```

The gist of this program is very simple; however, when the program is executed, we get an erroneous result.

division	REAL	2
dividend	REAL	4
divisor	REAL	2
notLessThan1	BOOL	FALSE

Figure 11.1: notLessThan1 program output

As can be seen, notLessThan1 is set to FALSE. This is not the expected output for the program. The expected output should be TRUE. This means there is a defect in the software. For a problem such as this, print debugging can be a good technique to find and eliminate the defect.

To start the print debugging process, the first thing we need to do is create a string variable. To do this, we'll need to include an extra variable, as in the following code snippet:

```
PROGRAM PLC_PRG
VAR
    debugMsg      : STRING(20); //remove when debugging is done
    division      : REAL;
    dividend      : REAL := 4;
    divisor       : REAL := 2;
    notLessThan1  : BOOL;
END_VAR
```

In this case, a variable called debugMsg was added to the variable list. This variable was declared as a string that can contain 20 characters. We could have also used a WSTRING, but I like to use a string with a fixed value to save on resources. I also use this type as a simple reminder to myself to remove it when I'm finished debugging as I personally use WSTRING more due to its increased flexibility.

STRING allows you to adjust the number of characters to compensate for the amount of information you want to display; however, if you are going to display more than 20 characters, you might as well use a WSTRING type. It is also a good idea to give the variable a name that is indicative of its purpose as a debugging variable. Another good idea is to add a reminder comment next to the variable that reminds both you and other developers to remove it when the debugging is done.

The next step in print debugging is setting up your print statements or, in the case of PLC programming, adding your message to the program. When it comes to PLC programming, this can be tricky. A good rule of thumb is to work inward. This means it is typically a good idea to place your debugging variable at the top of the code you want to analyze and at the bottom. You will then move them inward toward each other until you find the defect. When it comes to print debugging, you can and often should use multiple variables for complex programs, especially if your code has a lot of control statements such as IF statements or loops. This will help optimize the amount of data that you collect in each run.

For this example, we're going to put the message variable at the beginning and end of the program. This follows the best practice that we just established and will let us know where our code is causing the issue or whether we have a more complex problem, such as a corrupted file or corrupted cache. For this example, modify the code to match the following snippet:

```
debugMsg := 'start';
division := dividend / divisor;
```

```
IF division < 1 THEN
    notLessThan1 := TRUE;
ELSE
    notLessThan1 := FALSE;
END_IF


debugMsg := 'end';
```

This code has our debugging statements wrapped around our suspected erroneous logic. If there is a fatal problem that is not related to our code, the debugMsg variable won't say anything, and chances are you won't be able to press the **Play** button or run the program at all. If there is a problem with our logic and the program starts, the debugMsg variable will say start. Finally, if there are no problems and the program executes without error, debugMsg will say end.

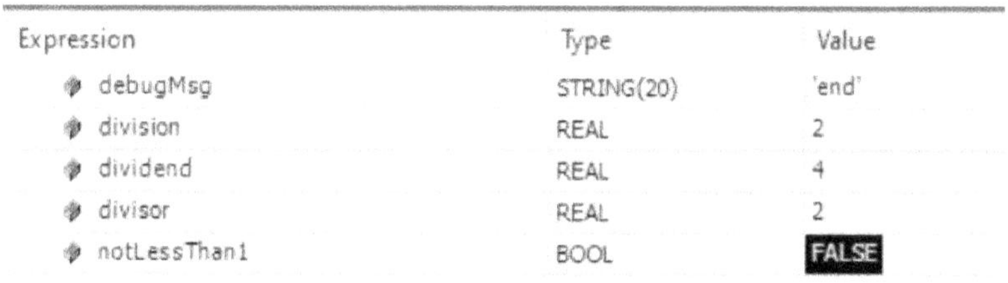

Expression	Type	Value
debugMsg	STRING(20)	'end'
division	REAL	2
dividend	REAL	4
divisor	REAL	2
notLessThan1	BOOL	FALSE

Figure 11.2: Print debugging output

The debugMsg variable is set to end. This means that our code is executing; as such, we can deduce that we have an issue with our logic. The next step would be to move the end message to another place in the code. This is where the strategy of moving the debug variables toward each other comes in.

If you study the code, you will notice that the notLessThan1 variable's state is changed in only one place—in the IF statement. There are also two branches in that conditional statement—the main IF statement and an ELSE statement. For situations such as these, it is a good idea to put a debugging statement in both conditions to see the path that the code is taking. To do this, you can modify your code to match:

```
debugMsg := 'start';
division := dividend / divisor;

IF division < 1 THEN
```

```
    notLessThan1 := TRUE;
    debugMsg := 'in main if';
ELSE
    notLessThan1 := FALSE;
    debugMsg := 'in else';
END_IF
```

The code will change the debugMsg variable to in main if when the division variable is less than 1 and in else when the division variable is greater than 1. When this code is run, you should get the following output:

Expression	Type	Value
🔷 debugMsg	STRING(20)	'in else'
🔷 division	REAL	2
🔷 dividend	REAL	4
🔷 divisor	REAL	2
🔷 notLessThan1	BOOL	**FALSE**

Figure 11.3: debugMsg output

For this run, the debugMsg variable came out as 'in else'. This means that the divisor is not getting set properly. Since there is only one place in the code where that is getting set, we can assume that we have values swapped in our variables. If you look at the variable list, you should see that by swapping the divisor and dividend, we'll get a value that is less than 1. After swapping the value, the output will be as follows:

Expression	Type	Value
🔷 debugMsg	STRING(20)	'in main if'
🔷 division	REAL	0.5
🔷 dividend	REAL	2
🔷 divisor	REAL	4
🔷 notLessThan1	BOOL	**TRUE**

Figure 11.4: Corrected variable assignments

In this iteration of the program, you can see that the debugMsg variable is set to 'in main if', and the notLessThan1 variable is set to TRUE. In other words, the program is working as intended.

Note

It is important to understand that this was a demonstration of print debugging. The true value of the divisor variable was shown in all the screenshots. Though the values are shown, when working with a non-trivial program or logic, it is not so straightforward. Print debugging is a very valuable technique to understand. In short, no matter what you're working on, you should know how to print debug. It is strongly recommended that you put together some trivial programs and debug them with the technique, as it is a vital skill to master.

Print debugging can be overkill sometimes. Sometimes you may get lucky, and you might be able to figure out a defect just by visually analyzing the code.

Understanding visual analysis

Major problems can often be caused by the smallest of issues. If your program is small, and you can easily flip through the code files and digest the code, with practice, you can sometimes spot the problem. For example, assume we have a program that will not run. Let's revisit a modified version of the last example:

```
division := dividend / 0;

IF division < 1 THEN
    notLessThan1 := TRUE;
ELSE
    notLessThan1 := FALSE;
END_IF
```

We can see that we have a division-by-zero issue on the first line. In this case, print debugging would be sheer overkill and do nothing but add overhead to our code and possibly introduce more defects or code rot. There are also times when there may be a defect so severe that the program will not start. In cases such as these, you will have to rely on your eyes to scan the often seemingly incoherent error information from the compilation system.

Fatal bugs that prevent the code from compiling or running will typically come from the following:

- Syntax errors
- Math errors
- Incorrect names
- Unhandled errors

Many advanced IDEs, such as CODESYS, can usually give you visual clues to issues such as these. These indicators usually come in the form of red squiggle lines. However, not all development systems have a feature such as this, so you need to get used to using your eyes, especially if you're integrating with new and advanced technologies that are prevalent in Industry 4.0.

What we have explored can be thought of as manual debugging. That is, we are not using any tools; we are only using certain techniques. For the most part, you will rely heavily on these; however, most IDEs have tools built in to help you debug your code. These tools are typically called debuggers. In the next section, we're going to explore how to use the debugger in CODESYS.

Exploring debuggers

A **debugger** is a program that allows you to debug your PLC code. Most modern IDEs have built-in debugging tools, and CODESYS is no different. So, when you experience an issue that is too complex to troubleshoot with print debugging, it is vital to use this tool.

Generally, most debuggers work similarly. That is, the rules that we're about to explore will apply to most IDEs. Though the general steps will be the same, the overall steps to use a debugger in a different system may vary.

Exploring breakpoints

The key to using a debugger is the concept known as **breakpoints**. The best way to think of breakpoints is as pauses in the code. Essentially, you use a breakpoint to halt the execution of the program at a certain line without terminating the program.

Note

Breakpoints will only pause the program when it is under development as they have to be enabled. If the program has a breakpoint in it and the program is uploaded to the PLC, it will generally be ignored and not affect the execution of the program under production conditions.

To create a new breakpoint, all you have to do is log in to (but not run) the application and right-click on the line you want your breakpoint to be at. Once you select that line, select **New Breakpoint**. This will open up a window where you create your breakpoint, as shown in *Figure 11.5*:

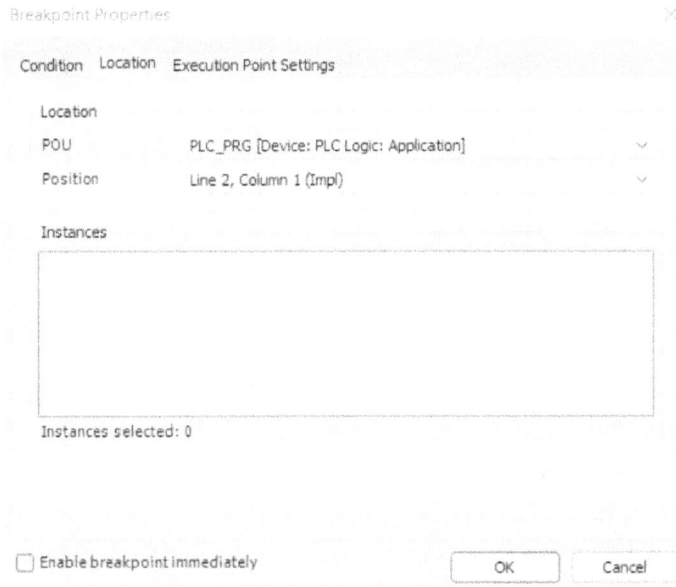

Figure 11.5: Breakpoint Properties window

The dialog shown in *Figure 11.5* is used to set the properties of the breakpoint. The default **Location** tab is where you select where the breakpoint is going to be. The **POU** section is the file in which the breakpoint will be placed, and the **Position** section is the line the breakpoint will pause.

To demonstrate breakpoints, we're going to troubleshoot the program that we debugged with the print debugging technique.

To do this, reset the variables in the program to their original state:

```
PROGRAM PLC_PRG
VAR
    division     : REAL;
    dividend     : REAL := 4;
    divisor      : REAL := 2;
    notLessThan1 : BOOL;
END_VAR
```

Similar to the original error, we will have the dividend and divisor swapped.

For this demo, we're also going to use the same logic as we did before. For this demonstration, the breakpoint is in the code with the red box around it, which is the line that does the computation. Now, log in, right-click that line, select **New Breakpoint**, and set **Position** to *line 1*. When you are finished, your code should look like the following.

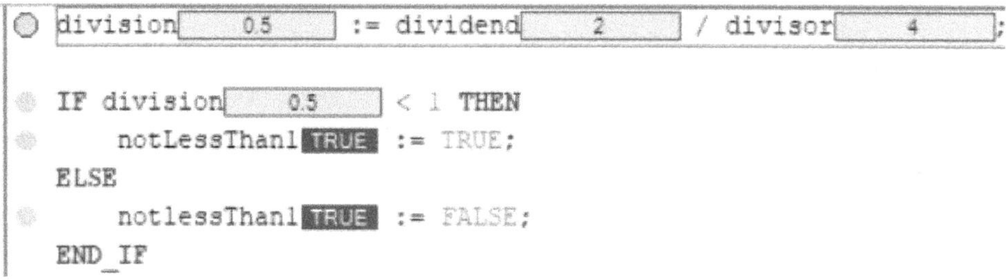

Figure 11.6: Breakpoint on line 2

Notice the red outline around the first line. This red outline is an inactive breakpoint. Essentially, there is a breakpoint there, but it is turned off. This means that when the program is run, it will run as if there were no breakpoint present. For the breakpoint to pause the program, log in and right-click on the line that is highlighted in red. You will see an option called **Toggle Breakpoint**, and depending on whether the breakpoint is enabled, you will either see **Enable Breakpoint** or **Disable Breakpoint**. The functionalities of these options are as follows:

- **Toggle Breakpoint**: Toggling the breakpoint will either enable or disable it. If the breakpoint is enabled when this button is pressed, it will disable it, and vice versa.
- **Enable Breakpoint**: This option will only be available when the breakpoint is disabled. When enabled, the breakpoint will act as a pause in the program. When a breakpoint has been enabled, the circle next to the line will be solid red.
- **Disable Breakpoint**: This option will only be available when the breakpoint is enabled. This option will disable the breakpoint, and when the program is run, it will be ignored. If the breakpoint is disabled, the circle next to it will be gray with a red circle around it.

To demonstrate a breakpoint, the first thing we're going to do is add a message variable and a message on *line 2*.

This is what the variable list should look like:

```
PROGRAM PLC_PRG
VAR
    message         : STRING(20);
    division        : REAL;
```

```
    dividend        : REAL := 4;
    divisor         : REAL := 2;
    notLessThan1    : BOOL;
END_VAR
```

For this code snippet, all we did was add a message variable. Unlike the debugMsg variable in the print debugging example, this variable will serve as an output to show off how the breakpoint works.

The following is the code to demonstrate the debugging tool:

```
message := 'before breakpoint';
division := dividend / divisor;
message := 'after breakpoint';

IF division < 1 THEN
    notLessThan1 := TRUE;
ELSE
    notLessThan1 := FALSE;
END_IF
```

This code is similar to the code we used for print debugging. However, to give a basic demonstration of the debugger, the message variable signals that the program is before the breakpoint, and a change in text signals that the program has moved past it.

Once the code is modified, you will want to log in to the program, add a breakpoint on the second line that contains the division operation, and enable it. When complete, your code should resemble the following screenshot:

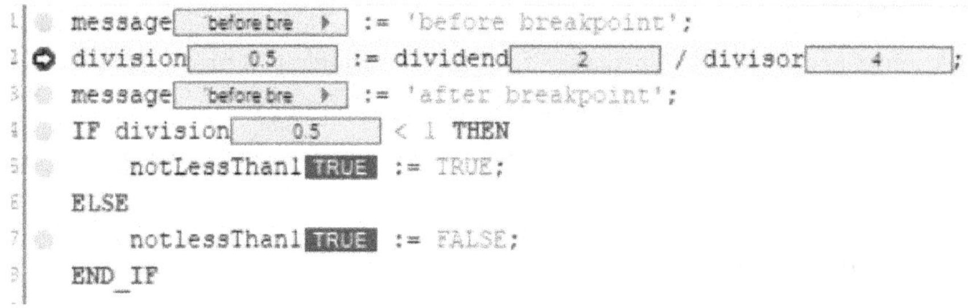

Figure 11.7: Enabled breakpoint

The code in *Figure 11.7* shows an enabled breakpoint. Depending on the state of your code, it may be a red line instead of a yellow one. When you press **Play**, you will get the following output for the variables and code:

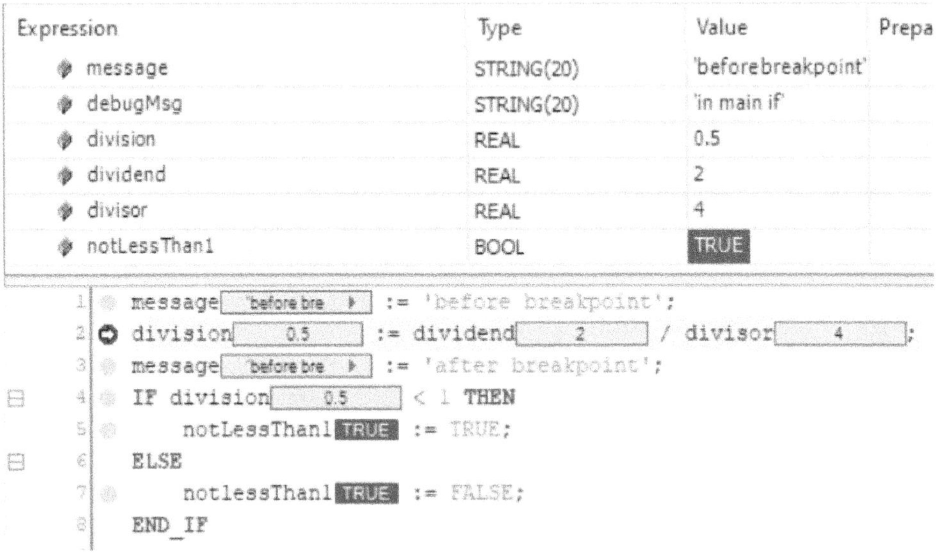

Expression	Type	Value	Prepa
message	STRING(20)	'before breakpoint'	
debugMsg	STRING(20)	'in main if'	
division	REAL	0.5	
dividend	REAL	2	
divisor	REAL	4	
notLessThan1	BOOL	TRUE	

```
1   message[ before bre  ▶ ]  := 'before breakpoint';
2 ◉ division[    0.5    ]  := dividend[    2    ] / divisor[    4    ];
3   message[ before bre  ▶ ]  := 'after breakpoint';
4   IF division[    0.5    ] < 1 THEN
5       notLessThan1 TRUE  := TRUE;
6   ELSE
7       notlessThan1 TRUE  := FALSE;
8   END_IF
```

Figure 11.8: Paused program

If your output does not match up exactly, don't worry as long as the message variable shows before breakpoint, and the second line is highlighted yellow in the code section. Also, notice that the division variable is set to 0. In other words, the computation did not run. It is important to remember that the line with the breakpoint will not run. Ultimately, this means that the program is paused at the second line and, as such, the breakpoint is working.

Exploring stepping

Inserting breakpoints is only half the process of using the debugger. A breakpoint will stop a program, which means that if the error comes after the breakpoint or we need to analyze what comes next, we need to be able to move to the next instruction. To do this, we can use what's called **stepping**.

Stepping is a tool used to manually control the flow of the program. There are a few stepping types, as follows:

- Step Over: The Step Over command allows the statement at the breakpoint to be executed and halts the execution again before the next command. A Step Over command can be called by pressing *F10* or the following button in CODESYS:

Figure 11.9: Step Over button

This button will perform the same operation as pressing *F10*. One point to note is that if the next command is outside the current **program organizational unit (POU)**—for example, a custom function block—that line will be executed as if it were one single command and not go into that code.

- Step Into: Step Into should be used when the next line of code is a POU, such as a subordinate POU, function block instance, function, method, or action. The Step Over command will treat the POU call as one command. This means that all the code in the POU will be executed as if it were one command. This is different from the behavior of Step Into, which will enter the POU and execute the first statement. After the first statement is executed, the program will then pause again. Essentially, you use Step Into when you're calling other POUs and you need to run those commands line by line.

Figure 11.10: Step Into button

The Step Into command can be called by pressing the shown button in *Figure 11.10* or by pressing *F8*. Unless you're trying to step into a specific POU, it is not necessary to use Step Into.

- Step Out: The Step Out command will run the POU code from the breakpoint to the end of the POU. Once the POU code is completed, the execution will return to the calling POU. This command is unique when compared to the other step commands because if this command is run in the main POU, it will execute to the end and will jump back to the first line of code in the POU. Once at the first line, it will pause there.

Figure 11.11: Step Out button

The Step Out command can be invoked by pressing the button or by pressing *Shift + F10*.

Other debugging commands exist in CODESYS. Each of these commands has its benefits and is used for different things. For now, you can experiment with the breakpoint program and Step Over command.

Since the first edition of this book came out, **large language models (LLMs)** have exploded in popularity. This explosion in popularity is actually starting to change the way some developers troubleshoot code. In the next section, we're going to explore how ChatGPT and similar models can be used to assist in the debugging of PLC code.

Debugging with ChatGPT

Although the title of this section is *Debugging with ChatGPT*, you don't have to use ChatGPT to get the most out of this section. There are many other excellent AI systems that can do just as good a job as ChatGPT. Some of these other AIs are systems such as Microsoft's Copilot and Google's Gemini. For this book, we're going to use ChatGPT, but you can use whatever you want. The key is not in the AI system that you use but the prompt you write and the way you interact with the system. Therefore, the first thing we need to do is understand how to construct a prompt!

Constructing prompts

This section is going to assume you know the basics of writing a prompt. If you're not familiar with prompts, you'll want to explore *Chapter 17* of the book *PLCs for Beginners*. However, assuming you have some idea of how to write a prompt, there are three things you want to work in:

- A description of the issues that is as detailed as possible
- Things that you know are not the problem
- Source code: Note that this will not always be possible; it is a nice-to-have piece of information but not always totally necessary

These are just some basic components. Typically, when writing the prompt, you want to have a solid question in mind. I have found that this is the optimal way of using an LLM in general. In fact, in my earlier writing, I used to solely refer to prompts as queries to promote the concept.

Understanding how to write a prompt is one thing. Using the prompt in practice is another. In the next section, we're going to apply these principles and create a prompt that can troubleshoot a piece of code.

Troubleshooting code with AI

To test ChatGPT or whatever AI system you're using, let's write a simple program that would have a division-by-zero error in it. For example, suppose we have the following code:

```
division := dividend / 0;

IF division < 1 THEN
    notLessThan1 := TRUE;
ELSE
    notLessThan1 := FALSE;
END_IF
```

In this case, assume we're getting a division-by-0 error. In a snippet this small, it is easy to see why; however, for larger code blocks, a simple error such as this could be easy to make.

To troubleshoot this using an AI such as ChatGPT, we could use a prompt such as the following:

```
Why is this code producing a division by 0 answer:
division := dividend / 0;

IF division < 1 THEN
    notLessThan1 := TRUE;
ELSE
    notLessThan1 := FALSE;
END_IF
```

In this prompt, we have a clear question (query), a clear issue with the code, and the source code itself. In other words, we have the backbone of a quality prompt. After entering the prompt, the system should output something such as the following:

```
The issue is that the code is trying to perform division by zero in the
expression dividend / 0, which is mathematically undefined. In most
programming languages, dividing by zero results in an error or produces
undefined behavior, and it may not behave as expected.
To handle this issue, you can add a check to ensure the divisor is not
zero before performing the division. Here's an updated version of the code
that prevents division by zero:
IF divisor <> 0 THEN
    division := dividend / divisor;
ELSE
```

```
    division := 0;  // or handle the case as appropriate
END_IF

IF division < 1 THEN
    notLessThan1 := TRUE;
ELSE
    notLessThan1 := FALSE;

END_IF

In this version, we check if the divisor is zero before performing the
division. If it is zero, we assign a default value to division (e.g., 0)
to avoid the division by zero error. You can customize the handling based
on the requirements of your system.
```

In this case, ChatGPT reworked the code to ensure that a division-by-zero error could not occur. It did this by wrapping the math equation in an `IF` statement and adding in some extra error handling.

Note

AI systems such as ChatGPT can give different answers. If the solution that you're given does not match the one that is presented here, don't worry. The system, prompt wording, and even your prompt history can influence the results.

The future of AI troubleshooting

Though it is impossible to clearly say what the future of AI as a troubleshooting tool will be, there are many programming systems that are being integrated with AI to allow for faster code troubleshooting and debugging. Due to the AI systems being in their infancy, their full impact has not been felt yet. However, these tools will be the future of debugging and software development in general.

AI pitfalls

It is important to note that many factors can affect an AI's output. It is also important to note that AI models are not infallible. Much like human programmers, systems such as ChatGPT, Copilot, and Gemini make mistakes. You should never rely wholeheartedly on the output of AI to complete a task. Though they typically produce quality results, they can, and often will, be wrong. When the model was trained, what dataset it uses, and even the feedback it was given can cause answers that are outdated or flat out inaccurate. When using an AI system, it is important to err on the side of caution. Now that we have explored AI, we can move on to our final project.

Troubleshooting: A practical example

When working with motors, it is sometimes necessary to incrementally stop a motor. Sometimes this is due to the process, while other times it is due to the motor or component. To demonstrate practical troubleshooting, we're going to create a state machine.

> **Note**
>
> This example will force variables. Forcing variables can be very dangerous in a real-world project. Forcing can be a great way to troubleshoot a project, but you need to exercise extreme caution when doing this!

The variables for the state machine will be structured like the following:

```
PROGRAM PLC_PRG
VAR
    machineState        : INT := 1;
    motorSpeedCutOff    : INT := 10000;
    runTime             : INT := 2;
    setSpeed            : REAL;
    numOfParts          : REAL := 8;
    motorOff            : BOOL;
    exc                 : __SYSTEM.ExceptionCode;
    motorSlowDown       : INT := 100;
    speed               : INT;
END_VAR
```

Once the variables are implemented, you can move on to the main logic of the program, which is as follows:

```
CASE machineState OF
    1:
        //machine off state
        motorOff := TRUE;
    2:
```

```
        //machine run state
   __TRY
        //set motor speed
        setSpeed := numOfParts / runTime;
        IF setSpeed >= motorSpeedCutOff THEN
            motorOff := TRUE;
        ELSIF setSpeed < motorSpeedCutOff THEN
            motorOff := FALSE;
        END_IF
   __CATCH(exc)
        //throw machine into error state
        machineState := 3;
   __ENDTRY
3:
        //error state
        runTime      := 0;
        machineState := 4; // go into motor slow down
5:
        //Motor Wind Down
        IF setSpeed <= 500 THEN
            FOR speed := 100 TO 500 BY motorSlowDown DO
                setSpeed := setSpeed - speed;
            END_FOR;
        END_IF;
        machineState := 1;
END_CASE
```

The purpose of this code is to gracefully stop the motor from running. However, this code block has a particular problem. According to the customer, the motor is not slowing down.

To troubleshoot this problem, the first thing that we need to do is reproduce it. We know that the only time the motor should be turned off is when there is an error of some kind. Therefore, we can throw a division-by-zero error.

The first thing we are going to want to do is run the application as normal. Therefore, we're going to set the machineState variable to case 2 which will put the motor in a normal state:

Expression	Type	Value
machin...	INT	2
motorS...	INT	10000
runTime	INT	2
setSpeed	REAL	4
numOf...	REAL	8
motorOff	BOOL	FALSE
exc	EXCEPT...	RTSEXCPT_NOEXCEPTION
motorS...	INT	100
speed	INT	0

Figure 11.12: Normal motor operations

Figure 11.12 shows that the motor appears to be on and operating as would be expected. Since the motor will only turn off when an error is thrown, we're going to set the runTime variable to 0.

After setting runTime to 0, we should be met with the following error:

Device.Application.PLC_PRG

Expression	Type	Value	
machineState	INT	4	
motorSpeedCutOff	INT	10000	
runTime	INT	0	
setSpeed	REAL	4	
numOfParts	REAL	8	
motorOff	BOOL	FALSE	
exc	EXCEPT...	RTSEXCPT_FPU_DIVIDEBYZERO	
motorSlowDown	INT	100	
speed	INT	0	

Figure 11.13: Abnormal behavior in motor

Figure 11.13 shows an error; however, our motor did not shut off. Upon examining the variable output, we can see that everything looks as if it should work. From here, we can either start with breakpoints or use print debugging. This is ultimately a matter of preference; however, for issues such as these, where we're not sure exactly why a case isn't transitioning, print debugging can often provide enough information for the amount of effort.

To carry out the print debugging process, we're going to add a variable called msg to the variable list. Essentially, just add the following code somewhere in the variable list:

```
msg: STRING(20);
```

At this point, we're going to want to use our visual inspection skills, as it is important to follow the flow of the program to get to the point where it should be executed as expected. We want to have a msg statement at the top of the catch block, error block, and in the last case statement.

After you run the program and set the value to the variables in *Figure 11.14*, you should get the same values in the **Value** column:

Device.Application.PLC_PRG		
Expression	Type	Value
machineState	INT	4
motorSpeedCutOff	INT	10000
runTime	INT	0
setSpeed	REAL	0
numOfParts	REAL	8
motorOff	BOOL	FALSE
exc	EXCEPTIONCODE	RTSEXCPT_FPU_DIVIDEBYZERO
motorSlowDown	INT	100
speed	INT	0
msg	STRING(20)	'in error case'

Figure 11.14: Program output

Here, we have something a little strange. We are held up in the error case, so we're in case 3. However, our machineState variable is set to 4. This means that we either have our case mislabeled or we have our state set to the wrong case. If we look at the source code for the state machine, we can see that our motorSlowDown case is labeled as 5. This would cause an error as the case is trying to go to case 4. Therefore, there is no case for it to go to, so it is hung up. We can fix this bug by either changing the case number of the variable or the case itself. Since it makes little sense to have case 3 followed by case 5, we will just relabel case 5 to case 4. Upon making the code change and running it, we will be met with the following proper case transition output:

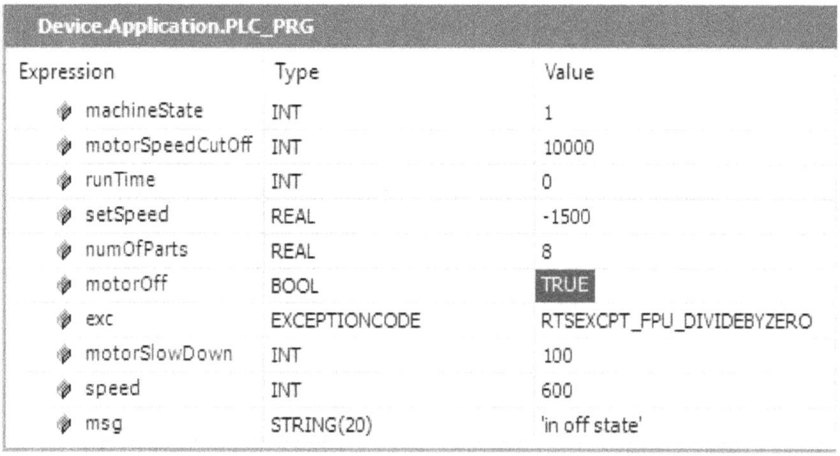

Figure 11.15: Properly transitioning the program

In the preceding screenshot, we can see that the motor is in the correct case.

Now, if we look at *Figure 11.15*, we can see something odd. The set speed is still way off. We have a negative number. When the setSpeed variable hits 0, it should simply cut off, so we should never have a value less than 0. This means we have a bug.

This bug can be found and remedied simply by looking at the code. If we have a for loop, it is going to run for a given number of intervals. For our program, this is not as desired. As soon as our variable is less than or equal to 0, we want the loop to break so that we can move on to the next statement. A more appropriate loop would be a while loop. Therefore, we can modify our program to match the following snippet.

This is the modified case 4 code:

```
//Motor Wind Down
msg := 'in motor case';

IF setSpeed >= 500 THEN
    WHILE setSpeed >= 0 DO
        setSpeed := setSpeed - speed;
    END_WHILE;
    IF setSpeed <= 0 THEN
        setSpeed := 0;
    END_IF;
```

```
END_IF;

machineState := 1;
```

You can simply replace the code in case 4 with this code. The WHILE loop will execute until the setSpeed variable—the variable that controls the motor speed—is either 0 or less than 0. Similar to the FOR loop, the WHILE loop can also produce a value less than 0; as such, we'll include an IF statement that will set the variable to 0 when the value is less than or equal to 0. Essentially, this is just a sanity check to ensure the value is set to 0.

To test the code, force the machineState value to 4 so that it will not leave the case, and you should be met with the following output:

Device.Application.PLC_PRG		
Expression	Type	Value
⊘ machineState	INT	Ⓕ 4
⊘ motorSpeedCutOff	INT	10000
⊘ runTime	INT	0
⊘ setSpeed	REAL	0
⊘ numOfParts	REAL	8
⊘ motorOff	BOOL	TRUE
⊘ exc	EXCEPTIONCODE	RTSEXCPT_FPU_DIVIDEBYZERO
⊘ motorSlowDown	INT	100
⊘ speed	INT	0
⊘ msg	STRING(20)	'in motor case'

Figure 11.16: while loop output

As can be seen, when we force the machineState variable, the setSpeed variable will always be 0. This is what we want. We have now fixed multiple bugs using forcing and print debugging, and the state machine is now working.

Chapter challenge

To get some practice with prompts, write a few prompts for a system such as ChatGPT to troubleshoot the code for you. Examine the outputs and see whether the AI system came to the same conclusion. This may take a few iterations to fully flesh out.

Summary

In this chapter, we explored various types of testing and troubleshooting techniques. By this point, you should have a decent background in troubleshooting code. Both testing and debugging are a bit of an art. It will take practice, and it will take some time to get good at it.

Overall, it should be noted that testing and troubleshooting are rapidly changing. The widespread adoption of AI is changing the way code is troubleshot. If you take anything away from this chapter, let it be the basics of how to write a prompt. As time passes, you will probably find yourself using systems such as ChatGPT and Copilot more often. For now, we're going to switch gears and look at SOLID programming!

Questions

1. Define print debugging.
2. Define interactive debugging.
3. Define the debugging process.
4. How can AI be used to debug code?

Further reading

- *CODESYS testing and debugging*: https://content.helpme-codesys.com/en/CODESYS%20 Development%20System/_cds_struct_test_application.html

12

Advanced Coding: Using SOLID to Make Solid Code

As the old saying goes, *"with great power comes great responsibility,"* and in terms of OOP, this could not be truer. Object-oriented programming is an excellent methodology to program literally anything. However, it can easily end up a huge mess. In my formative years as a developer, I often put little to no forethought into how my code would work and be used; I just coded. This flaw in thinking led to a few programs being sent to the cyber-trash heap way before their time. When implemented incorrectly, OOP can produce a chaotic mess of a program.

Until this point, we have explored the power of OOP and how it can allow us to reduce the amount of code that we have to write. However, what we have not explored is how to keep codebases maintainable. By default, OOP does not necessarily translate into code that is easy to maintain, expand upon, or, for that matter, understand. Even when using proper relationships between function blocks, proper design patterns, and the like, code can still easily become a mess. This mess can be a major issue for industrial automation, where systems are constantly adapted to an ever-changing world. So, if OOP is the way of the future for industrial automation programming, how can we ensure that our OOP code will last?

Enter the world of SOLID programming. SOLID programming is a set of rules that will help you drastically improve your code. In this chapter, we are going to learn about SOLID programming by exploring the following concepts:

- A soft introduction to SOLID
- How SOLID benefits code
- The principles of SOLID

Lastly, we are going to use SOLID principles to design a simulated industrial painter.

Technical requirements

For this chapter, all you will need is a copy of CODESYS installed. The examples for this chapter can be found at the following URL: `https://github.com/PacktPublishing/Mastering-PLC-Programming-Second-Edition/tree/main/Chapter%2012`.

Some of the code for this chapter will be a bit different than the other chapters. Much of the code in this chapter will be more akin to pseudocode that follows the IEC 61131-3 Structured Text syntax/structure due to the examples being used to merely demonstrate concepts. The goal of the examples is to provide a very familiar code structure so the principles can be applied to other languages and non-PLC projects that will be common in Industry 4.0. It cannot be stressed enough that these principles are language- and platform-agnostic. With that, the general structure and flow of the pseudocode are mostly compatible with CODESYS, and with minor modifications, can be run as examples. Nonetheless, the code is not designed to run as-is unless stipulated.

Introducing SOLID programming

If you're a traditional developer, you may have heard of SOLID programming before. SOLID is a very common set of principles that are used across the IT industry to produce well-architected code. If you're an automation programmer, you probably haven't heard of the concept before, and that's due to the immature implementation of OOP in PLC programming.

When I was first introduced to SOLID programming, I was incredibly confused about its purpose. My young, inexperienced self simply could not fathom that OOP did not ensure quality code. After all, as long as you're following proper OOP principles, you should be producing quality code, correct? Well, the answer to that is "Wrong." Quality code stems from well-architected code.

A quality program is a program where things can be easily added or removed, bugs can be easily found, and code can be easily changed without the risk of breaking other code. This is where SOLID comes into play. SOLID programming is a set of general rules that, when followed, will drastically improve the quality of your program's architecture.

So, what is SOLID programming? SOLID programming is a set of five **object-oriented design (OOD)** principles. SOLID programming is the brainchild of Robert C. Martin, a.k.a Uncle Bob. What Uncle Bob devised are five principles that, when implemented properly, can allow your program to become flexible enough to be maintained so it can stand the test of time. With the introduction of OOP into the industrial automation world, having flexible code is a necessity. Before we start diving into SOLID programming, we need to first understand its benefits.

Benefits of SOLID programming

As every automation engineer knows, automation systems can stay in production for decades on end. As every automation engineer also knows, during that time, the process will change, which will require new software, hardware will become obsolete and will need to be replaced, and so on, which, as you can guess, will require software modifications. As someone who has spent countless hours sifting through thousands of lines of code at a customer site for hours on end with multiple different employers, I can say that when it comes to architecture, the extra effort is worth it. Even when you're working on well-organized and well-architected codebases, you'll find that tracking down a single error can be quite daunting. When the codebase is poorly designed, tracking bugs can become a Herculean task.

When implemented properly, SOLID can produce code with the following qualities:

- Easier to debug
- Cheaper to maintain
- Easier to scale (add or remove features)
- Performs better
- Runs more efficiently

In general, SOLID will greatly improve the quality of your codebase and make it much more scalable and easier to fix in the field.

It is best to think of SOLID as a series of general rules as opposed to hard standards. These general rules will produce code that is easy to maintain, expand, and, if necessary, modify. In other words, most think of SOLID as a set of best practices. When implemented correctly, these principles will allow you to build very robust code that is easy to maintain in the future.

Since SOLID is more of a set of best practices, it's important to understand that the context in which it is thought of and how it is implemented will vary, like OOP. However, much like how the general rules don't change from language to language, neither will the rules of SOLID. With that in mind, let's explore the principles that govern SOLID programming.

The governing principles of SOLID programming

The principles that govern SOLID programming are as follows:

1. **S**: Single-responsibility principle

2. **O**: Open-closed principle

3. **L**: Liskov substitution principle

4. **I**: Interface segregation principle

5. **D**: Dependency inversion principle

> Note
>
> These principles are widely used in general-purpose programming. This means that if you're using a general-purpose programming language such as C# or Java to create your HMIs or other automation software, you can use these principles to keep that codebase clean as well!

As a PLC developer, you will use some of these principles more than others. For me personally, the principle that I use the most is the **single-responsibility principle** (**SRP**). This principle can be applied to almost any code module, such as a function, interface, struct, function block, or anything else.

The single-responsibility principle

The SRP is, in my opinion, the most important of the five principles to implement. The SRP states that a code module should do one thing and one thing alone. This goes back to the one-sentence rule. If you have to use the word *and* to describe your module, you have violated the SRP, and you should break the component out. Generally, this is a trick that many experienced developers use to ensure that code components are properly broken out.

Now, what qualifies as a code module? Well, that is kind of an open-ended question. However, as an everyday developer, the code modules that you will interact with the most will generally consist of the following:

- Functions/methods
- Function blocks or classes (in a general-purpose programming language)
- Structs

- Interfaces
- Microservices

Any time I'm developing something that uses a code module like the ones just mentioned, I recommend trying to summarize its responsibility in a complete sentence without the word *and*. If the word *and* appears in the sentence, it usually means the module is doing more than one thing. This is especially true for methods/functions. Even in the realm of traditional software development, it is not uncommon to see functions/methods that do multiple things. This will usually lead to serious trouble when one of the module's responsibilities must be changed. The moment you change a line of code in a module, that module should be treated as new, untested, and potentially defective. If the SRP is ignored, you've effectively broken multiple components of the program by altering that single method, function, or whatever else it might be.

The key to using the SRP properly is the sentence that describes the code modules. This can be very tricky when you first start using the technique. For many who are first starting out, there is a big gray area as to how to format the sentence. For example, if you have a machine with two different motor brands, a lot of developers will simply group the logic for both motors together. In some cases, this could work; other times, it won't. When you're trying to use the one-sentence rule, you need to be fair with your sentence, and you need to have a clear understanding of what you're trying to do. If you have two or more functionalities that could be broken if there were a change to one, then you should break that module out.

Implementing the SRP

For some reason, developers (both experienced and non-experienced) love to bunch several different responsibilities in methods or functions. When things inevitably have to be changed, you'll end up having to retest several different behaviors. To demonstrate this with code, consider the following function with the following variables:

```
FUNCTION motors : BOOL
VAR_INPUT
    pos        : INT;
    motorPos   : INT;
END_VAR
VAR
    turnOnMotor : BOOL;
END_VAR
```

The body of the function is composed of the following:

```
//turn on motor
IF turnOnMotor = FALSE THEN
    turnOnMotor := TRUE;
END_IF

//home motor
IF motorPos <> 0 THEN
    motorPos := 0;
END_IF

//set the new pos
MotorPos := pos;
```

> **Note**
>
> The code in this example is designed to reflect a function that does not follow the SRP. The operations of the function are simple examples to demonstrate ignoring the concept.

In this case, we have a function called `motors`. Already, we have a clue that this function is doing too much due to the plural name. After exploring the body of the function, we can see that this component is responsible for turning the motor on if it's off *and* homing the motor if it's not homed *and*, finally, setting the position of the new motor.

Now, suppose that our boss wants to modify this so that the function toggles the motor state as opposed to just enabling the motor. This request means there's a possibility for new bugs to be introduced, and on top of all that, we can't reuse any of this code. As such, in the best of cases, we will get unneeded and redundant code that may or may not have defects in it. In this case, the modifications to the code will require extensive rework and testing of non-relevant features. In other words, instead of only testing the new feature (in this case, the toggle function), we have to test all three responsibilities.

> **Note**
>
> Though redundant code in automation engineering isn't necessarily seen as bad by many practitioners, it actually is and should be avoided.

If we were to summarize this function in a sentence, we would get something like the following:

This function turns on the motor and homes the motor and positions the motor.

As can be read in the sentence, the word *and* appears twice. This means the function is performing at least three responsibilities. In these situations, it is usually a good idea to break out each responsibility that comes after the word *and* into a function of its own, and call those functions from another. In other words, we need to create a façade or orchestrator function. A better solution to this problem would be to create three functions, such as the following:

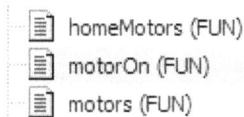

    homeMotors (FUN)
    motorOn (FUN)
    motors (FUN)

Figure 12.1: Functions

Note

The `motors` function name does not follow the verb rule and is plural. The name `motors` was kept as a means to help keep track of the original function from the last example. It is also not uncommon for a function such as this to be named a plural noun in the automation industry because it is a façade function that can be used to control multiple motors. When implemented correctly, having a façade method or function responsible for controlling multiple like-items, such as different motors, will not violate the SRP; however, in most real-world situations, it would normally be preferred to change the `motors` name to a singular verb.

The `homeMotor` function will consist of the following code:

```
FUNCTION homeMotor : BOOL
VAR_INPUT
END_VAR
VAR
    motorPos     : INT;
END_VAR
```

The body will consist of the following:

```
//This function homes motor
IF motorPos <> 0 THEN
    MotorPos := 0;
END_IF
```

Next, the `motorOn` function will consist of the following:

```
FUNCTION motorOn : BOOL
VAR_INPUT
END_VAR
VAR
    turnOnMotor : BOOL;
END_VAR
```

The body of the function will be as follows:

```
//This function turns on motor
IF turnOnMotor = FALSE THEN
    turnOnMotor := TRUE;
END_IF
```

Finally, what consists of the `motors` function will now be reduced to the following:

```
FUNCTION motors : BOOL
VAR_INPUT
    pos         : INT;
END_VAR
VAR
    motorPos    : INT;
END_VAR
```

The body of the function will now consist of the following:

```
motorOn();
homeMotor();
//set the new pos
MotorPos := pos;
```

Though this is little more than a fancy pseudocode example, we demonstrated how we can architect the SRP into our system. In this case, all we did was break the logic into separate functions and call those individually with a façade. By doing this, we can now modify the functions without directly affecting other functionality. We can also call the new functions individually, which means no redundant code. This design will also make our code more robust because we now have more granular control over its behavior.

Note

Though you are not directly changing or at risk of breaking outside code when using the SRP, it is still generally a good idea to test the dependent functionality.

Now that we have a basic understanding of the SRP, we need to move on to the next principle of SOLID, the **open-closed principle (OCP)**.

The open-closed principle

The general rule of thumb in an ideal world is that once a stable code module is implemented, you don't want to modify it to add functionality. This kind of leaves us in a pickle, as, at some point, we are going to need to add new features; in short, we are going to have to eventually scale the program.

If you perform a Google search on the open-closed principle, you will usually find a definition that states the following:

Objects or entities should be open to extension but closed to modifications.

This essentially means that instead of modifying a class/function block to add new functionality, it is better to extend it using principles such as inheritance, polymorphism, or advanced design patterns. This means that the OCP is ultimately a design principle. The OCP is something that needs to be baked into the design phase, and it will usually take practice to properly implement and plan for.

This principle also goes hand-in-hand with the SRP, in my opinion, as you will need to ensure your modules are only doing one thing to properly pull off the OCP. To properly implement this principle, you must have a clear understanding of the relationships between the objects, and each object should closely follow the SRP. Put bluntly, properly implementing the principle starts in the design phase of the SDLC.

Implementing the OCP

Implementing the OCP takes practice to master, and like many other things in programming, it is often subjective as to what constitutes a properly open-closed architecture. However, consider that we have a function block that drives a motor, as in the following example.

In this pseudocode example, we are going to examine a function block that consists of two methods. The function block will be called MotorControl, and it will have a motorOff and a motorON method. Both methods will use a series of function block-level variables that will look like the following:

```
FUNCTION_BLOCK MotorControl
VAR_INPUT
END_VAR
VAR_OUTPUT
END_VAR
VAR
    motor : INT;
    wait  : WSTRING;
    state : BOOL;
END_VAR
```

The motorON method will look like the following:

```
CASE motor OF
    1:
        //company 1
        wait := "10 ms";
        state := TRUE;
    2:
        // company 2
        wait := "2 ms";
        state := TRUE;
END_CASE;
```

The motorOff method will consist of the following:

```
CASE motor OF
    1:
        //company 1
        wait := "10 ms";
        state := FALSE;
    2:
        // company 2
        wait := "2 ms";
        state := FALSE;
END_CASE;
```

If you look at the code, you will see that both motor functions have a case for different types of motors. The code shows that each type of motor will have a unique pause before the motor is either turned on or shut down. In other words, this code will work for these two specific motors. However, suppose we want to add a third motor with a waiting period of 5 milliseconds. To implement this, we would have to modify two different methods. This means we will need to break our golden rule and change the existing code. A change like this would include an extra case with a state change and a wait time. A change like this isn't that big of a deal in theory. However, a simple change like this can lead to broken code for motors that are not relevant to the upgrade, and enough of these small changes can add up over time to morph the codebase into something unmaintainable, unstable, and messy. At the very least, to be thorough, you will have to retest each of the motors. As we established before, in the productivity and profit-driven world of industrial automation, this can lead to extra downtime, which, in turn, will lead to an extra loss of money.

To summarize the code thus far, we created a program that does not follow the OCP; in other words, we created code that is not scalable. Our program cannot be modified without manipulating old code, which can lead to maintainability issues later on in the product's lifespan. Put bluntly, our program is not architected well, and sooner or later, we will end up with issues. In all, if this code is deployed to a customer site, it will eventually end up costing the owner downtime and money.

With that in mind, how can we alter the code to keep it well-architected and create a more flexible and SOLID architecture? The answer to that will reside in the `motorControl` function block. Conceptually, the `motorControl` block does follow the SRP because if we were to describe it with a sentence, it would read like the following:

The function block controls the state of the motors.

In this case, it can be argued that the motors being plural could mean that it is not fully in compliance with the SRP even though it is. This type of semantics is where the one-sentence rule can get gray. As we design the software, this is one area that needs special attention, as we are kind of in compliance with the SRP, but we're still experiencing issues with flexible code.

To fix these scalability issues, let's redesign the program using UML.

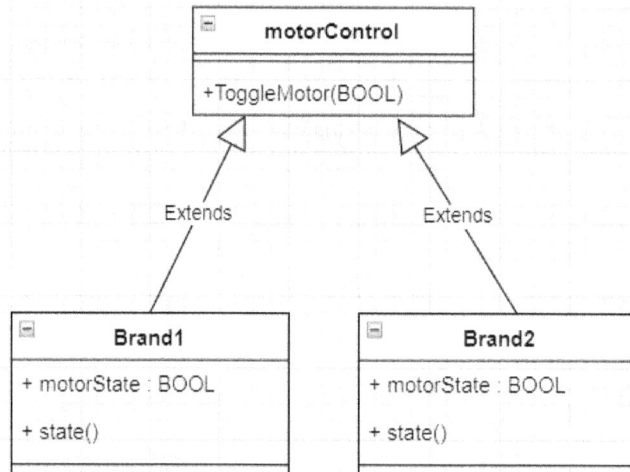

Figure 12.2: Open-closed motorControl function block

With this design, the motor brand function blocks will inherit from the motorControl function block. In this design, the only thing the Brand function blocks will keep track of is the state of the motor. The way this program is designed, if a new brand had to be added, all we would have to do is create a new function block that extends motorControl. This is in contrast to the old design, where, if we wanted to add a new motor, we would have to modify at least two different blocks of code.

If we were to turn the UML design into code, it would have the following project tree:

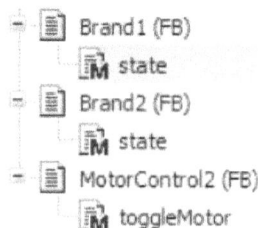

Figure 12.3: Open-closed project tree

The pseudocode for the `toggleMotor` method in the `MotorControl2` function block should match the following:

```
METHOD PUBLIC toggleMotor : BOOL
VAR_INPUT
    state : BOOL;
END_VAR
```

The body of the method should look like this:

```
IF state = TRUE THEN
    toggleMotor := FALSE;
ELSE
    toggleMotor := TRUE;
END_IF
```

The variable logic for the `Brand1` and `Brand2` function blocks will be simple, like the following:

```
FUNCTION_BLOCK Brand2 EXTENDS MotorControl2
VAR_INPUT
END_VAR
VAR_OUTPUT
END_VAR
VAR
    motorState : BOOL;
END_VAR
```

The logic for the `Brand1` method will resemble the following:

```
//turn motor off
motorState := toggleMotor(TRUE);
```

By the same extension, the `Brand2` method will resemble this:

```
//Turn motor on
motorState := toggleMotor(FALSE);
```

As can be seen in the two methods for this example, both methods are very similar and only pass an argument to the `toggleMotor` method.

This design is much more flexible and scalable. With this design, scaling the functionality of the program is as simple as adding a new function block. This is a major improvement over the original design, which required many different changes to existing code to simply add support for another motor.

With all that being said, it is unrealistic to think that you are never going to modify the old code. Old code will eventually have to be changed – there is simply no way around that. You will eventually have to update a function block or method to add a new base feature, such as a speed control method or whatever else. There is no way of knowing the future and, by extension, there is no way of working everything into a design. Now, what the OCP is getting at is that you don't want to have to modify existing code to add support for new things, such as extra motors or the like. Generally, if it is a functionality that is reflected across all child function blocks or targets a specific function block, it is, in my opinion, okay to modify. However, if you are constantly having to modify code to add support for new components, you probably need to revisit your design with the OCP in mind.

In summary, the OCP is a design principle. This is not something that you can bake into your code on the fly, as it will need to be addressed during the design phase. Mastering this rule will take practice to implement properly, and people may disagree with the overall best course of action. However, with practice, you will learn how to implement this rule, which, in turn, will increase the quality of your code. With all that being said, we can now move on to the *L* in SOLID, which stands for the **Liskov substitution principle (LSP)**.

Liskov substitution principle

The LSP is, without a doubt, one of the hardest SOLID concepts to understand and implement. The concept is usually defined with something akin to the following:

Objects of a parent class/function block should be replaceable with objects of a child class/function block without affecting the behavior.

In short, this idea means that the child function blocks should not restrict or change the behavior of the parent function blocks. This is a simple idea, but it can be hard to understand or implement.

What the LSP boils down to is that you should be able to swap a parent reference for a child reference, and all code that relies on the parent's contract should still behave correctly. To properly implement this principle, you must have developers who are knowledgeable on the topic. This isn't a principle that can normally be enforced by outside software; instead, you usually must find violations by testing and, in the case of automation programming, code reviews. It is my experience that when an organization tries to implement SOLID, this principle can be either overlooked or warped, since enforcing it is usually a matter of style and ensuring behavioral correctness.

Implementing the LSP

To demonstrate the concept, we're going to explore a violation of the LSP first. It is common to use a square and a rectangle as an example. This is a common example, and it is a good example to research to grasp the LSP. In practice, both a square and a rectangle have an area equal to the following equation:

$$Area = length * width$$

In mathematics, a square can be defined as a type of rectangle; therefore, we have an "*is-a*" relationship. Essentially, a square will use the same area equation and can inherit from the rectangle function block. We can code this out with the following code:

The rectangle function block

The variables for the main function block will be as follows:

```
FUNCTION_BLOCK rectangle
VAR_INPUT
END_VAR
VAR_OUTPUT
END_VAR
VAR
    width  : INT;
    height : INT;
END_VAR
```

This function block will have three methods named setWidth, setHeight, and getArea. All the methods will have a return type of INT and an access specifier of PUBLIC where applicable.

- The getArea method will have no variables but will have the following logic:

  ```
  getArea := width * height;
  ```

- The setWidth method will have one variable, w : INT, that will be embedded in the VAR_INPUT block. The logic for the file will be as follows:

  ```
  width := w;
  ```

- Finally, setHeight will also have one variable, h : INT, embedded in the VAR_INPUT block. The logic for the method will be as follows:

```
height := h;
```

Essentially, the function block will be responsible for setting the dimensions of the shape and retrieving the area.

The square function block

The square function block will inherit from the rectangle block, but will have no logic or variables; it will only have two methods named setHeight and setWidth that will have a return type of INT.

- The setHeight method for the square function block will have one variable, h : INT, in the VAR_INPUT block. The logic for this block will match the following:

```
Width  := h;    // force square
Height := h;
```

- The setWidth method will have one variable, w : INT, in the VAR_INPUT block, and the logic will be as follows:

```
width  := w;    // square keeps sides equal
height := w;
```

Once the two function blocks are implemented, the PLC_PRG variables should be as follows:

```
PROGRAM PLC_PRG
VAR
    rRect       : rectangle;
    rSquare     : square;
    errorRect   : BOOL;
    errorSquare : BOOL;
    areaRect    : INT;
    areaSquare  : INT;
END_VAR
```

While the logic for the POU will be as follows:

```
(* Test with Rectangle *)
rRect.setWidth(w := 5);
rRect.setHeight(h := 4);
```

```
areaRect := rRect.getArea();        // 5 * 4 = 20

IF areaRect <> 20 THEN
    errorRect := TRUE;              // should stay FALSE
END_IF;

(* Test with Square, used "like" a Rectangle *)
rSquare.setWidth(w := 5);
rSquare.setHeight(h := 4);
areaSquare := rSquare.getArea();   // 4 * 4 = 16, not 20

IF areaSquare <> 20 THEN
    errorSquare := TRUE;           // becomes TRUE → LSP is violated
END_IF;
```

When the code is run, you should be met with *Figure 12.4*:

Device.Application.PLC_PRG		
Expression	Type	Value
+ ⬦ rRect	rectangle	
+ ⬦ rSquare	square	
⬦ errorRect	BOOL	FALSE
⬦ errorSquare	BOOL	TRUE
⬦ areaRect	INT	20
⬦ areaSquare	INT	16

Figure 12.4 – Non-compliant LSP output

You should be able to swap a parent reference for a child reference, and all code that relies on the parent's contract should still behave correctly. In this example, even though a square is a rectangle, they are not compatible enough for inheritance to be properly applied and for everything to still make sense. In other words, a square is a rectangle, but a square is not enough of a rectangle for the square to inherit from. In this case, the implied contract is that setWidth and setHeight set their respective dimensions independently. The PLC_PRG POU can reasonably expect SetWidth(5); and SetHeight(4); to yield an area of 20. In square, those same calls produce an area of 16 instead, so substituting a square where a rectangle is expected breaks the caller's assumptions and violates the LSP.

If this was a violation of the LSP, what does a compliant example look like? In this demo, we're going to rework the example to create a compliant LSP program. First, we need a more generic structure for both function blocks to inherit from. If we think about it, a square is a rectangle, but they are both shapes. This means we can rework our example, to have a structure similar to the following.

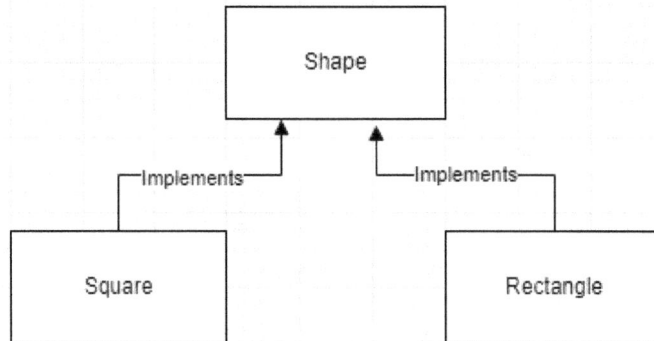

Figure 12.5 – Compliant LSP structure

In this program structure, we are going to have both function blocks derive from something more generic, a general shape.

In this example, we're going to implement a shape interface called IShape with a method called GetArea with the following code.

```
METHOD getShape : INT
VAR_INPUT
    METHOD GetArea : REAL
END_VAR
```

We're going to rework our rectangle function block to implement IShape and have a method called setSize with the following variables.

```
METHOD PUBLIC setSize : INT
VAR_INPUT
    w : INT;
    h : INT;
END_VAR
```

The method block will have the following logic.

```
width  := w;
height := h;
```

This function block will also automatically implement the GetArea method, which will consist of the following:

```
GetArea := width * height;
```

The rectangle function block itself will have the following variables.

```
FUNCTION_BLOCK rectangle IMPLEMENTS IShape
VAR_INPUT
END_VAR
VAR_OUTPUT
END_VAR
VAR
    width  : REAL;
    height : REAL;
END_VAR
```

The square will also implement IShape and have one unique method called setSide that will have the following variables.

```
METHOD PUBLIC setSide : INT
VAR_INPUT
    si : INT;
END_VAR
```

The line for the logic will be as follows:

```
side := si;
```

The GetArea method will also have one line of logic, which will be as follows:

```
GetArea := side * side;
```

The PLC_PRG POU will consist of the following:

```
PROGRAM PLC_PRG
VAR
    rect    : rectangle;
    sq      : square;
```

```
    shape    : IShape;   // parent reference
    areaRect : REAL;
    areaSq   : REAL;
END_VAR
```

The main logic for the POU will be as follows:

```
// Use rectangle via IShape
rect.SetSize(w := 5, h := 4);
shape := rect;
areaRect := shape.GetArea();    // 20.0

// Use square via IShape
sq.SetSide(si := 5);
shape := sq;
areaSq := shape.GetArea();      // 25.0
```

We created a reference to IShape, and we call GetArea using that reference. The magic with this example is that all we're doing is changing what object the IShape reference is pointing at.

When this code is run, we should get what's shown in *Figure 12.6*:

Device.Application.PLC_PRG		
Expression	Type	Value
+ ⬦ rect	rectangle	
+ ⬦ sq	square	
+ ⬦ shape	IShape	16#000...
⬦ areaRect	REAL	20
⬦ areaSq	REAL	25

Figure 12.6 – Compliant example

In this example, IShape is the parent and the two function blocks are the children. In this case, all code that relies only on the IShape contract keeps working. In other words, if you see shape, you should be able to use rect or sq.

If you contrast this with the previous, non-compliant example, you may notice that the overall code structure is slimmer and more maintainable. Barring the IF statements that were used as error checks, this version of the code has, overall, fewer and more streamlined methods. Though the methods for the non-compliant example were short, you have to consider what that would be like if this were a production environment. You would have more methods, which would probably be more complex.

The LSP is, in my opinion, the most confusing of the SOLID principles. What constitutes compliant LSP code is often up for debate in practice. However, a simpler principle that we're going to explore in the next section is the **interface segregation principle (ISP)**.

Interface segregation principle

As we have explored, anytime we implement an interface, we have to implement all the methods that come with it. This can be good and bad in a way. In a sense, when we implement an interface, we never have to worry about accidentally missing a method implementation. On the other hand, if we are not careful and our interfaces are not designed well, we can end up with methods that are not used in the function block, which is a bad practice. In programming, we don't want unused code in our programs, as it can clutter and bloat the codebase. Conversely, if you do remove an interface's method in a function block, the code typically won't compile.

This is where the ISP comes into play. Essentially, the meaning of the principle can be summed up with the following:

A function block should not have to implement an interface it does not use, nor should it depend on methods that it does not use.

The best way to think of this principle is to *use it or lose it*! It is a very sloppy but very common practice to leave unimplemented code in your program due to interfaces. If you implement a general interface, you will likely end up with methods that your project does not need. As a result, you will have useless methods floating around your codebase. Even though a system such as CODESYS will implement the methods automatically, from a purely organizational point of view, your codebase will become more cluttered and one of those useless interfaces will eventually get deleted and cause havoc with compilation. Though it is common to use *fat*, general interfaces, it should be avoided whenever possible. If you see your function block is dependent on methods that it does not use, it is a signal that it's time to redesign your interfaces.

The ISP recommends that it is better to use multiple smaller and more specific interfaces than one *fat*, general one. It is important to remember that interfaces are models. If your function block is a hybrid of multiple things, you can use multiple models to craft it. For example, consider a checking and savings account. Both accounts will allow you to withdraw and deposit money; however, a savings account will build up interest over time. Though this type of program would not normally be utilized in a PLC, for the sake of example, let's look at some code to explore this concept.

Implementing the ISP

An inexperienced programmer may do something such as the following to implement accounts:

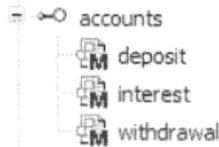

Figure 12.7: Accounts interface

In this case, we have three methods: deposit, interest, and withdrawal. When we implement this interface, we get the following:

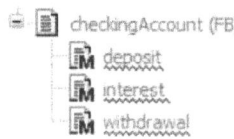

Figure 12.8: Checking account function block

The interest method was automatically added when we implemented the accounts interface. This means that we have unused code and, as we stated before, this is not good. We violated the interface segregation principle and, as such, we are now dependent on unused code. So, with that in mind, how can we fix this?

The cleanest way would be to create a savings accounts interface that extends a checkingAccount one. To do this, we are going to create a savings and checking account function block, as well as a savings and checking account interface, similar to the following:

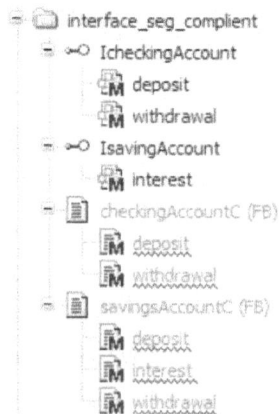

Figure 12.9: Interface segregation compliant

In this example, the savingsAccount function block implements the IsavingAccount interface. The IsavingAccount interface itself extends the IcheckingAccount interface, and, as such, the savingsAccount function block that implements IsavingAccount will consist of the deposit, withdrawal, and interest methods. With this implementation, the savingsAccount function block will have all three necessary methods. You can think of IsavingAccount as a composite interface. Though, in this case, we only implemented one interface in the savingsAcccount function block, we actually implemented multiple smaller ones through interface inheritance.

On the other hand, the checkingAccount function block will only implement the IcheckingAccount interface. This means that it will not require the interest method, which, in turn, means that the function block will not have unnecessary code. As a result, the program is now in compliance with the ISP. Now that we have the ISP under our belt, we can look at the final principle, the **dependency inversion principle (DIP)**.

Dependency inversion principle

When developing modern-day software, you want your code to be as loosely coupled as possible. It is common, especially in automation, to have to work with various types of libraries and APIs. The kicker to this is that this software will change especially when parts are swapped out. For example, it is not uncommon for the customer to opt to put a different brand of hardware in the machine. This means that if you have something such as a motor drive that requires a certain library, your software will need to be changed to accommodate the modification. As has been the whole point of this chapter, you don't want to change existing software and, if you do, change it as minimally as possible. Academically, the DIP can be defined as:

High-level modules should not depend directly on low-level modules; both should depend on abstractions.

This is where loosely coupled architecture comes into play. When it comes to consuming low-level software components such as **application programming interfaces (APIs)** or the like, you don't want your software closely tied to it. To use something such as a drive library, you want to create a middleman component to act as a go-between. In other words, you don't want to talk directly to the API; you want to talk to a function block that will talk to the API for you.

The middleman API is like a façade function block. The middleman will always have a set of stable methods that can be used. For example, you could have an on, off, and ready method that your high-level code will call. The middleman function block will be responsible for determining which API method(s) to call to accomplish the task. In terms of a block diagram, the DIP can be viewed as the following:

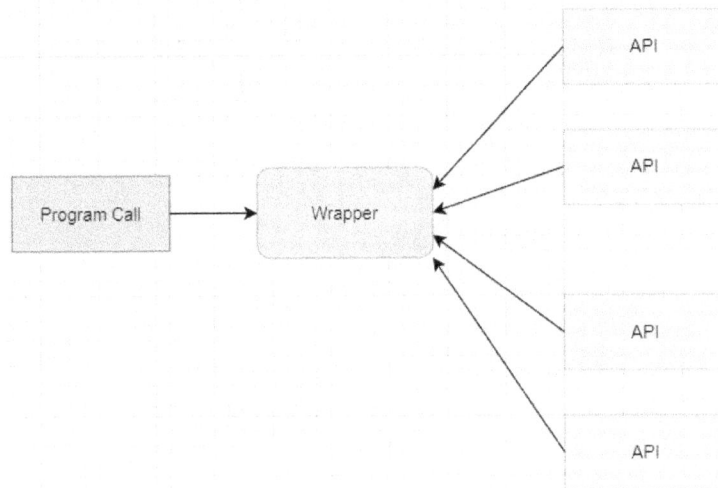

Figure 12.10: Dependency inversion diagram

This design will, for the most part, allow you to keep your program intact. This means that if the lower-level modules ever change, your program won't have to. All you'll need to do is modify the wrapper to support the API changes. We can demonstrate this with the following example.

Implementing the DIP

Suppose we have two APIs, called Api1FB and Api2FB, that provide the logic for turning a device on. To promote a loose design, we are going to create a DriveFB function block that will act as a middleman or façade for the CartFB function block, which will be the high-level code that the user is manipulating with calls from a control panel. In other words, for this example, CartFB will be the **Program Call** represented in *Figure 12.10*. The DriveFB will serve as a wrapper that will prevent the higher-level CartFB block from having to know about the APIs.

Note

The focus of this project is to show separation of high-level and low-level logic. The DriveFB function block is designed to act as an example of a wrapper. Though this is a common real-world example of how DIP is often implemented in practice, in the strictest sense of the principle the function block should be abstracted out more, using structures such as interfaces. Nonetheless, the example will be similar to how DIP-style wrappers are often implemented in the field.

For this example, we are going to create a GVL file and the following function blocks:

- CartFB
- DriveFB
- Api1FB
- Api2FB

All function blocks with the exception of CartFB will have a PUBLIC method named on that has a return type of BOOL.

When we are done, the structure of the project should look like the following:

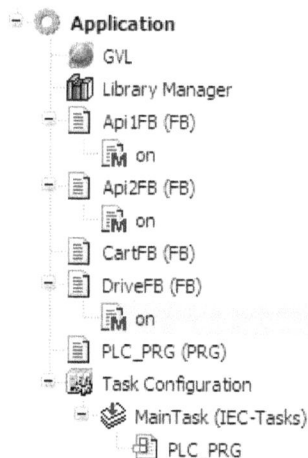

Figure 12.11: Dependency inversion project

Once you have this in place, you can start to set up the code.

First, the GVL will only consist of a single variable. The code for this structure will be as follows:

```
{attribute 'qualified_only'}
VAR_GLOBAL
    msg : WSTRING;
END_VAR
```

The `Api1FB` function block will only have a single line of code in the on method, which will be the following:

```
GVL.msg := "api1";
```

The same can be said for the on method for the `Api2FB` function block, which will consist of the following:

```
GVL.msg := "api2";
```

The `DriveFB` method will consist of the following:

```
FUNCTION_BLOCK DriveFB
VAR_INPUT
END_VAR
VAR_OUTPUT
END_VAR
VAR
    a1 : Api1FB;
    a2 : Api2FB;
END_VAR
```

The on method for the `DriveFB` block will consist of the following variables:

```
METHOD on : BOOL
VAR_INPUT
    api : INT;
END_VAR
```

While the body of the `DriveFB` on method will consist of the following:

```
IF api = 1 THEN
    a1.on();
ELSE
    a2.on();
END_IF
```

The next function block to set up is the `CartFB` block, which will comprise the following:

```
FUNCTION_BLOCK CartFB
VAR_INPUT
    in : INT;
END_VAR
VAR_OUTPUT
END_VAR
VAR
    d : DriveFB;
END_VAR
```

The body of the `CartFB` function block will comprise the following:

```
d.on(in);
```

To call the `DriveFB` function block and utilize our mock APIs, we will use the following variable in the `PLC_PRG` file.

```
PROGRAM PLC_PRG
VAR
    c : CartFB;
END_VAR
```

`Api2FB` will ultimately be invoked with the following line:

```
c(in:=2);
```

When the program is run, you should see the following output in GVL:

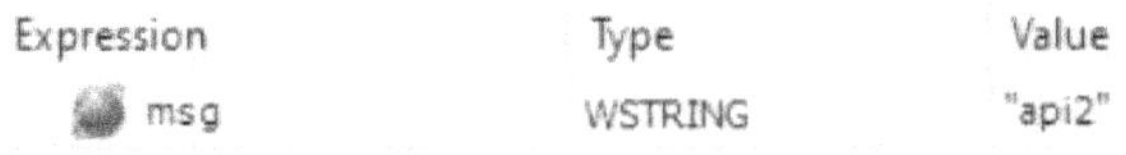

Expression	Type	Value
msg	WSTRING	"api2"

Figure 12.12: GVL output for the program

If you replace 2 in the final line of the `PLC_PRG` file with 1, you should see the other API be invoked.

In this project, we created a loosely coupled program that requires minimal to no programmatic modifications to the high-level logic (`CartFB`) if either of the APIs is ever changed. Now, this code was merely for demonstration. The `IF` statements are not the most effective or robust way of implementing this program. It was done as a means to easily demonstrate the concept of using a wrapper. Overall, this is a very simplified version of DIP. A much better solution would include creating a factory function in the `DriveFB` function block, using more abstraction for the wrapper, and passing around a reference to the objects; these suggestions would promote a better and more

SOLID solution for the architecture. Regardless of how you do it, the goal of DIP is to promote loosely coupled software. By doing this, you will need fewer code changes to adapt and have an overall more flexible and stable design.

By this point, we have explored all of the SOLID principles. SOLID is not an approach that can be easily mastered. SOLID is more of a mindset that must be practiced to fully understand. Now that we have a basic understanding of SOLID, let's try to design a simulated painting machine.

Final project: design a painting machine

Painting machines are often complex devices that have many moving parts. For our final project, we are going to design a simulated device that can move a part on a conveyor belt and paint a sentence on it. For this project, we are going to set the following requirements:

- Drive the conveyor belt (belt on/off)
- Select between two paint APIs
- Paint a message on a part

With these requirements, we can formulate a design like the following:

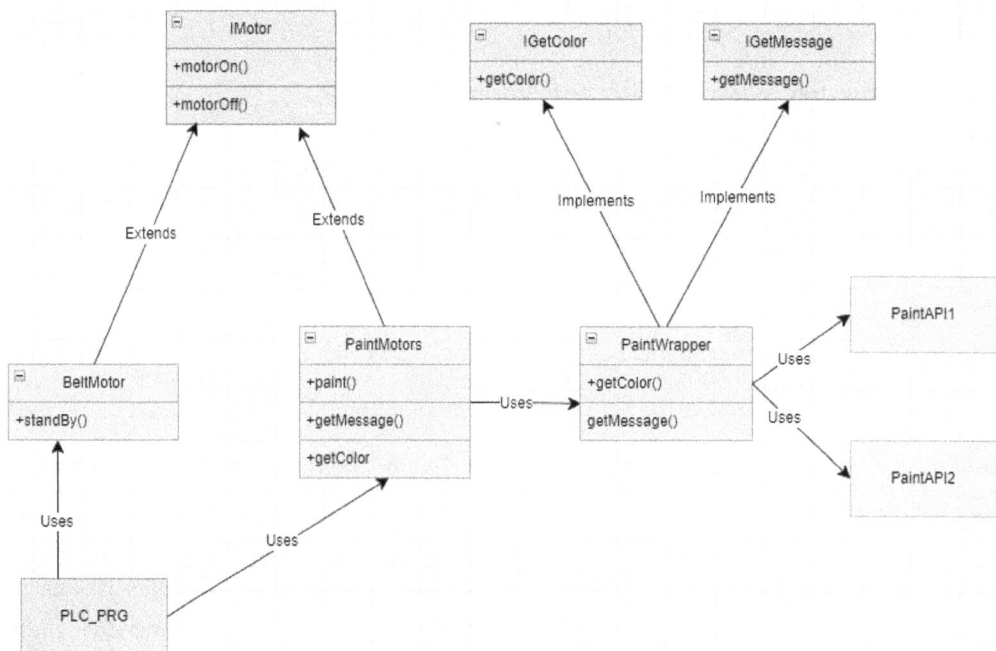

Figure 12.15: Painter design

In this case, the `PLC_PRG` file is going to act as our orchestrator. The file will control when the belt is running and when the machine is painting. In this design, we'll use a `beltMotor` function block and a `paintMotors` block, which will implement `IMotor`. The `paintMotors` block will use the `paintWrapper` function block as a wrapper; in this case, this will be our mock API. The `paintWrapper` function block will use two different vendor paint APIs, and it will also implement two interfaces, called `IGetColor` and `IGetMessage`.

In short, this example is SOLID. All our components, such as our methods, interfaces, and function blocks are describable with one sentence. This means we are following the SRP. Our components are defined singularly enough that if we needed granular control over them, such as calling a singular method or creating a coherent façade, we could.

In this design, if more motors need to be added (added features), we can add them without needing to change existing code in `beltMotor` or `paintMotors`. Therefore, as the machine is upgraded and new functionality is added, we never have to worry about inadvertently breaking `beltMotor` and `paintMotors`. This means that our motor design follows the OCP.

We also have the basis for the LSP. Since we have the `IMotor` interface, any place in the code that works with an `IMotor` instance can use either a `beltMotor` or `paintMotors` instance instead. The key is that, as long as `beltMotor` and `paintMotors` satisfy the contract defined by `IMotor`, the LSP is satisfied.

The `paintWrapper` function block will enforce the interface segregation principle, as the two interfaces only provide the relevant methods for the function block to use. If, at some point, we decided that we no longer need a `getColor` method, we could remove this interface and the method with no leftover code.

The `paintWrapper` function block also serves as a façade wrapper that provides a buffer between the APIs and the `paintMotors` function, which tells `paintWrapper` to start. Essentially, this function block is just there to ensure that the API function blocks are loosely decoupled from the rest of the program. As with the DIP example, we could abstract this out more, but even with this implementation, the code will be more loosely coupled than most non-SOLID PLC projects. In fact, many PLC programs that implement the DIP will usually use a pattern similar to this.

In this diagram, the two APIs can be thought of as generic vendor APIs. In this design, the high-level logic of the PLC program does not need to know how these APIs work or even what they do. All the high-level logic needs to do is funnel through our wrapper, and that logic will utilize the vendor APIs.

Overall, this design is about as SOLID as a PLC program is going to get. Realistically, you're not going to implement all these principles in a single project. However, even if you implement just a few of these principles, your PLC code will likely be light-years ahead of your competitors.

Summary

As was seen in this chapter, SOLID can take some extra effort during the design phase. However, when done correctly, these five principles can ensure that your code is easy to fix and expand upon. In the fast-paced world of industrial automation, this is a must. You need to be in and out of a customer site as quickly as possible, and SOLID can foster this.

Each of these principles will take some time to master, but once you do, the payback will be well worth it. At this point, you should have a good enough background to start expanding your knowledge and experience on these concepts. Hence, with all this in mind, we can now move on to another very important aspect of automation programming: HMI design.

Questions

1. What are common code modules?
2. What should you do if the word "and" appears in the summary of your module?
3. What is the interface segregation principle?
4. Name the five principles of SOLID.

Further reading

- *SOLID: The First 5 principles of Object-Oriented Design*: https://www.digitalocean.com/community/conceptual_articles/s-o-l-i-d-the-first-five-principles-of-object-oriented-design

- *SOLID*: https://en.wikipedia.org/wiki/SOLID

- *Exploring the Liskov Substitution Principle*: https://www.infoworld.com/article/2971271/exploring-the-liskov-substitution-principle.html

- *LSP: The Liskov Substitution Principle*: https://medium.com/@gabriellamedas/lsp-the-liskov-substitution-principle-e43910b638bc

Join our community on Discord

Join our community's Discord space for discussions with the authors and other readers:

`https://packt.link/embeddedsystems`

Part 3

HMI Design

In this section, you'll shift your focus to creating intuitive, functional human-machine interfaces (HMIs). You'll learn how to design user inputs and outputs, build clean and effective layouts, and craft interfaces that guide operators naturally through complex systems. This section also covers the critical topic of alarms: how to design them, prioritize them, and implement them safely to prevent catastrophic mistakes. By the end of this part, you'll understand how thoughtful HMI design improves usability, safety, and overall system performance.

This part of the book includes the following chapters:

- *Chapter 13, Industrial Controls: User Inputs and Outputs*
- *Chapter 14, Layouts: Making HMIs User-Friendly*
- *Chapter 15, Alarms: Avoiding Catastrophic Issues with Alarms*

13

Industrial Controls: User Inputs and Outputs

The cold reality is that the customer isn't going to care about your codebase. You could have the most eloquently written program ever produced, and the machine it's on may still be considered a failure by the customer. When it comes to the customer or end user, all they are going to care about is how they interface with the device. This means that a well-written UI is key to the success of a machine. If the UI is bad and hard to use, the machine is also going to be hard to use, and vice versa, no matter how well the software is written.

The key to the success of any machine is what's known as the **human–machine interface (HMI).** Though often delegated to the junior developer of the group, the HMI is a vital component of the device, and it will dictate whether the machine will be in operation for 20 years or 20 minutes. HMI development is as much an art as it is a science.

In this chapter, we're going to explore HMI development by covering the following topics:

- Introduction to HMI design
- Switches
- Buttons
- LEDs
- Potentiometers
- Sliders
- Spinners

- Measurement controls
- Control properties

Once we've explored all of the common controls, we're going to build a sample HMI panel, as well as some PLC code for the controls to interface with.

Technical requirements

If you opted to use another development environment, you need to download and use CODE-SYS for this chapter with the visualization tool installed. The project that will be developed in this chapter can be downloaded from `https://github.com/PacktPublishing/Mastering-PLC-Programming-Second-Edition/tree/main/Chapter%2013`.

> CODESYS Visualization
>
> You may need to separately install the visualization kit separately. You can do this by launching the CODESYS installer under the tools tab. Once there you can click on the Browser tab and search for Visualization to install the kit.

HMI development is as much an artistic endeavor as it is an engineering practice. I strongly recommend that you pull down the code and modify it using the principles that will be covered in this chapter.

Introduction to HMI design

In the old days of automation, the way an operator interacted with a machine was with a physical control panel that consisted of physical components such as switches or buttons. This could easily lead to bottlenecks and restricted progress as modifying or fixing the panel would cost time and money. In the modern automation era, the physical panels of old gave way to programs that live on a touchscreen computer. These programs are called HMIs.

HMI is industrial jargon for a specialized piece of software that allows machine operators to easily interface and control the machine. In other words, an HMI can be thought of as an industrial **UI (User Interface)**. The HMI will be the point of contact for the operator. This means that an HMI must be easy to use and well laid out. Often, the HMI will determine whether the end user considers the machine a success.

HMIs offer much flexibility and reliability over their physical counterparts. Since an HMI is a software program, if new features are added to the machine, updating the user panel is as simple as adding the necessary controls to the screen and setting up the necessary logic. Also, unlike their physical counterparts, the controls of an HMI will not break. When it comes to an HMI, the only thing that can fail, and it will sooner or later, is the computer on which the HMI program lives or is displayed.

How are HMIs made?

Since HMIs are programs, there are many ways to produce them. There are various software packages that are specifically designed for HMI development such as C-more and Red Lion. These types of packages usually use simple drag-and-drop software components and are uploaded to specialized computer hardware that is designed to run the HMI. You can also use a SCADA system such as Movicon for HMI development.

If you're particularly savvy with programming, you can also use a general-purpose programming language such as Java, C#, C++, or some other high-level language. Though not commonly used anymore for traditional programming applications, Delphi and VB.NET are still commonly used in the automation world to create HMIs. Developing an HMI with a traditional language such as Java, C#, or a series of web-based technologies will offer much more flexibility and power, but it will take much more time and technical knowledge to develop. However, more complex architectures and processes can be used with these HMIs. For example, technologies such as containerization, cloud computing, and AI/ML can be easily integrated into the system.

When using a general-purpose programming language to develop an HMI, you will typically use a graphical framework like the following:

- WinForms
- WPF
- QT
- JavaFX

HTML5/CSS3/JavaScript

These technologies are not frameworks, as they are used to create frontend web applications. These technologies are becoming the norm in automation, especially in places that have adopted concepts that are found in Industry 4.0

CODESYS and similar systems typically have some type of simple HMI development system that can be used to create basic soft control panels. However, before we start digging into creating an HMI, we need to understand a few basic principles.

Basic principles for designing an HMI

Designing an HMI is like designing any other UI; the first step is to wireframe it. Wireframing involves creating a rough sketch of the way the UI controls should be positioned. An example of this is shown in *Figure 13.1*:

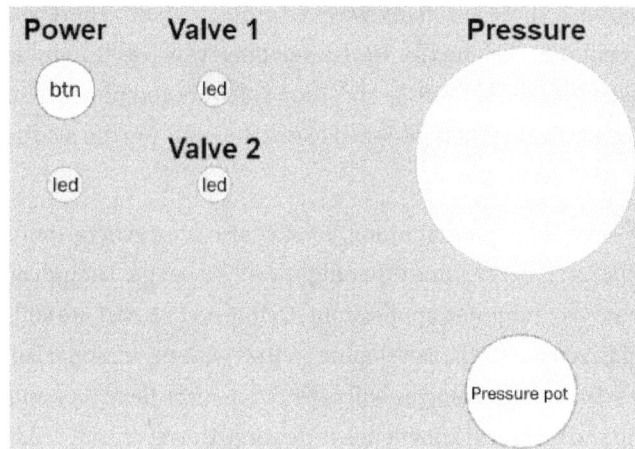

Figure 13.1: Wireframe example

When it comes to layouts, controls with similar responsibilities should be stacked vertically, with their label either on, to the left, or to the right of the control. You can also center the label above or below the control. Ideally, you want to label the controls similarly to what's shown in *Figure 13.1*; however, that is not always possible.

The responsibility of an HMI

How much should an HMI do? This is a great question, and it is very important. In short, an HMI should be dumb. This means that its primary—and hopefully only—responsibility is to take user inputs from the operator and display data from the PLC. Typically, you do not want any complex logic in the HMI. It should not do any computations, process any data, or make any decisions related to the PLC's operations.

With that, the notion of a dumb HMI is evolving. As new technologies are adopted, what's considered basic operations is also changing. For this book, we're only going to use an HMI as a means to input and display data. For high-tech facilities such as smart factories, this may differ as the HMI may be responsible for basic security, grouping users, and so on; however, that will vary. The best rule of thumb is to delegate as little responsibility to the HMI as possible.

These are basic design principles that will be explored throughout the rest of this book. With that, we need to learn the steps involved in creating an HMI.

Adding an HMI

The easiest way to create an HMI project is to simply create a standard project, something we have done throughout this book. Once you've created a project, you will want to right-click **Application**, navigate to **Add Object**, and select **Visualization...**, as shown in *Figure 13.2*:

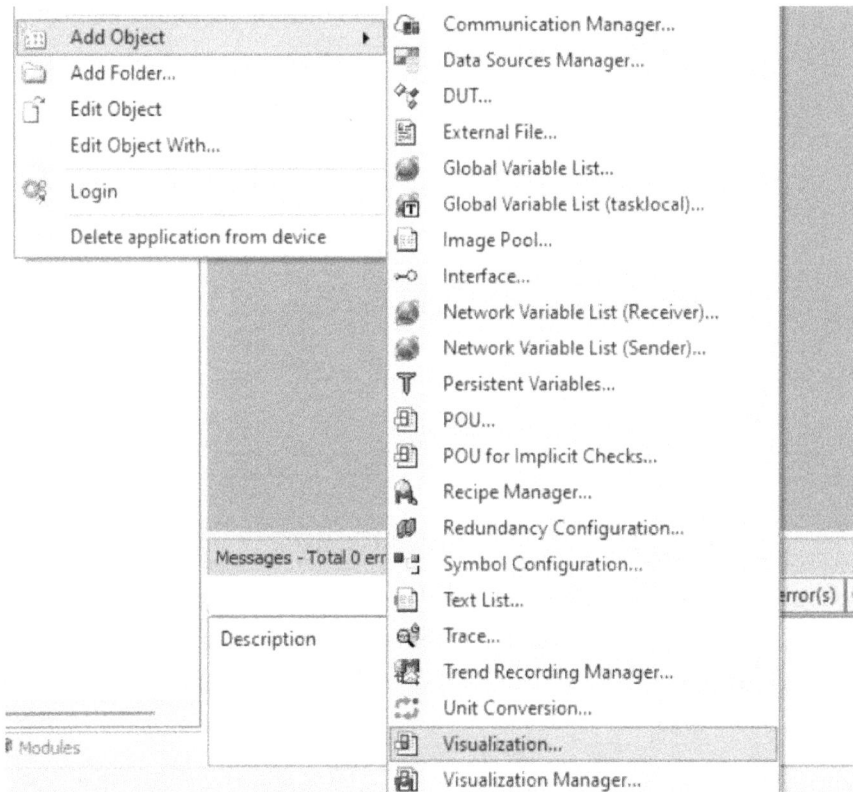

Figure 13.2: Adding an HMI to a project

> **Note**
>
> Depending on your version of CODESYS you may have to install the visualization tool separately!

Once you've done this, you'll see a window similar to the one shown in *Figure 13.3*:

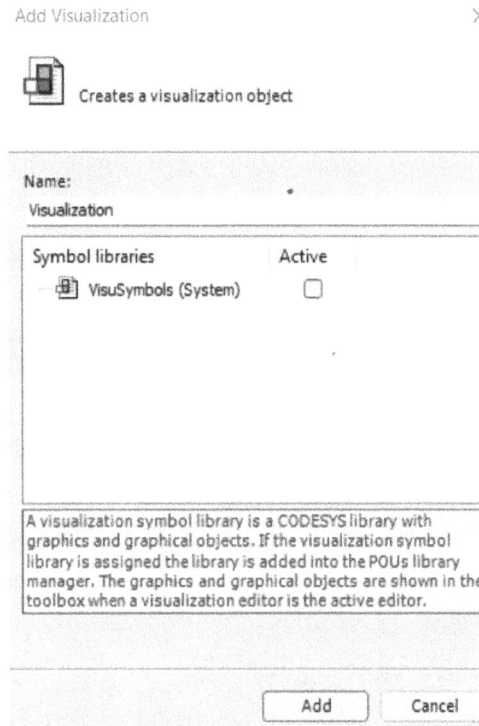

Add Visualization   ×

Creates a visualization object

Name:
Visualization

Symbol libraries	Active
VisuSymbols (System)	☐

A visualization symbol library is a CODESYS library with graphics and graphical objects. If the visualization symbol library is assigned the library is added into the POUs library manager. The graphics and graphical objects are shown in the toolbox when a visualization editor is the active editor.

Add   Cancel

Figure 13.3: Add Visualization

Click **Add** and wait a few minutes for the controls to render. Upon completion, you will be met with a new area to the right of the screen that contains HMI controls; see *Figure 13.4*. As can be seen in the figure, there are many different controls to choose from, including LEDs and switches.

Figure 13.4: HMI controls

The preceding figure shows all the controls you can use. Each tab will contain more controls, so each tab is worth exploring. Now that we can add an HMI to the screen, we can start exploring!

Exploring common HMI controls

All systems need some way for the operator to send input signals and receive feedback. For purely physical systems, switches, buttons, and so on are used for the inputs, while control elements such as LEDs and gauges are used for the outputs. This can be costly, and in the modern computer-driven world, this can also be unnecessary as we can simply program in our controls. As such, the remainder of this section will explore software-based controls.

Flip switches

As we all learned in high school, a **switch** causes a break in a circuit that will essentially cause the flow of electricity to stop when it reaches the switch. In other words, with the switch closed, the electricity is free to flow in the circuit, which will cause the equivalent of a TRUE condition. If the switch is open, the electricity will not be allowed to flow throughout the circuit, which will cause a FALSE condition. In terms of HMIs, a switch can be thought of in a similar sense. A digital HMI switch will behave the same way as a physical switch will. When the switch is *on*, the variable that it is attached to will be set to TRUE, and when the switch is *off*, the variable will be set to FALSE.

Two common types of HMI switches are the rocker and dip switch, both of which can be seen in *Figure 13.5*.

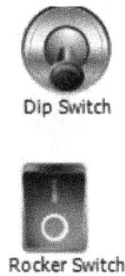

Dip Switch

Rocker Switch

Figure 13.5: Flip switches

These are flip switches, and they behave similarly to a common wall switch. When you flip the switches up, they will produce a TRUE condition, and when they are flipped down, they will cause a FALSE condition. The only way to toggle a switch's state is to toggle the switch itself. Switches like these are commonly used to put things into a given state until the operator decides to change that particular state. For example, the operator can flip a switch up to turn a fan on and flip it down to turn a fan off. Now that we understand what flip switches are, let's look at push switches.

Push switches

Another type of switch is the push switch. A push switch will behave the same way as a flip switch, but instead of flipping it, you will push it. In their non-pressed state, push switches will produce a FALSE state; while pressed, they will produce a TRUE state. The CODESYS HMI builder offers a few different types of push switches. Two common push switches can be seen in *Figure 13.6*:

Push Switch   Push Switch LED

Figure 13.6: Push switches

Now that we've explored switches, we can move on to **buttons**, which are their derivatives.

Buttons

As with flip switches, buttons behave the same way as their physical counterparts do. An HMI button will only change states while the button is pressed. In general, when the button is released, the state will change back to whatever it was before. However, many HMI development systems will allow you to configure the button so that it latches a bit or pulses for a single scan. Typically, you will want to use a button to perform operations such as jogging an axis into position, starting a process, inputting data, changing/opening an HMI screen, and so on.

For most HMI systems, buttons are much more customizable in terms of appearance. You can customize their color, the text that appears on them, whether they will be normally on or normally off, and so on. A typical button in its default state—in other words, when you add a button to the screen—can be seen in *Figure 13.7*:

Button

Figure 13.7: Button

However, after customization, you could make it look like the button shown in *Figure 13.8*:

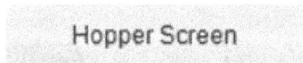

Hopper Screen

Figure 13.8: Customized button

Buttons are very important components. In my experience, I have always found myself using buttons more than switches. There is no set rule for when to use one over the other, so long as you get your desired results. However, as a general rule, I would use buttons in the following situations:

- Changing an HMI screen
- Starting a process that will end and needs to be restarted
- Entering data
- Jogging components into place

Switches and buttons are very common, powerful HMI components. At their core, they are inputs for Boolean variables. With that being said, if switches and buttons are inputs, LEDs are their outputs. Let's take a closer look.

LEDs

One of my favorite HMI components is LEDs. Nothing screams advanced technology more than blinking lights! Ever since I was a child, I have loved playing with LEDs, and as an adult, not much has changed. I still have to constantly keep my love of adding lights to my HMI control panels in check.

In CODESYS, LEDs are referred to as **lamps**; however, in everyday speech, any light indicator is normally referred to as an LED. So, for this book, we will use the term LED to refer to lamps. LEDs are, in my opinion, one of the simplest and most powerful status indicators an HMI developer can use. An LED, or lamp, in CODESYS will look like what's shown in *Figure 13.9*:

Figure 13.9: CODESYS LED/lamp

LEDs can serve multiple purposes in an HMI. They can determine the status of a device, indicate whether a component is on or off, whether there is an issue, whether there is about to be an issue, and so on. In CODESYS, LEDs can be set to five different colors: red, yellow, green, blue, or gray.

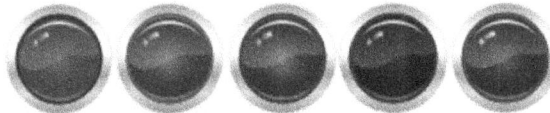

Figure 13.10: The five LED colors

The primary way LEDs are used to relay status information is via their color. Colors such as red, yellow, and green all have meanings that can signify a certain status.

LEDs will be used a lot in your day-to-day life and throughout the rest of this book; however, for now, we're going to switch gears and talk about **potentiometers (pots)**.

Potentiometers

Outside of switches, arguably the other most common HMI control is the potentiometer or, as it is most commonly referred to, pot. In a nutshell, pots allow you to input a numerical value in a range. For example, in the same way a physical pot will allow you to adjust the resistance in an electrical circuit, an HMI pot will allow you to adjust a value in the software.

Pots have many different uses and are very common in HMI development. They are commonly used in applications that require temperature control, such as ovens, and speed control input, and they are even used as inputs for things such as part counters. A pot is depicted in *Figure 13.11*:

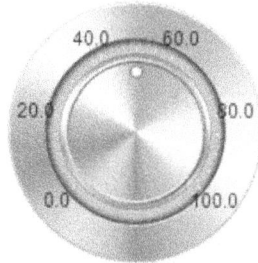

Figure 13.11: Potentiometer

By default, a pot will have a range from 0 to 100; however, this can be adjusted. When working with pots, it is very important to remember to adjust your range. This is a common mistake that even the most experienced HMI developers make. So, when you're working with pots, it is worth keeping that in mind. Similar to switches and buttons, pots also have cousins in HMI development called **sliders**. Now that we have a grasp on pots, we are going to switch gears again and take a quick look at sliders.

Sliders

Sliders can best be thought of as pots that are depicted as straight lines. Sliders work similarly to pots. For the most part, anywhere you can use a pot, you can use a slider, and vice versa. The differences between pots and sliders are mostly aesthetic. Sliders in CODESYS are depicted in *Figure 13.12*:

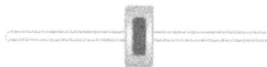

Figure 13.12: Slider

Sliders can be customized in much the same way pots can. For example, you can customize attributes such as the range in the same way you can with pots.

In my experience, both sliders and pots can be difficult for operators to work with. A large part of this is due to operators usually wearing work gloves while interacting with the screen. In my opinion, sliders are a bit harder to work with than pots due to them being smaller, and typically not having a range on them. However, this is just my opinion, your experience and your customers' experience may be different. Either way, pots and sliders are excellent input controls, and any HMI developer should have a basic understanding of both.

In terms of the data type of the variable, it is common to use an integer type for both sliders and pots. You can use a REAL type, but your ability to control the full resolution of the value will be greatly affected. For example, it will be hard to hit a specific number after a point. If you need a floating number, you may want to opt for a numerical input such as a keypad. Generally, I never like using pots or sliders for any inputs that need a value of less than 1. In other words, if you need a high level of precision, you probably don't want to use either of these types of controls. With all that in mind, we can now explore **spinners**.

Spinners

Spinners are another input control. Different from pots and sliders, spinners have an up-and-down button that allows operators to adjust the value. Essentially, spinners are very simple and handy. Usually, for many applications where a slider or pot may be inefficient, a spinner can be an excellent alternative. A simple spinner can be viewed in *Figure 13.13*:

Figure 13.13: Spinner

If you use a spinner, you should set a maximum and a minimum value. Essentially, these values will act as the range for the spinner. The buttons that can be used to set the value can be seen in *Figure 13.13*. When the *up* button is clicked, the value in the spinner will increase, whereas when the *down* button is clicked, the value in the spinner will decrease. The one downside of spinners is that the buttons on the spinner can be hard to click when they are small. As such, when you do use spinners, it is important to consider the button size to aid the operator's ease of use.

In all, spinners are very simple but very powerful controls. Similar to pots and sliders, spinners are excellent for operator inputs.

Now that we have explored spinners, we can focus on **measurement controls**, which can display data from the PLC or input controls.

Measurement controls

While an LED is a common readout for a control, such as a switch or a button, measurement controls, such as gauges and bar graphs, are used as readouts for controls such as pots and sliders. Readout controls usually vary the most between HMI development systems. In CODESYS, the readouts are mostly analog. These controls are handy for displaying things such as temperature or pressure. CODESYS offers several different styles of readouts to choose from.

The following control readouts can be used for many different applications. First, let's look at *Figure 13.14*, which shows a bar graph.

Figure 13.14: Bar graph

A bar graph is a simple straight line with a green bar inside that points to a value. As the value of the variable increases, the green line will as well. A bar graph is great for things such as showing the percentage of a job that is complete, and so on.

Figure 13.15: 90° gauge

The 90° gauge works the same way as the bar graph does, but it has a more compact and different style.

Figure 13.16: 180° gauge

The preceding figure depicts a 180° gauge while the following figure depicts a 360° one. Both function similarly to the other gauges, with the only difference being the style.

Figure 13.17: 360° gauge

For the most part, the only real differences between the different displays are their shapes. As with pots and sliders, you will need to set the range on these as well. As shown in *Figures 13.14* to *13.17*, gauges range from 0 to 100 by default. This means that if you're planning to have input values that are more than 100 and you don't set these values, you're going to peg out the gauge before you reach the upper limit of the control. So, the moral of the story is that whenever you use pots, sliders, or gauges, you need to remember to change your ranges for the controls!

Histograms

Though it is not a gauge, another measurement control that can be used is a **histogram**. Histograms show the current values in an array. They are excellent for applications such as reading temperature from multiple thermal couples or pressure from multiple pressure gauges.

Histograms are great controls, have many different use cases, and excel at displaying real-time data. They are great for monitoring:

- Real-time temperature readings
- The statistics of a machine run
- Monitoring the status of multiple production lines
- The voltage/current that is being drawn from different parts

Essentially, the ideal use case for a histogram is for monitoring real-time data. An example of the output for a histogram can be viewed in *Figure 13.18*:

Figure 13.18: Histogram

As stated previously, histograms work off an array. Unlike gauges or LEDs, which read a single variable, histograms read an array. The following code was used to generate the graph data shown in *Figure 13.18*:

```
PROGRAM PLC_PRG
VAR
    hist : ARRAY [1..4] OF INT;
END_VAR
```

The variables were set with the following code:

```
hist[1] := 10;
hist[2] := 30;
hist[3] := 25;
hist[4] := 42;
```

As can be seen from the histogram and its respective code, the value of each element in the array will correspond to a bar in the graph.

Text fields

Arguably, one of the most important controls that you will use is a text field. Text fields will generally spin up a keyboard or require the use of an external keyboard, depending on the system. Most HMI development systems will generate a keypad of some type when the text field is interacted with; however, if you use a general-purpose programming language, you will either have to use a physical keyboard, the keyboard provided by the computer that the software lives on, or code one up.

A text field provides many benefits over controls such as sliders or pots. For instance, they allow your operator to enter data more easily as they can press numbers or letters, and they provide them with a better level of control. For example, if the operator must enter a decimal value, they can with relative ease by simply inputting the value. This differs from trying to manipulate a slider or pot, which can be difficult with a glove on. Here are some common use cases for text fields:

- Inputting decimal values
- Entering raw text for things such as job run metadata, which can include the operator's name, the date, the customer's name, and more
- Inputting the number of parts to make for a job run
- Inputting precise temperatures
- Inputting precise voltages
- Programming machine movements

There are many more use cases; the only real limit is your imagination as a programmer.

Adding a text field to the screen is as easy as dragging and dropping it in a desired location. Once you've done that, you need to configure it. To configure the control, you will need to add the following to the Texts property:

```
Input: %s
```

Once you've done that, you should have a text field similar to the one shown in *Figure 13.19*.

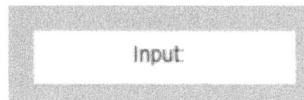

Figure 13.19: Partially configured text field

A text field will usually be connected to a PLC variable. This variable will hold the data that the user inputs; therefore, you will want to add a variable to hold the input, similar to what we did with the other controls. Once you've done that, navigate to **Inputs** and **OnMouseClick**. Thereafter, you will need to click the field with the three dots, as we did previously, and configure the pop-up so that it matches what's shown in *Figure 13.20*:

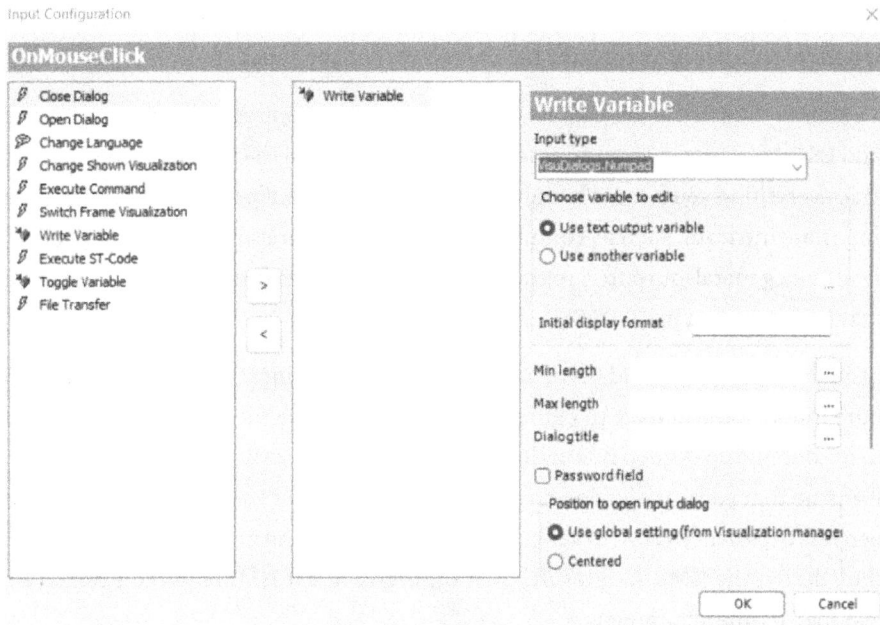

Figure 13.20: Text field configuration

At this point, click **OK** and run the program. When you click the text field, you should see a keypad appear, as shown in *Figure 13.21*:

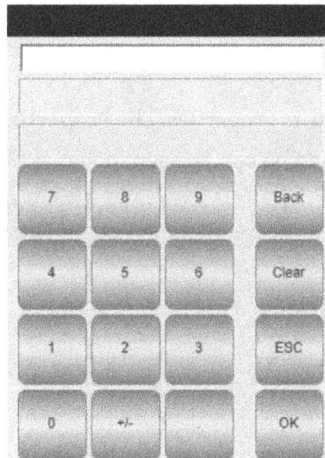

Figure 13.21: Input pad

Now, enter a number and click the **OK** button on the pad. Go back and check the variable that you assigned to it; notice how the variable matches what you typed in.

This is a common numerical keypad. As can be seen, it has numerous buttons, such as a **Clear** button and **ESC**. There are even buttons for changing the sign of the value you enter. In short, the keypad has everything you need. Though the keypad has several powerful features, if you need something more intricate, such as complex mathematical operators, you will typically need to code one up in a general-purpose programming language or use one of the more complex key inputs that are supported in CODESYS.

Typically, built-in keyboards and keypads are great for basic things; however, if your application needs more input, you will have to figure out either how to use the operating system's built-in keyboard or code one up. A good rule of thumb is to design your HMI around your expected inputs. This will ensure that your application has all the necessary input controls for the operator to use. I have seen less experienced HMI developers fall into this trap and not fully think out their inputs before they design the HMI. As a result, their otherwise great HMI required extensive rework, which cost lots of time and money.

At this point, we have explored many of the controls that CODESYS offers. CODESYS offers many more tools, but, for the most part, these are the basic controls that you'll need to know about to get you through a project. Now that we know what the various controls do, we need to know how to customize them via the **Properties** menu.

Control properties

The core of any control is setting up its properties. The **Properties** screen will vary for each control. An example of such a **Properties** screen can be seen in *Figure 13.22*:

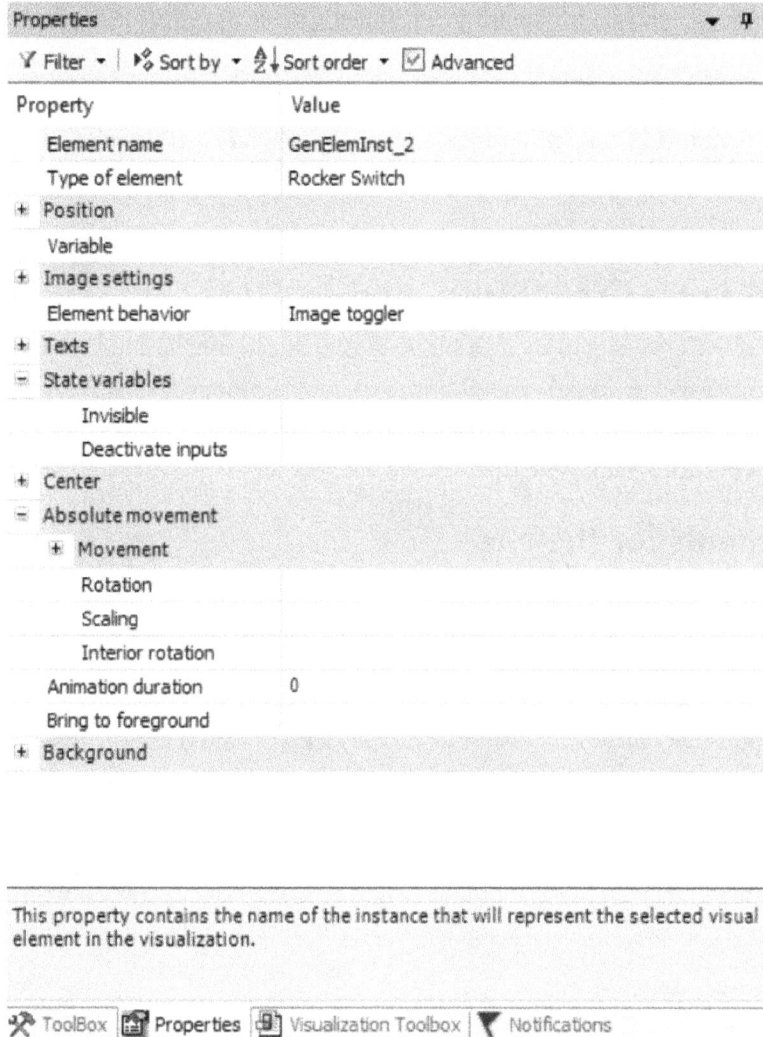

Figure 13.22: Rocker switch Properties menu

Figure 13.22 shows the **Properties** screen for a **Rocker Switch** element. You will need to set the properties for each component you have on the HMI, even if the controls are the same. For example, if you have 20 rocker switches, you will need to set the properties for each one to ensure they behave properly.

As stated previously, it is very important to remember that each control type will have a different set of properties that need to be configured. There will be common fields for each control, but each control will ultimately have different fields that need to be set up. It is important to become familiar with the controls that we have studied, as they will vary. At this point, we have explored many common controls and covered the basics of how to configure their behavior.; therefore, we're going to move on to our final project.

Final project: Creating a simple HMI

For this project, we are going to create a simple HMI that can control a histogram. The HMI we are going to create will be straightforward: when a switch is flipped, an LED is going to turn on, and a pot will become visible. When the pot appears, we will be able to turn the pot to adjust one of the lines on the histogram. With that in mind, let's set up some basic requirements.

Requirements for the HMI

The HMI will need the following:

- Four rocker switches that will control the visibility of four different pots
- Four LEDs that indicate when the rocker switch is on
- Four pots that will only be visible when the rocker switch is on
- Each pot will control exactly one bar on the histogram
- Both the pot and the histogram will have a range of 0 to 100 (default range)

With these requirements in mind, minimal code will be required to make the HMI function as intended. These requirements also dictate that there will be the following controls:

- Four rocker switches
- Four LEDs
- Four pots
- One histogram

Designing the HMI

From these requirements, we can move on to the design phase and layout of our HMI. Much like coding, there is no set way of laying out an HMI. So, we're going to use a layout that looks similar to this:

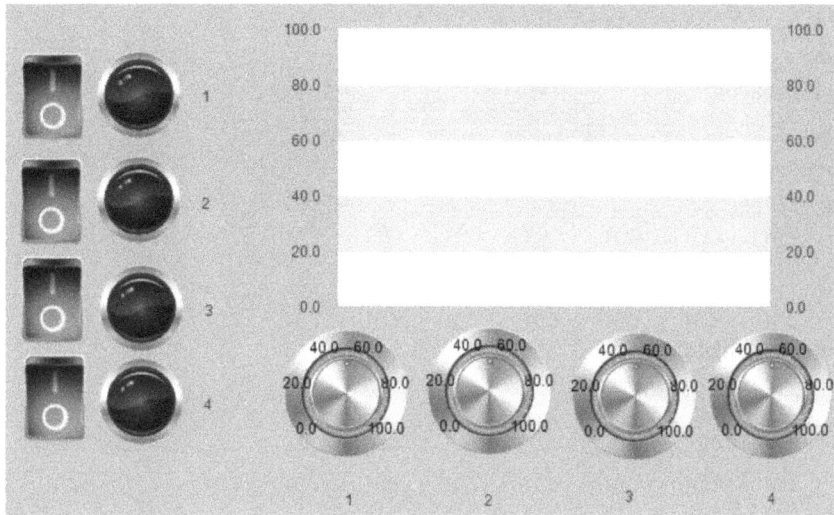

Figure 13.23: HMI layout

In this layout, we have four rocker switches next to their corresponding LEDs. The focal point of the HMI is the histogram in the middle. Underneath the histogram reside the four pots that will control the bars in the output. In short, this is a condensed design that will get the job done. However, this isn't the only design that can accomplish the job. For fun, I recommend that you play around with the design to see whether you can improve the layout.

Now that we have a design in place, we can start building the HMI.

Building the HMI

The code for the HMI will consist mostly of variables. There should be four switch variables that are tied to LEDs and an **Invisible** field. However, to make the pot visible when the switches are on, we will have to add four additional variables that will be the inverse of the state of the switch. This may seem to be counterintuitive, but a True variable in the **Invisible** field for the pot will cause the pot to be invisible. This will create a counterintuitive situation. For this example, for learning purposes, we will remedy this problem with PLC logic but note that there is a better method of accomplishing the same task, something we will explore later on.

The variables for the HMI will look like this:

```
PROGRAM PLC_PRG
VAR
    hist : ARRAY [1..4] OF INT;
    sw1  : BOOL;
```

```
    sw2  : BOOL;
    sw3  : BOOL;
    sw4  : BOOL;
    pot1 : BOOL;
    pot2 : BOOL;
    pot3 : BOOL;
    pot4 : BOOL;
END_VAR
```

The logic will look like this:

```
pot1 := NOT sw1;
pot2 := NOT sw2;
pot3 := NOT sw3;
pot4 := NOT sw4;
```

This logic will simply invert the Boolean state of the rocker switch. These inverted variables will be responsible for causing consistent visible/invisible behavior when used with the current state of the switch.

If you take a closer look at the variables, you will notice that an array has been declared. This array will be assigned to the histogram. Each pot will be assigned an element of the array. Therefore, the graph will dynamically update when the pots are turned.

Once the variables are in place, you can start to add the controls to the screen. This is a simple task as all you have to do is select the different controls and, with your mouse button held down, drag them to the screen. Once they are on the screen, you can move them around as needed and resize them. Lay out the elements in a similar fashion to what can be seen in *Figure 13.23*.

With all this set up, we need to start assigning variables to the HMI components. The first controls we are going to hook up are the switches and the LEDs:

1. All we have to do is select the control and find the **Variable** field, as shown in *Figure 13.24*:

Figure 13.24: Empty Variable field

2. To assign a variable, click on the button with the three dots on the right. You should see a pop-up similar to the following:

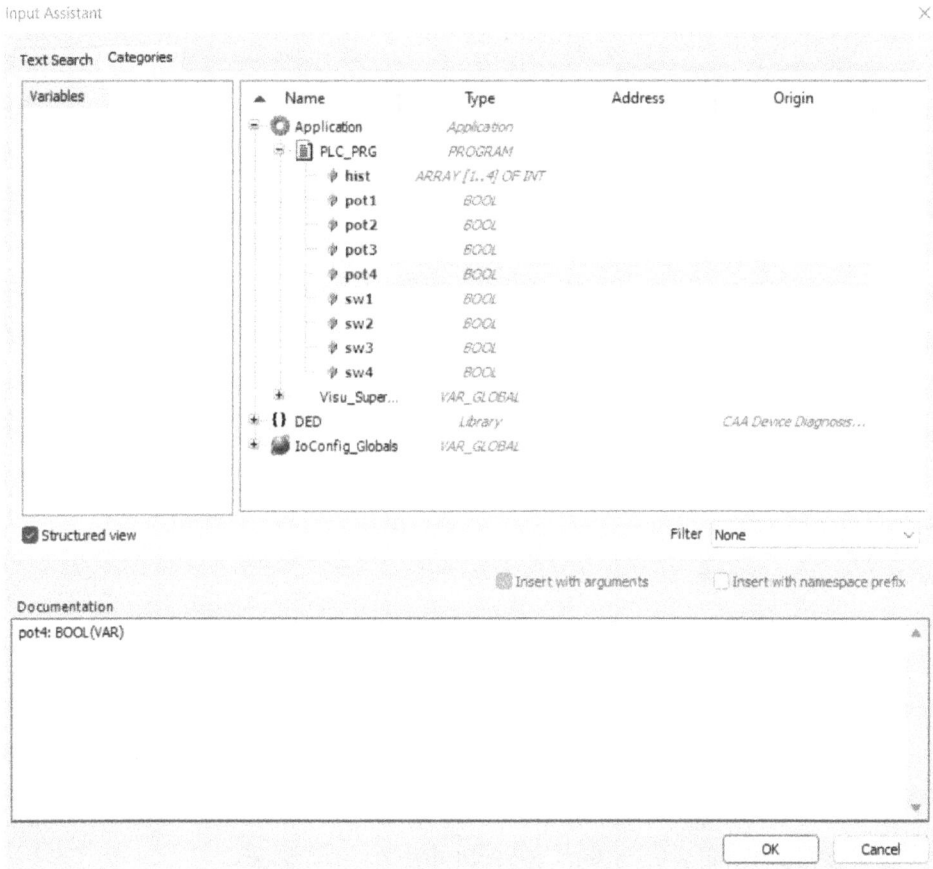

Figure 13.25: Input Assistant

3. Select the corresponding variable for each switch and LED. In short, you will assign a switch variable to both the rocker switch and the LED that is next to it.

Once you've done that, you can move on to setting up the pots. These will require a little more setup than the switches and LEDs.

4. For the pots, you will need to set up the **Variable** and **Invisible** fields.

Figure 13.26: Pot Invisible field

5. Assign the corresponding pot variable to this field using the same process that we used with the LEDs and switches. When you are finished, your field should look like this:

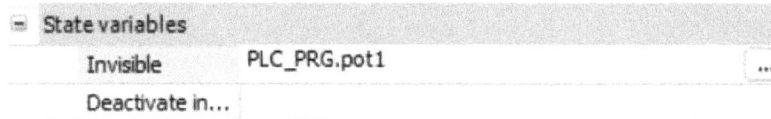

Figure 13.27: Pot Invisible field with variable

6. Repeat this process for each pot.

7. At this point, assign the `hist` array element to the pot's **Variable** field.

Variable   PLC_PRG.hist[1]

Figure 13.28: Pot variable

8. You will have to add the square brackets and the element number to the end of the variable.

9. Repeat this process for each pot variable.

10. This will be very simple: all you will have to do is assign the array to the **Data array** field in the histogram's **Property** menu. Adding this array will be the same as adding a normal variable. When you are done, the field should look like this:

Data array   PLC_PRG.hist

Figure 13.29: The Data array field for the histogram

With this final component set up, you can now start the HMI by performing the same set of steps that we used to run the PLC code.

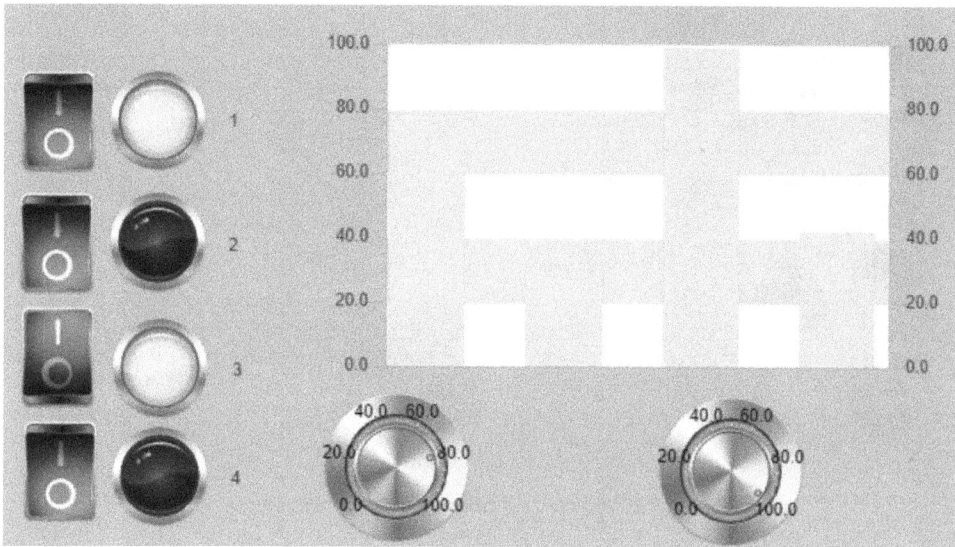

Figure 13.30: Working HMI

In this case, we have two pots turned off and two turned on. We can move the pots and watch the chart change.

Now, what we have works, but it is not the best solution. For this HMI to work, we have to have code that inverts the switch's state. To accomplish this, we created PLC logic. This is a bit unnecessary and somewhat bad, as we now have the PLC doing a menial task. For our purposes, a better solution would be to get rid of the pot variables altogether and simply put the NOT keyword in front of the switch variable, similar to what's shown in *Figure 13.31*.

Figure 13.31: Switch variable with the NOT operator

When the variables are set with the **NOT** operator and run, we can observe the same behavior.

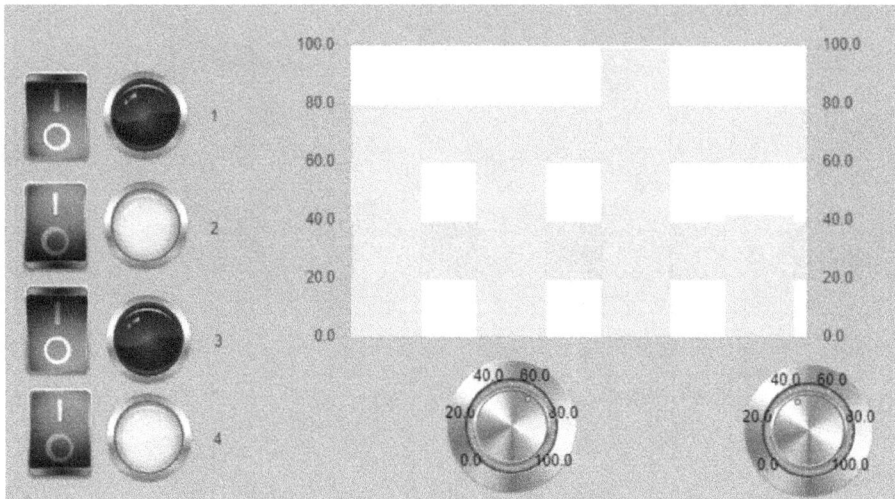

Figure 13.32: Inverted variable in the Properties menu

The main takeaway is that we can manipulate an HMI via the PLC code or through the HMI properties. For our purposes, adding the PLC code was not the optimal solution. However, there will be times when manipulating the HMI via the PLC code will be the optimal solution. A general rule I like to go by is to try to keep the HMI control manipulation on the HMI side. This isn't always possible, but it should be strived for as it will cause less code bloat and free up your PLC to complete more important tasks.

With that, we have created a functioning HMI. At this point, I recommend modifying the current HMI so that you get acquainted using the low-code tooling that is provided in CODESYS. Most of the major HMI and SCADA development packages implement a similar method for creating HMIs, so you're going to want to get used to using the tools as they all work similarly.

Summary

In this chapter, we explored common HMI components such as switches, buttons, LEDs, pots, sliders, and spinners. We also learned how to hook up HMI components and how PLC code can manipulate the controls. Then, we explored how simply using commands in the **Properties** field can allow us to manipulate the controls without the need for the PLC.

As I have stated previously, HMI development is as much an art as it is a science. The next chapter will be dedicated to best practices of laying out an HMI so that your operators can use it effectively. For now, I strongly recommend getting used to the controls and the layout of the **Properties** menu.

Questions

1. What is a button?

2. What kind of data structure do histograms take?

3. Can we add keywords to a property field?

4. Can we manipulate an HMI via PLC code? If so, when should we?

Further reading

- CODESYS visualization: `https://content.helpme-codesys.com/en/CODESYS%20 Examples/_ex_visualization.html`

Get This Book's PDF Version and Exclusive Extras

UNLOCK NOW

Scan the QR code (or go to `packtpub.com/unlock`). Search for this book by name, confirm the edition, and then follow the steps on the page.

Note: Keep your invoice handy. Purchases made directly from Packt don't require an invoice.

14

Layouts: Making HMIs User-Friendly

HMI development has a lot in common with graphic design. Much like with graphic design, there are a few rules that should be followed as closely as possible to ensure that the HMI is user-friendly. There is a difference between laying out an HMI and something akin to a website. I usually like to consider HMIs as the cousins of traditional user interfaces. Both types of interfaces have certain things in common, such as a logical layout and efficient coloring.

Though these types of user interfaces are cousins to one another, an HMI will have a person staring at it much more often. As such, certain factors must be considered that would normally be ignored when developing something such as a website.

Due to operators using the HMI more frequently and in a much more high-paced and mission-critical environment, HMIs need to be easy to use, easy to look at, well organized, as consistent as possible with other machines, and provide just enough information for the operator to do their job without overloading them with too much information. This means that things as simple as color selection are vital to the success of the HMI. To create a successful, functional HMI, we are going to explore the following concepts:

- Colors
- Grouping/positions
- Blinking
- Organizing the HMI into multiple screens

To round out the chapter, we are going to create a simulated carwash HMI screen that would be found at a kiosk.

Technical requirements

Like in the previous chapters, the only technical requirement for following along is a working copy of CODESYS that has the visualization package installed. The code for this chapter can be found at the following page: `https://github.com/PacktPublishing/Mastering-PLC-Programming-Second-Edition/tree/main/Chapter%2014`.

The importance of colors

Believe it or not, colors can utterly sink an HMI. Choose the wrong colors and your HMI will literally hurt your operator's eyes. A general but not normally followed rule is that you want to use dark, pastel colors for your HMI. This will reduce the contrast of the HMI screen and make it easier to operate. As a rule of thumb, you want to avoid bright colors. Normally, HMI developers will opt for colors such as black or gray for backgrounds and different shades of gray for control colors. To start the color discussion, let's look at backgrounds.

Backgrounds

In terms of backgrounds, most industrial guidelines recommend shades of gray. Outside of the guidelines, I like to stick with shades of gray as a personal aesthetic choice unless specified otherwise. However, some organizations I have worked for have primarily used black or shades of dark blue as backgrounds to great success.

Black backgrounds are excellent; however, they do require a bit more work when there is heavy use of labels, and if you're not careful, it can cause eye strain in the wrong environment settings. To put that in perspective, you'll probably have to adjust label colors for any background, but in my opinion, black requires a more drastic change. Consider *Figure 14.1*; upon studying the figure, the first thing you may notice is that there is a heavy contrast between the components.

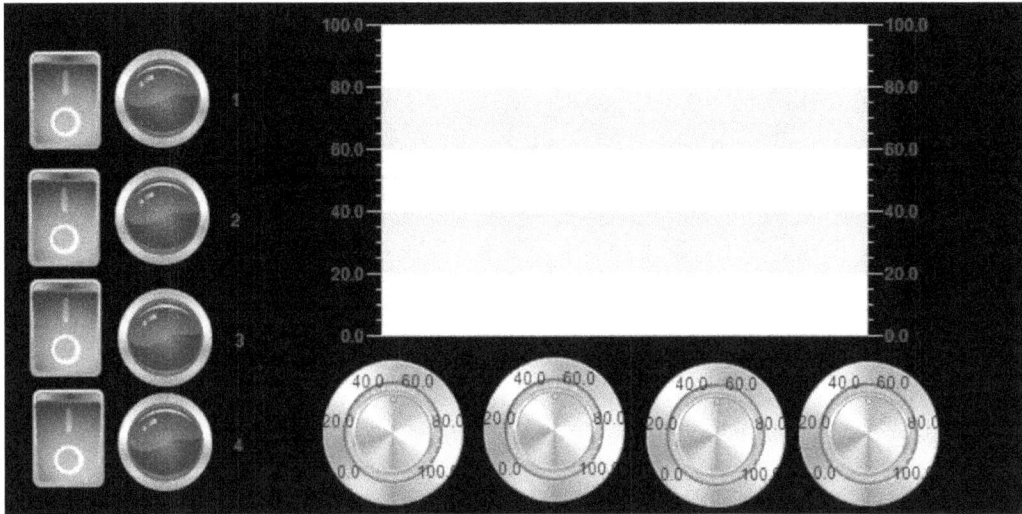

Figure 14.1: Black background HMI

Comparing *Figure 14.1* to *Figure 14.2*, the dark gray creates less contrast, and the labels are easier to see; however, the lighter background will be a little harsher on the eyes compared to the black background.

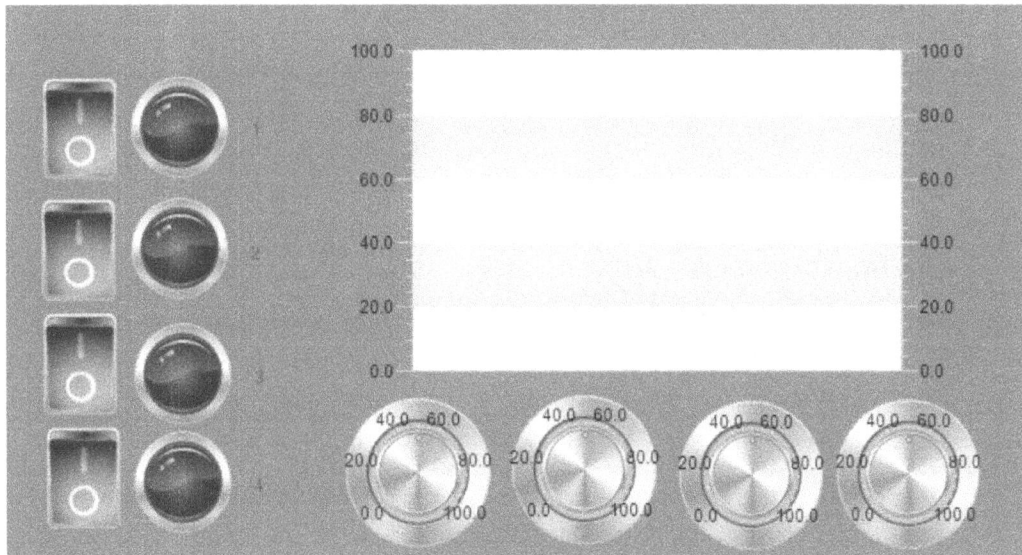

Figure 14.2: Dark gray background HMI

> **Note**
>
> Neutral colors, such as shades of gray, blue, and black, are typically preferred by many organizations. However, it is not uncommon for people to ignore that rule and use whatever colors look good. So, keep in mind that as long as your operators can easily see the device and are not going home with migraines, you can use what you want.

In short, the black background is a bit easier to look at, especially in low-light environments, but the gray background is easier to work with in terms of contrast and is closer to compliance with industry guidelines. In all, you're going to have to do some color matching with both, but the gray background is generally favored.

The meanings of red, yellow, and green are pretty much universal across the world. In the next section, we're going to explore how to leverage these colors in our HMIs.

Red, yellow, and green

The colors red, yellow, and green are everywhere, from streetlights to industrial machinery. They are so common that their meaning is almost ingrained in us. However, the following list is going to explore the three colors and how to use them to great effect in an HMI:

- **Red:** The color red is usually an indicator that something is wrong or stopped and is typically associated with an alarm. You usually reserve red for alarms, controls such as buttons that can go into an erroneous state, stop/off LEDs, critical warning LEDs, or controls that are indicating that something is reaching its operational limits. By instinct, anytime the operator sees red, they will assume that there is something wrong so you must use the color cautiously.

- **Yellow:** Yellow is an odd color. Yellow is in-between green and red in terms of meaning. Yellow will usually signal that your machine is still in an operational state, and all systems are still functioning, but you need to be cautious because something could go wrong at any time. In other words, yellow is a warning color.

  Yellow is used a bit more liberally in HMIs. You can generally get away with using yellow without confusing anyone. The only thing that you need to be mindful of is color consistency. If you're using yellow to mean warning, you don't want to color a control a similar shade of yellow, as that will cause confusion. Though you can be a little more liberal with yellow than you can with red, it is advisable to use this color sparingly as well.

- **Green**: Green means *go*. The color green as an indicator means everything is running and working as intended and operating within its intended operational limits. Generally, green can be used liberally as well; however, consistency also matters with this color. If you're using green in your HMI to signify a working state, you still want to be careful when using it for non-indicator components. Though it is common to see green controls and LEDs, you generally want to reserve this color to indicate the following: on, working, ready, or normal operation.

Control colors

Choosing the correct coloring for controls is a bit of an art. Many of the best-practice documents will tell you to pick shades of gray; however, this rule is rarely followed. Most of the time, people will select colors for controls based on aesthetics. I have personally seen controls colored in every color of the rainbow. Usually, this isn't a problem. The only time that color matters is when you are using red, green, or yellow due to it possibly being confusing to the operator. However, it is usually a good idea to make things such as buttons and controls a shade of gray, with the colors green, red, or yellow acting as a status indicator. Regardless, no matter what color you choose, you will want to gray out controls that are not active.

Colors are never enough on their own. Without proper labeling, no matter what color you choose, confusion can still set in, and the only way to remove the confusion is with proper labeling.

Labeling colors

Regardless of the color that you choose, you always want to clearly state the status of the machine on the HMI. If you look at *Figure 14.3*, you will notice that there are no labels under the LEDs. In this case, an operator who is not familiar with the machine may not know what each LED is supposed to signify. They could assume any number of things, such as the machine is in shutdown mode or there is a broken part.

Figure 14.3: Unlabeled LEDs

However, in *Figure 14.4*, the LEDs are labeled in such a way that the operator will know exactly what they mean.

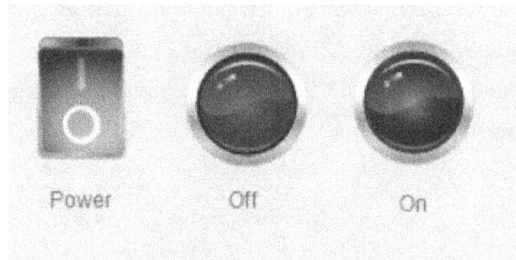

Figure 14.4: Labeled LEDs

There is a lot to the art of color selection. This short tutorial is just to give you an idea of selecting the proper color and why it is important to label the controls, no matter what color they are.

Understanding grouping/position

Another key aspect of HMI design is grouping. Controls and readouts need to be logically grouped so the operator can easily control the machine and take the necessary readings. When it comes to grouping, I have seen two schools of thought. The first one is to stack the controls vertically, as in *Figure 14.5*:

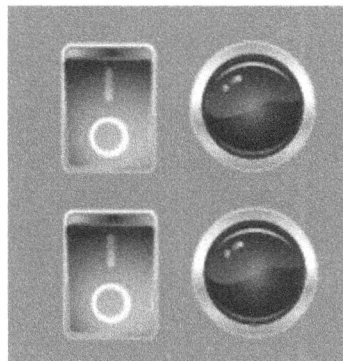

Figure 14.5: Vertical stacking

With the controls laid out as they are in *Figure 14.5*, the operator scans the controls in a top-to-bottom motion. This configuration is known as **side navigation**. Normally, the side navigation is on the left of the screen. Left navigation is considered more efficient and faster for operators. The key to this layout is that each component gets equal weight. This means that, in theory, visually, the bottom switch is as important as the top. There is an important caveat to this. Though each switch should have equal weight in this layout, people will typically default to assuming the top switch is the most important. Therefore, when you're designing an HMI, it's a good idea to put what could be considered more important controls at the top.

Left-side layouts are common for things such as selecting submenus and homing different parts. This layout is also handy for configurations such as the one seen in *Figure 14.5*. Since all the controls technically have the same visual weight, the operator is less likely to overlook one.

Another type of layout is where you place the controls on either the top or bottom of the screen in a horizontal pattern. In terms of HMI development, it is common to have fields for data input toward the top of the HMI or data entry page. Typically, data entry fields will be laid out in a horizontal pattern at the top of the screen; however, it's not uncommon to stack these on the side when there are many of them. In terms of controls such as buttons, I prefer to put these along with their indicators toward the bottom of the layout. This will typically leave the middle of the screen reserved for data output displays. Consider *Figure 14.6*:

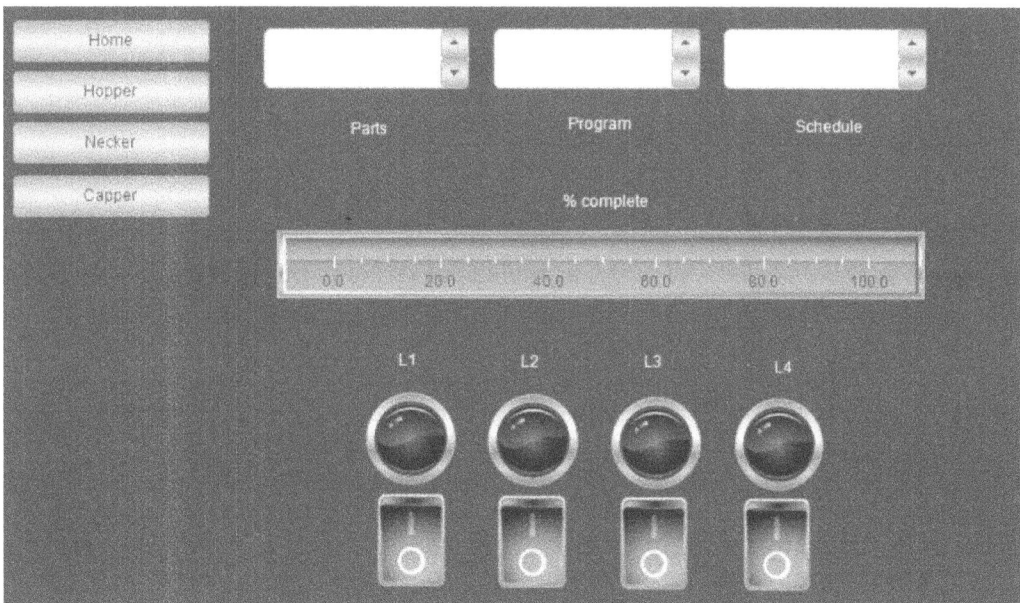

Figure 14.6: Example HMI

In the example HMI, we have four buttons to the left of the screen. In a live HMI, these would allow us to navigate to those submenus. These are to the left so the operator can efficiently scan them. This layout will allow the operator to navigate to the menu they need without having to look around the screen.

Moving to the right, we can see we have three input fields labeled **Parts, Program**, and **Schedule**. These fields are input fields. They are positioned so the operator can select their menu, then easily scan to the right and start inputting the data for the run.

In the middle of the screen, we have a **% complete** bar. This bar provides pivotal information to the operator, mainly how far along the run is. This readout is placed squarely in the center of the screen. The reason for this is that it draws the operator's attention. In the case of the operator needing a readout, they simply have to look at the middle of the screen to get their data. This reduces the extra scan time and, ultimately, makes the HMI easier to use.

At the bottom of the screen, we have four switches and LEDs. These controls will turn on **L1, L2, L3**, and **L4**. They have an LED placed on top so the operator can easily scan to see whether the light, and by extension process, is on.

Figure 14.6 is by no means a perfect HMI. Nonetheless, it does demonstrate some basic layout principles. When developing an HMI, it is very important to make it as easy on the eyes as possible. In other words, you don't want the operator to have to search for their controls or readouts. With that in mind, let's switch gears and talk about blinking.

Best practices for blinking

Nothing says hi-tech and advanced like blinking lights. Everyone loves blinking lights. However, much like many other features that we have seen, blinking can be as much a curse as it can be a blessing. When used properly, blinking can be used to indicate an emergency (such as an issue that could cause harm to personnel or property), or it could mean that a job is loaded and ready to go. In either case, blinking is distracting.

If you blink a component such as an LED, button, or popup, you need to be aware that this action will take the operator's attention away from the controls and put their focus on the blinking component. For some things, such as issues or emergencies, this is what you want. However, blinking components for the sake of it is bad. Generally, I will only blink a component under the following conditions:

- Machine malfunctions
- Safety-related issues (open door, safety sensor tripped, etc.)
- E-stop has been engaged

For the most part, you can blink any component you want as long as it has an **Invisible** field, as shown in *Figure 14.7*:

Figure 14.7: Invisible field

However, just because you can blink a component, it does not mean you should. Generally, you should only blink components such as LEDs. On the contrary, you never want to blink one of the following:

- A switch
- A button
- An input field
- A popup
- Anything with text

Blinking something in the preceding list can create not only an annoying situation for the operator but also a potentially dangerous one. For example, if you were to blink a popup with an error message, you could potentially make it so the operator cannot tell whether there is a safety issue or a component failure. You could also make it difficult for the operator to acknowledge the popup. The same can be said for something akin to a blinking switch. If the switch has to be flipped, the blinking could interfere with the operator's ability to do so.

In cases of popups, input fields, and so on, it is okay to have a blinking element to them. This element might change the color of the control without disabling it, adding a blinking border, or something along those lines. For example, if the operator needs to press a button; it is okay to maybe have it rhythmically change colors. In the case of blinking colors, you can still get the operator's attention without the need to make the control hard to use. Regardless of what you blink, you must remember to do it tastefully and ensure that it is assisting, as opposed to distracting, the operator.

Blinking a component

To blink something, you can use a series of timers, or you can use the **Util library**. Util is a library packed with a lot of different function blocks. One such function block is `Blink`. As the name suggests, the `Blink` function block is an abstraction layer that can be used to easily blink a component.

To demonstrate blinking a component, the first thing we will need to do is import the Util library. To do this, double-click **Library Manager** in the tree and click the **Add Library** button. Once you do this, you will be met with the following screen:

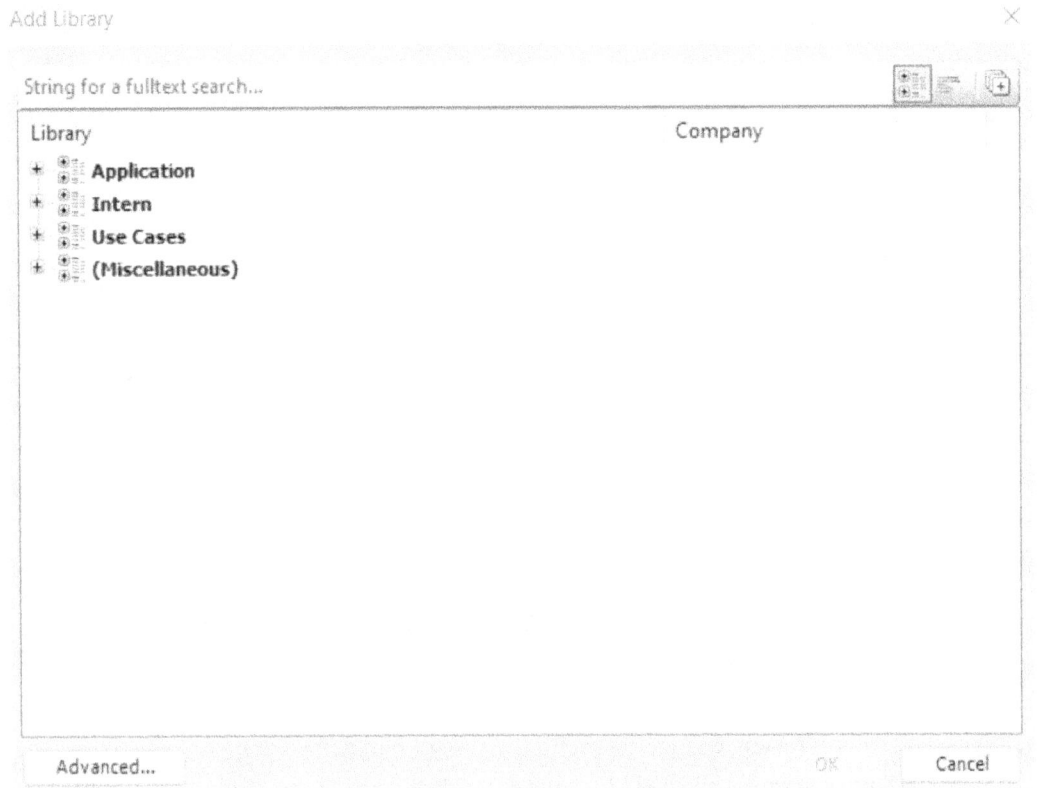

Figure 14.8: Library search screen

From here, you will want to click on the *expand* button next to **Application**, and then you will need to expand **Common**, as shown in *Figure 14.9*:

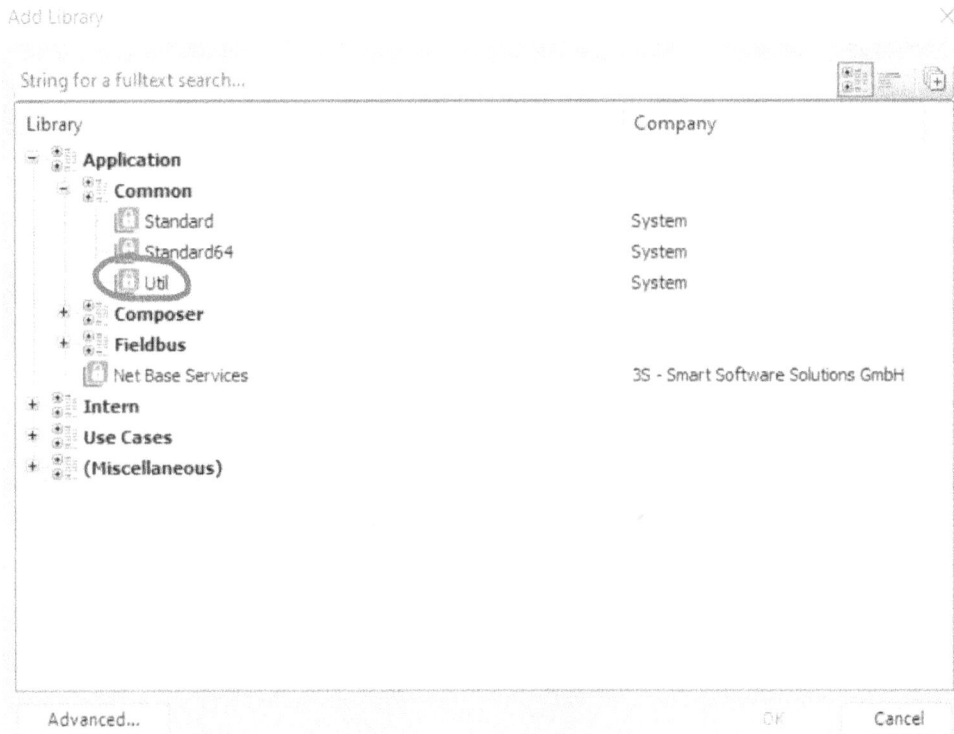

Figure 14.9: Util library

Double-click on **Util**, and the library should be imported. From there, add an HMI screen to the project and drop in an LED and switch, as in *Figure 14.10*:

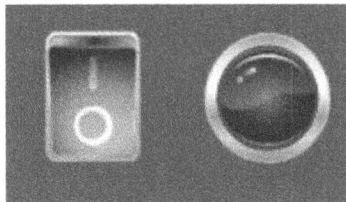

Figure 14.10: Blinking LED HMI setup

Next, navigate to the PLC_PRG POU file and add the following variables:

```
PROGRAM PLC_PRG
VAR
    blink  : Blink;
    led    : BOOL;
    enable : BOOL;
END_VAR
```

These three variables are all that are needed to blink the LED. The blink variable references the Blink function block, the led variable will be assigned to the LED in the HMI, while enable will be tied to the switch and will dictate whether the LED is blinking or not.

The logic to flash the LED is as follows:

```
blink(ENABLE:=enable,TIMELOW:=T#500MS, TIMEHIGH:=T#500MS, OUT => led);
```

This line of code has multiple inputs and a single output. The last argument is the output (OUT) that will dictate whether or not the LED is on. TIMEHIGH will determine how long the LED is on, while TIMELOW will determine how long the LED is off. Finally, enable will determine whether the blink functionality is active or not.

To see the blink function in action, start the code and flip the switch. Observe that the LED is blinking at a steady rhythm. The LED will be on and off for 500ms respectively, before repeating.

As stated before, you should never blink a component such as a button; however, it is common to blink the color of the control. You'll do something like this whenever the control goes into an error state. To demonstrate this, we are going to add a simple button to the HMI so that it resembles *Figure 14.11*:

Figure 14.11: Button

To alternate the button color, we need to set three settings, as in the following screenshot:

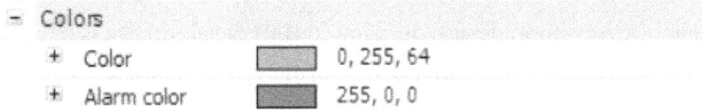

Figure 14.12: Button color and alarm color

In the **Colors** dropdown, you will see the **Color** and **Alarm color** fields. The **Color** field will set the default color, similar to *Figure 14.11*. The **Alarm color** field will set the secondary color, or in other words, the color that the button will be toggled to.

The third field that we need to set is the **Toggle color** field, as in *Figure 14.13*:

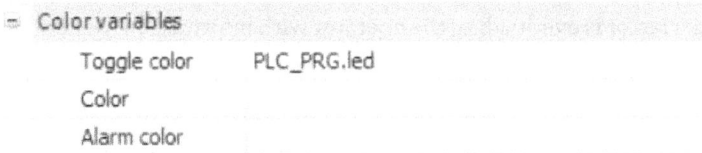

Figure 14.13: Toggle color setting

For this example, we are going to use the same code that we used to blink the LED. To alternate the color, all that is needed is to set the led variable in the **Toggle color** field.

When you run the example, observe that the button will change between red and green. You can also use this feature to simply change the color of the button. Either way, you will set the **Color**, **Alarm color**, and **Toggle color** fields in the same manner. This means that by simply setting the **Toggle color** variable in the PLC logic without the blink code, you can dynamically change the color without blinking.

Animation

A cousin to blinking is animation. It is common to have animation on HMI screens. However, much like blinking, this must be used wisely. Animation is used quite often to simulate a process in as close to real time as possible. This can be handy as operators can easily track the process. Much like blinking, animation can be very distracting.

In all, blinking and animation can be used to great effect; however, both animation and blinking must be used wisely.

Now that we have explored blinking and have touched on animation, we need to switch over to one of the most important concepts in HMI development: screen organization.

Organizing the screen into multiple layouts

One very common, but very poor, design decision in HMI development is to group multiple different screen responsibilities or way too much information on a single layout. There are many reasons why this is bad. Some reasons are as follows:

- Screen disorganization
- Cluttered appearance
- Poor usability
- Overloading the operator with irrelevant information

These are just a few reasons why screen organization is very important. Arguably, the most important organizational factor is overloading the operator with information. Generally, you only want to display the information that is relevant to the operator. If you include too much information, the operator can easily become confused, or they can tune the information out. Ultimately, they can end up ignoring important developments. One common way to combat this is to split an HMI application into multiple screens.

Generally, screen organization can be determined with the one-sentence rule. You usually want to be able to describe the layout's responsibility in one sentence without the word *and*. Much like with functions or methods, if you have to use the word *and*, you will want to split everything after the *and* into a layout of its own. HMIs that follow the one-sentence rule will generally produce cleaner layouts that are easier for the operator to use.

In CODESYS, HMIs are broken out into what are called visualizations. Essentially, visualizations are individual screens packaged into a single HMI. It is common to have a home screen that is the main entry point for the HMI, and the user can navigate to other screens from there.

Creating a visualizations screens

Adding a new screen is quite simple in CODESYS:

1. You simply right-click on the **Application** manager, then select **Add Object**, and, finally, select **Visualization...**, as in *Figure 14.14*:

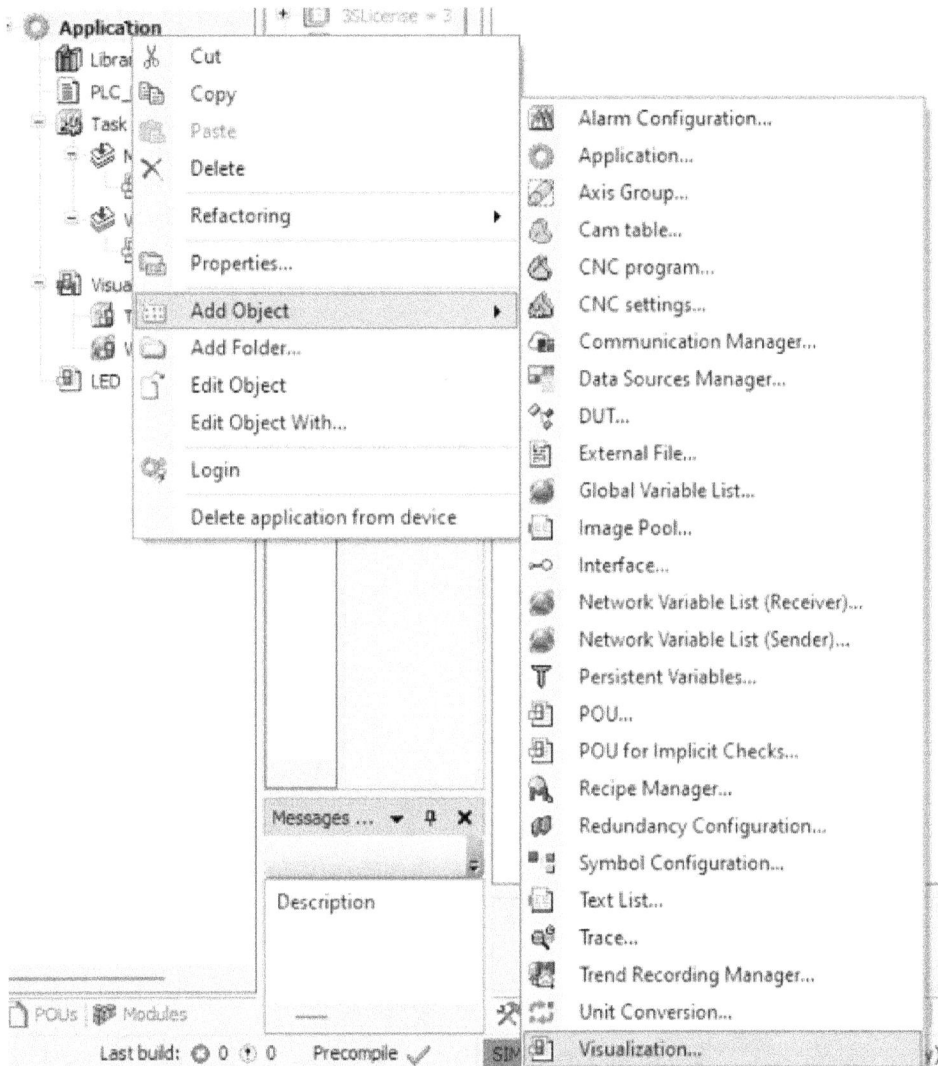

Figure 14.14: Add visualization

Once you complete this step, you should be met with a screen similar to what is shown in the following screenshot:

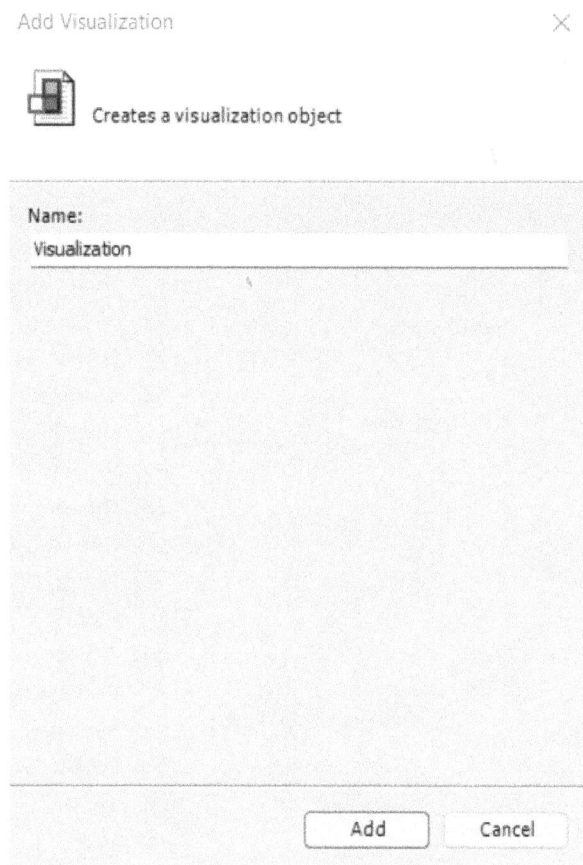

Figure 14.15: Add Visualization screen

2. In the case of this example, we are going to name the screen pumps. Hence, change the name to pumps and click **Add**.

Once you have followed these steps, the new HMI screen will be added to the project. An example of this can be seen in *Figure 14.16*. By default, the first screen that is added to the project, which is the one that is generated with **Visualization Manager**, will be the first screen that appears when the HMI is run. It is common to make this default screen the landing screen—or, as it is most often called in automation, the home screen.

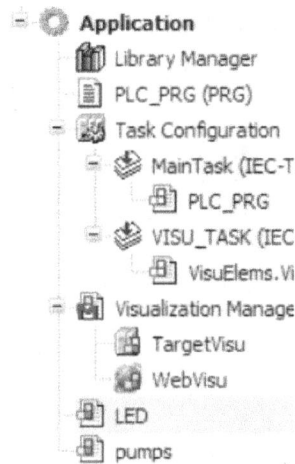

Figure 14.16: The pumps screen added to the HMI

Figure 14.16 shows that a new screen has been added to the project tree. As it stands right now, the screen is blank, and even if we did add something to it, it would never load. As stated before, by default, the screen that is generated with **Visualization Manager** is the default screen that will be loaded when the program runs. Usually, this is fine, as the default screen will simply serve as the program's home screen; however, there are times when we need to change the default screen. The following section will be dedicated to setting the default screen when the program loads.

Changing the default screen

Luckily, changing the default HMI view in CODESYS is simple. To demonstrate this concept, follow these steps:

1. Ensure that there is a visualization manager attached to your project. If you have an HMI already set up, you should have a visualization manager. Regardless, once configured, add the following controls to it:

Figure 14.17: Visualization screen

2.   Now, add another screen called **V2** and add the following controls to the layout:

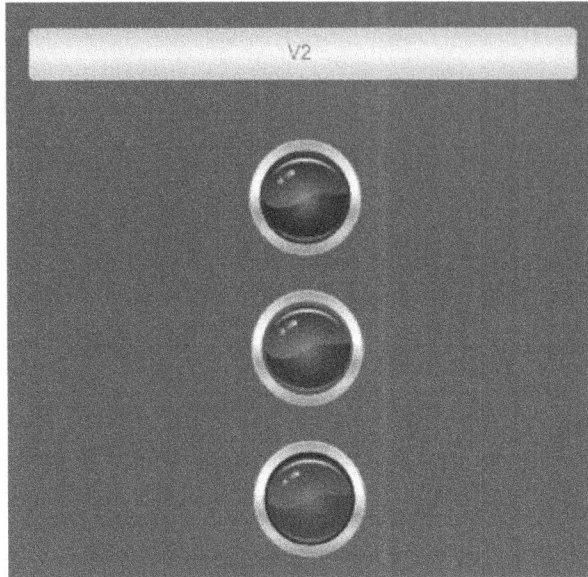

Figure 14.18: V2 screen

When you run the application, you'll see the **Visualization** screen by default.

3.   If you want the **V2** screen to load, all you need to do is click on **Visualization Manager** and navigate to the **TargetVisu** tab. The option to set the right screen as the default resides in that tab.

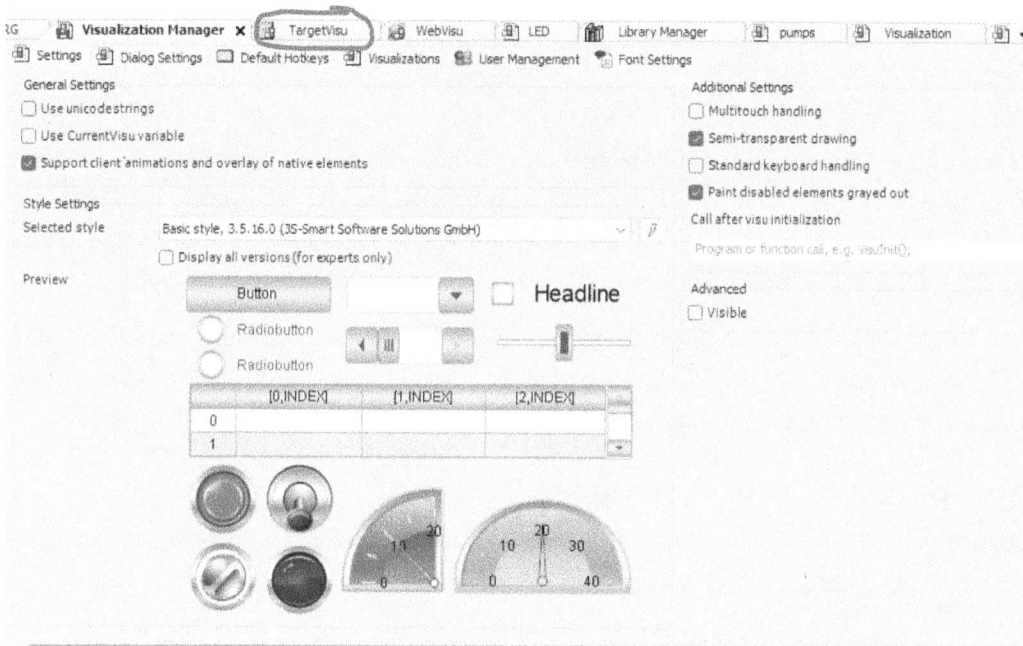

Figure 14.19: Visualization Manager

4. Once you click on the tab, you should see a screen similar to what is in *Figure 14.20*. If you study that figure, you will see a button with three dots in the **Start visualization** row. To select the right screen, click on that button and you will see a selection menu. Select the screen that you need to run at startup.

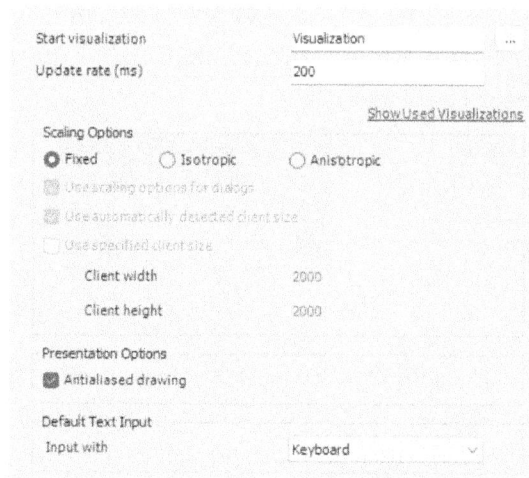

Figure 14.20: Visualization selection

5. For this example, change the screen to **V2** and run the program. Run the program as normal, and you should see that you are met with what is shown in *Figure 14.18* as opposed to *Figure 14.17*.

It is hard to get the full effect of this feature using a simulated PLC environment as we're doing here. You can simply click on the HMI screen you want to view and, with modern versions of the development environment, that screen will load. You will typically need an actual HMI device to see the full benefits of this feature.

In normal HMI development, you will need to navigate between screens quite often. As it stands right now, we cannot do that, and we are only able to choose the startup screen. After you run this example, we are going to explore the process of screen navigation.

Navigating between screens

Usually, you will navigate between screens using buttons; therefore, this tutorial will consist of a series of buttons that will enable our navigation features:

1. Add the following layout to your default **Visualization** screen:

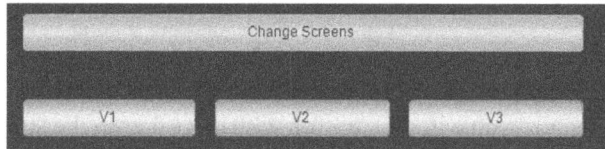

Figure 14.21: Default layout

2. Next, add three visualizations called **V1**, **V2**, and **V3** to the project. For demonstration purposes, simply add a button called **Home** on each of the screens, as well as a label to signify the layout, as in *Figure 14.22*:

Figure 14.22: V2 HMI

3. We now need to configure the button to navigate to the right screen. There are many ways to do this, but the easiest, in my opinion, is to set up the navigation as a button click. The first set of buttons we're going to configure will be the button on the home screen. To do

this, double-click the button, enable **Advanced** at the top of the properties menu, and scroll down to **Input configuration**, and then click the **OnMouseClick** field, as in *Figure 14.23*:

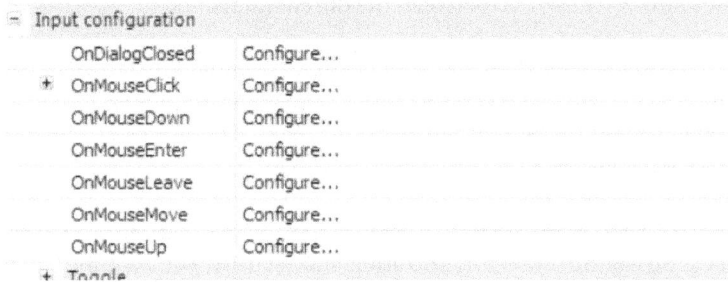

Figure 14.23: Button menu

4. Once you click on that, you should be met with what is shown in *Figure 14.24*. Click on **Change Shown Visualization** and double-click the button with three dots next to the **Assign** field. You will be met with another window; select the corresponding visualization for the button you are working on.

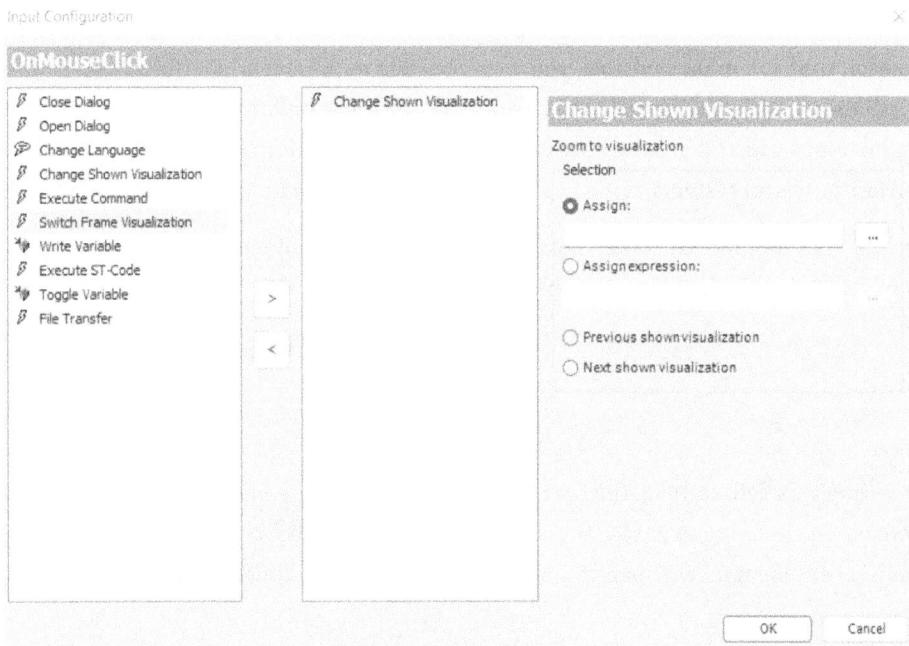

Figure 14.24: OnMouseClick

5. Once these steps are complete, you can set the button in the **V1**, **V2**, and **V3** screens to the default screen. After you run the program, press the button on the default screen and notice how it will switch between the different layouts.

Sometimes adding extra layouts can be overkill. Depending on the functionality and the complexity of the HMI, a common technique is to simply hide components.

Hiding components

Though it is very easy to navigate between screens, many HMI developers will typically hide components as an easy means of organization and security. For example, suppose there is a calibration button on the HMI that only the technicians and other authorized personnel should have access to. A control like this would normally be hidden behind a passcode control mechanism. That is, the user would have to enter a password to access the button. From there, the button would appear on the screen; when the button is pressed, other fields, such as calibration menus, would appear on the layout. In CODESYS, and many other platforms, this is typically done by making the controls invisible.

By making the controls invisible, you're also rendering them inactive. This means that if you have a button that you make invisible, you won't be able to press it while the button is hidden. This technique essentially allows you to create different screens without needing extra layouts. This methodology can and usually does make the system difficult to maintain; however, for cheaper and less powerful systems, it may be the only option to give the illusion of multiple layouts.

For the **Visu** tool in CODESYS, most control elements can be made visible or invisible with a simple **BOOL** variable in the **Invisible** field.

Figure 14.25: State variable

By setting this field to TRUE, the control element will become invisible and inactive. Conversely, by toggling the value to FALSE, the element will become visible and enabled. For simple HMIs, this is an excellent mechanism to control the layout of your touchscreen.

> Note
>
> Most systems will have some type of method to render controls invisible. However, the way it works and the setup will vary from system to system.

Final project: Creating a user-friendly carwash HMI

One application for HMIs is carwashes. Carwashes are excellent use cases for PLCs in general, and when you think of how a modern carwash works, they're also a great use case for HMIs. In this scenario, we're going to create an HMI for a kiosk that controls a self-service carwash.

HMI goals

Creating an HMI for anything that's meant to be used by the public can be tricky. On one hand, the HMI needs to be able to absorb all the information needed for the application to function; however, at the same time, it needs to be easy enough for anyone to use. From a purely economic point of view, if the HMI is too hard to use, potential customers will go elsewhere, which means a loss of revenue for your business. Therefore, it is important to have a very simple interface that anyone can use.

For this carwash, any user who signs up will get a secret code, in this case, 123. Once the user logs in, an LED will turn on, and the user can select which wash they want. For this application, we need a simple interface that allows the user to press a button to select their wash. Also, we need authorized personnel to be able to log in to the device and perform tasks such as calibrating the carwash. This means that we need hidden controls or extra HMI layouts. Due to the simplicity of this application, it will probably be easier to simply hide controls as opposed to making a whole new screen. Therefore, let's start setting up the code.

PLC code

To begin, we're going to need the following variables,

```
PROGRAM PLC_PRG
VAR
    led        : BOOL;
    num        : INT;
    admin      : BOOL := FALSE;

    basicWash : BOOL;
    deepWash  : BOOL;
    bestWash  : BOOL;
END_VAR
```

In this case, the admin variable will make the admin controls visible, the led variable will turn on an LED, and num will hold the value inputted from the built-in keypad.

The main logic for this code will look like the following:

```
IF num = 123 THEN
    led := TRUE;
    admin := FALSE;
ELSIF num = 111 THEN
    admin := TRUE;
    led := FALSE;
ELSE
    led := FALSE;
    admin := FALSE;
END_IF
```

The screen mode is dictated by the number the user enters. If they enter 123, they will be able to wash their car; however, if they enter 111, they will be able to use more advanced maintenance controls. With the code laid out, we can now move on to the actual HMI layout.

HMI layout

To begin, lay out the HMI to match the following:

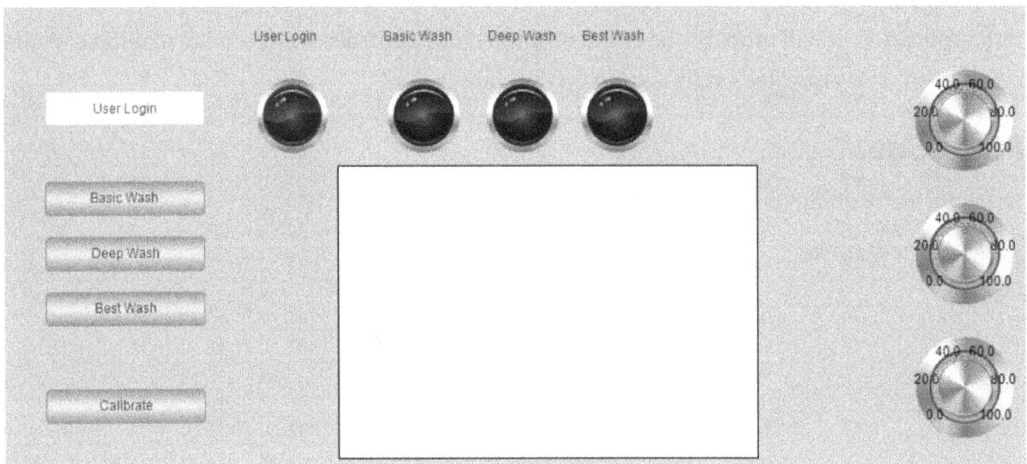

Figure 14.26: HMI layout

Note

For drag-and-drop systems, you can sometimes skip wireframing. These systems make it easy to move controls around on a whim, which can make wireframing redundant.

In this case, we have a text field, the white box in the top-left corner, which will open up a number pad. The configuration for this control will match *Figure 14.27*:

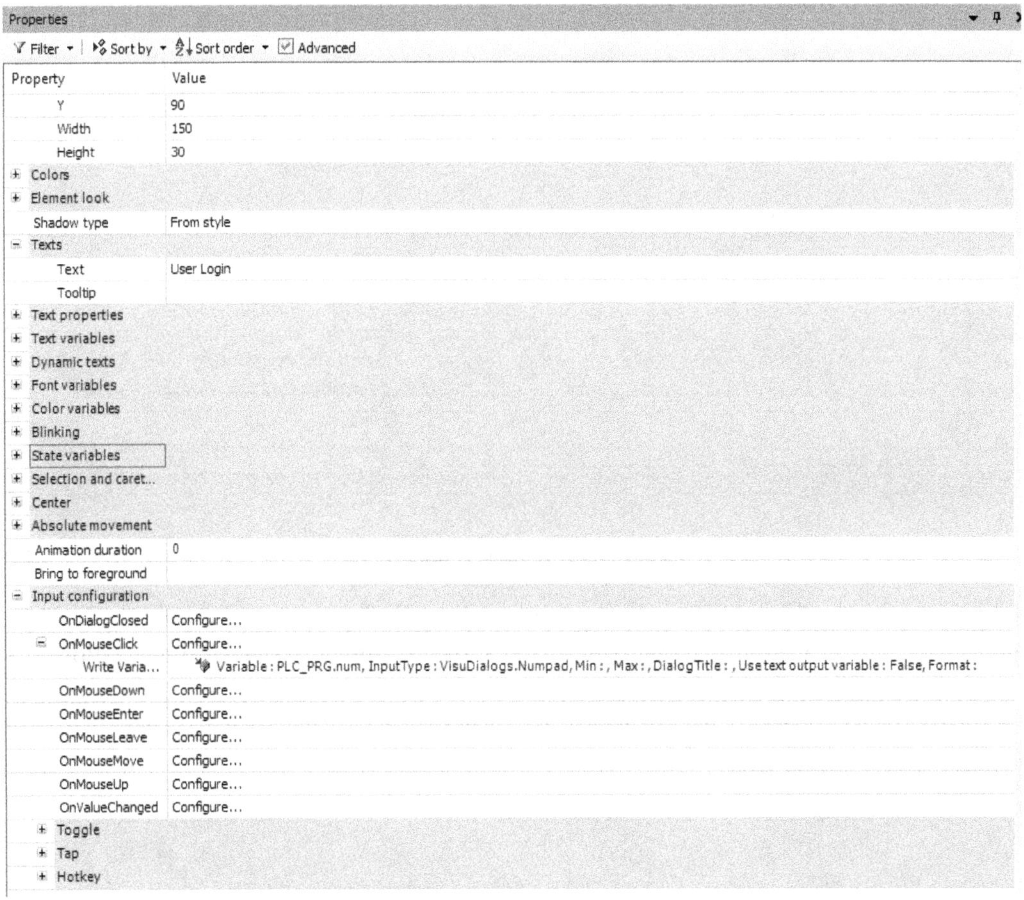

Property	Value
Y	90
Width	150
Height	30
⊕ Colors	
⊕ Element look	
Shadow type	From style
⊟ Texts	
Text	User Login
Tooltip	
⊕ Text properties	
⊕ Text variables	
⊕ Dynamic texts	
⊕ Font variables	
⊕ Color variables	
⊕ Blinking	
⊕ State variables	
⊕ Selection and caret...	
⊕ Center	
⊕ Absolute movement	
Animation duration	0
Bring to foreground	
⊟ Input configuration	
OnDialogClosed	Configure...
⊟ OnMouseClick	Configure...
Write Varia...	⟲ Variable : PLC_PRG.num, InputType : VisuDialogs.Numpad, Min : , Max : , DialogTitle : , Use text output variable : False, Format :
OnMouseDown	Configure...
OnMouseEnter	Configure...
OnMouseLeave	Configure...
OnMouseMove	Configure...
OnMouseUp	Configure...
OnValueChanged	Configure...
⊕ Toggle	
⊕ Tap	
⊕ Hotkey	

Figure 14.27: Text field configuration

Once you have the control on the screen, double-click it and navigate to the **Input Configuration** property. For this one, we want to set **OnMouseClick**, as can be seen in *Figure 14.27*. Once you're there, click on the button with three dots and set up the popup to match what is shown in *Figure 14.28*:

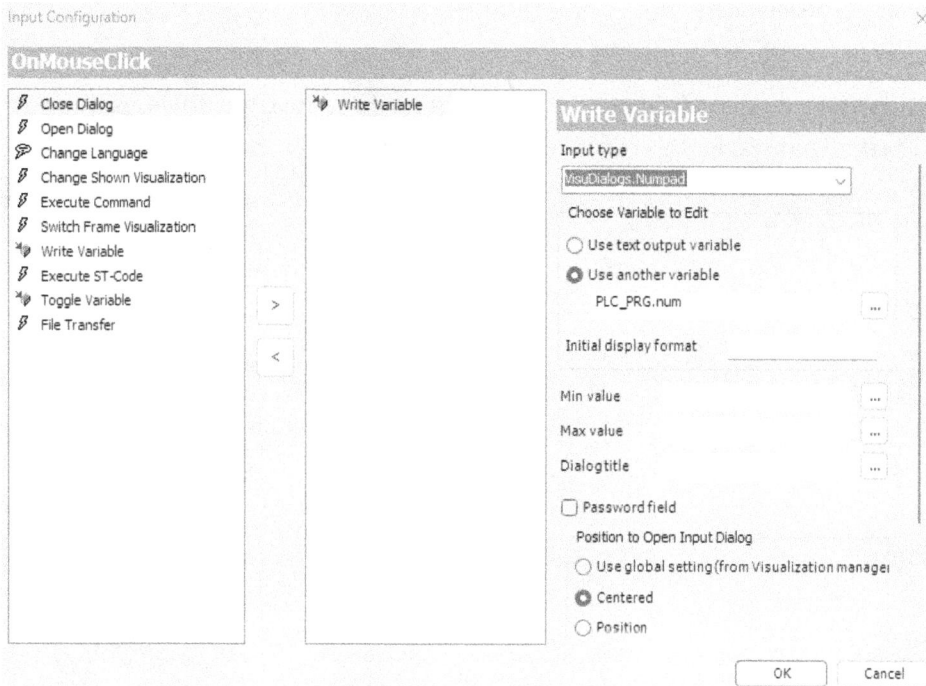

Figure 14.28: Numpad configuration

In terms of the user login LED, set the properties to match *Figure 14.28*. The only setup that is required is to set the **Variable** property to PLC_PRG.led and **Image** at the bottom to **Yellow**.

Property	Value
Element name	GenElemInst_9
Type of element	Lamp
− Position	
X	250
Y	78
Width	70
Height	70
Variable	PLC_PRG.led
− Image settings	
Transparent	
Transparent color	■ Black
Isotropic type	Isotropic
Horizontal alignment	Left
Vertical alignment	Top
− Texts	
Tooltip	
− State variables	
Invisible	
− Center	
X	535
Y	45
− Absolute movement	
− Movement	
X	
Y	
Rotation	
Scaling	
Interior rotation	
Animation duration	0
Bring to foreground	
− Background	
Image	Yellow

Figure 14.29: User LED setup

The wash LEDs should be configured like the following:

Property	Value
Element name	GenElemInst_17
Type of element	Lamp
− Position	
X	373
Y	78
Width	70
Height	70
Variable	PLC_PRG.basicWash

Figure 14.30: LED configuration

For each of these, set the **Visible** field to match their respective variable and set the color to green using the **Image** field as in *Figure 14.29*.

The placeholder box, calibrate button, and pots should have their **Invisible** field set to match what is shown in *Figure 14.31*.

State variables

Invisible   NOT PLC_PRG.admin

Figure 14.31: Maintenance controls

The wash buttons are going to be a bit trickier to set up. For these buttons, we're going to set their **Invisible** field to NOT PLC_PRG.led and we're also going to configure them to run a small ST script. To do this, we need to navigate to **OnMouseClick** and add **Execute ST-Code** as an event. This will give us a small box to write the script in, like the following:

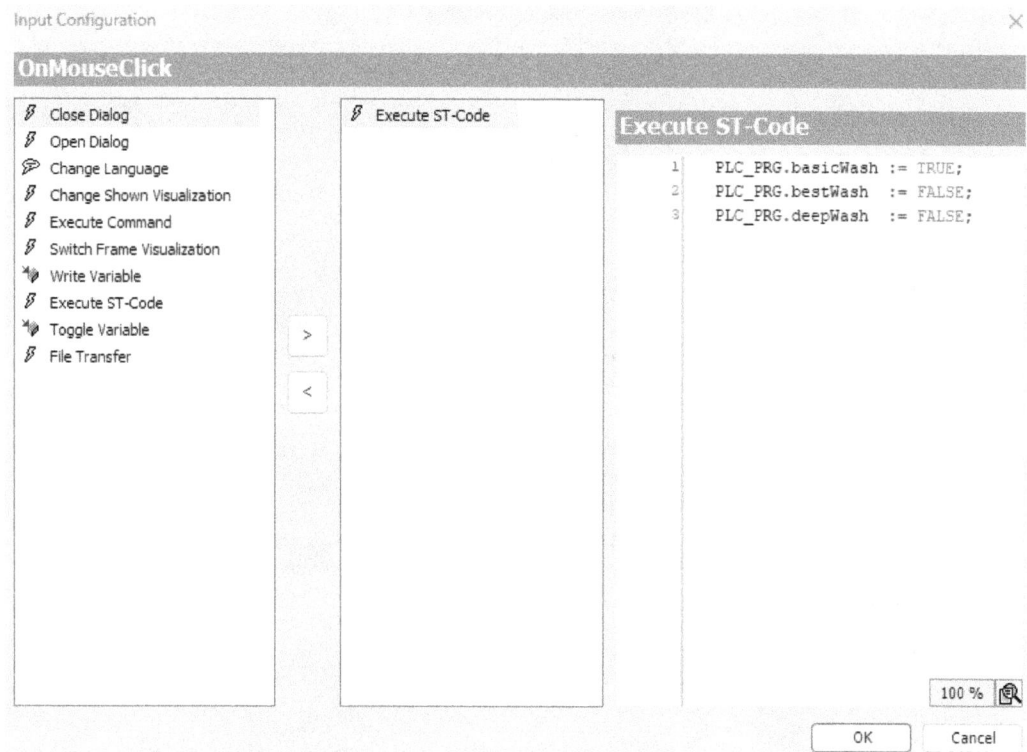

Figure 14.32: ST script

This feature allows us to write custom code and manipulate any globally available variable. This is a very powerful feature and is very handy for buttons like these.

The code for the basic wash button can be viewed in *Figure 14.32*. However, the code for the deep wash button should match the following snippet:

```
PLC_PRG.basicWash := FALSE;
PLC_PRG.bestWash  := FALSE;
PLC_PRG.deepWash  := TRUE;
```

The code for the best wash button should be as follows:

```
PLC_PRG.basicWash := FALSE;
PLC_PRG.bestWash  := TRUE;
PLC_PRG.deepWash  := FALSE;
```

These scripts will essentially turn on the LED for the corresponding wash while turning off any other LED that is on.

At this point, the code and HMI are set up, and we can now test!

To begin running the project, you should start at the screen shown in *Figure 14.33*:

Figure 14.33: HMI startup

As can be seen, in our startup mode, all we have are the wash and user login LEDs, as well as the **User Login** field. From here, click the **User Login** field, and the number pad should appear.

> **Note**
>
> The number pad may appear off-screen. You may have to search for it!

On the number pad, enter 123 and press the **OK** button. When the operation is complete, you should be met with what is shown in *Figure 14.34*:

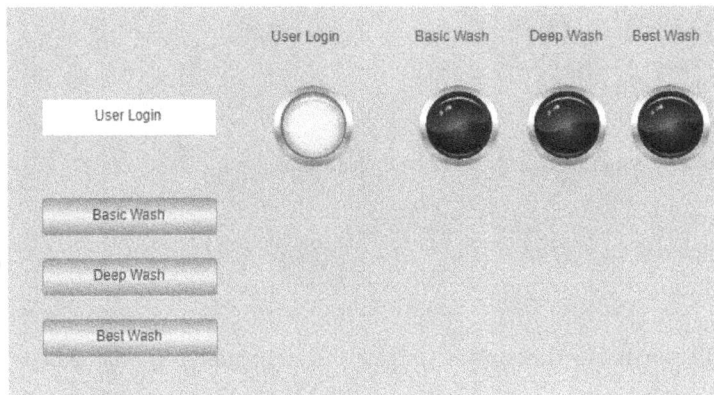

Figure 14.34: User HMI view

You should be able to click on a wash button and watch the corresponding LED light up while the others turn off.

Next, click **User Login** again and enter 111 on the keypad. Follow this up by pressing the **OK** button. When you do this, you should be met by what is shown in *Figure 14.35*:

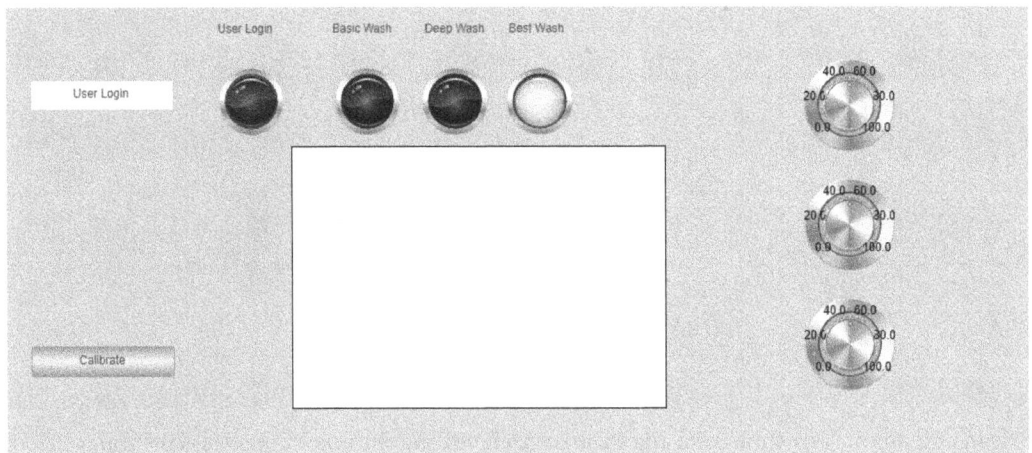

Figure 14.35: Admin panel

Notice that the wash buttons vanish as expected. Also, notice that an LED is on. The LED may not be on when you enter this mode. The reason why the LED is on in the screenshot is because **Best Wash** was selected when this mode was entered. If you have a wash selected while you enter this mode, the wash's LED will stay on.

Now, we can give ourselves a pat on the back because we created a working carwash HMI. The next thing to do is make it better!

Improvement challenges

Try to implement the following improvements to the PLC/HMI program:

- Clear the wash modes when the user is changed.
- Use a loop to lock the HMI out when the user inputs an invalid code more than three times.
- Break out the admin mode into its own screen.
- Add an emergency stop button.

Summary

Creating HMI panels is as much an art as it is a science. You need to ensure the colors are correct, the controls are logically laid out, and so on. In this chapter, we explored many of the basic principles of creating multiple windows, laying out switches, configuring LEDs, and more. We also learned how to hide certain controls when it is convenient for us. **Visu** is a great tool to create basic HMIs such as the one we created here. However, it can be used for more than simple controls. **Visu** can also be used to alert the operator to dangerous situations. In the next chapter, we're going to explore how to set up and use alarms!

Questions

1. What do the colors red, green, and yellow typically represent?
2. What color backgrounds should you primarily use?
3. How many responsibilities should each HMI screen have?
4. What should your default screen be?
5. How do you make a control invisible?

Further reading

- *Standard colors on HMI*: https://www.mesta-automation.com/standard-colors-on-hmi/
- *Design Tips to Create a More Effective HMI*: https://blog.isa.org/design-tips-effective-industrial-machine-process-automation-h

Join our community on Discord

Join our community's Discord space for discussions with the authors and other readers:

`https://packt.link/embeddedsystems`

15

Alarms: Avoiding Catastrophic Issues with Alarms

In many SCADA and HMI systems, alarms are dedicated controls that are specifically designed to warn operators about the status of the machine. Normally, alarms will allow you to change colors, display text, log issues, and more. Each HMI or SCADA package that offers alarms will offer different alarm types, styles, functionality, and more; however, the core principles that govern most alarms are universal.

Much like HMIs, developing and properly implementing an alarm is as much an art as it is a science. This chapter is dedicated to implementing alarms logically and effectively. To do so, we are going to explore the following concepts:

- What are alarms?
- Where to use alarms
- Alarm configuration: info, warning, and error setup
- Alarm HMI components
- Alarm PLC code
- How to acknowledge alarms

Alarms for motors are very common as the devices can easily overheat, draw too much current, or have other issues that need to be logged or reported. After exploring alarm concepts, we are going to round out the chapter by creating an alarm for a motor.

Technical requirements

As per every other chapter in the book, a full version of CODESYS will need to be installed. The visualization tool will need to be installed as well. As usual, the examples can be downloaded from the following URL: https://github.com/PacktPublishing/Mastering-PLC-Programming-Second-Edition/tree/main/Chapter%2015.

What are alarms?

Alarms in automation programming are designed to get the operator's attention and alert them to some type of situation. The most common type of alarm is meant to alert operators to a safety issue; however, they can also be used to warn of a machine malfunction. The goal of an alarm is to prevent harm to a device, machine, or, more importantly, a person.

In automation programming, an alarm will usually consist of an HMI component, PLC code, and sometimes physical hardware. Though not every machine will use hardware with their alarms, some common physical components used in conjunction with software are as follows:

- LED strobes
- Lamp strobes
- Loud buzzers
- Speakers that play prerecorded messages

Though technically not an alarm, I like to count things such as automatic locks in this category. This is because they are a safety mechanism that can get an operator's attention under the right circumstances and prevent damage or harm.

In terms of the HMI, many systems support alarm components that are specifically designed to give warning and error messages. However, as an HMI developer, you are not restricted to these components alone. It is not uncommon to use things such as blinking lights, pop-up boxes, and more for alarms. In fact, it is quite common, especially in systems that don't have alarm elements, to use things such as the following:

- Digital LEDs that blink
- Gauges
- Text fields that change colors
- Popups
- Buttons

For weaker development systems, a common technique for an alarm is to make a very large button invisible and set the button's text dynamically. When the alarm is triggered, the code can be rigged to make the button visible with a message indicating the status of the machine. This makes it very easy for the operator to acknowledge the alert by simply tapping the popup.

When should you use an alarm?

Much like with anything else, picking when and where to use an alarm is often up to the discretion of the developer. Not all alarms have to warn of danger; however, if danger or damage could result from the condition, it is absolutely necessary to implement one. Typically, common applications for alarms are as follows:

- E-stops
- Out-of-range values such as voltage or temperature
- Network issues
- Cybersecurity issues (hackers/unauthorized users)
- Misbehaving electrical components
- Parts that are damaged or nearing their end of life
- Warnings for fire, gas pollution, or some other dangerous conditions

Cybersecurity alarms

The machine or environment will dictate what type of alarm you need. Though not as common as alerts for things such as out-of-range components, with the rise of Industry 4.0 and the networked machines that it supports, alarms for cybersecurity are becoming a major factor in the development of a machine/network. For networked systems such as SCADA systems, it is advisable to dedicate some alarms to warn of intruders. For this type of application, the core alarm would typically live in some type of custom system and not necessarily the machine's HMI or PLC code. However, with the proper networking, external software, and PLC coding, one can easily rig the PLC to receive a warning and pass that on to an HMI for the operator to see. This is advisable for networked machines that can inflict major damage or bodily injury.

In terms of cybersecurity, there are many third-party security/monitoring systems that can be installed on advanced PLCs that are powered by Linux, Windows, or are deployed to a centralized computer that monitors a network. While many of these systems can usually give alerts (alarms) to the user without interrupting the HMI operations, others can send information using some type of network protocol, such as TCP/IP, that can be intercepted by the PLC or other computer systems and trigger an actual alarm in the HMI.

What should an alarm say?

For an alarm to be of any use, you will need the alarm to logically reflect the issue/warning at hand. An alarm should, at a minimum, include attributes such as the following:

- The issue that was detected
- Whether the alert is an info, a warning, or an error alarm
- Timestamp (if possible)
- Whether the alarm was acknowledged (if possible)

Most HMI systems will give you options for these four attributes. However, some alarm systems do not give these options, especially if they are built using a traditional programming language or weaker systems. Regardless of the options available, you want to give the operator as much information as you can to pinpoint the cause of the message, especially if the alarm relates to a warning or an error message.

Logging alarms

You will usually want to log your alarms if possible. This is especially true if your system is monitoring for intruders. You will log alarms when they need to be kept long-term or used for analysis. This can be easier said than done. Some systems do allow you to store the alarms in either a database or something such as a CSV file. However, not all systems support this. This type of behavior is more commonly associated with complex SCADA systems and HMI systems that are built using a traditional, general-purpose programming language.

Ultimately, what your alarms need to detect, and what they should say, is going to be up to the end user, your organization, and you. Much like HMIs, creating a decent alarm is a bit of an art. The next step in our journey is to learn how to configure alarms.

Alarm configuration: info, warning, and error setup

To do anything with alarms, the first thing we need to do is set up an alarm configuration. An alarm configuration is the configuration setup, such as the colors, the fonts, and so on, that will govern the info, warning, or error alarm. In CODESYS, this is a relatively easy task. To add an alarm configuration, you will simply need to right-click **Applications** and follow the path in the following screenshot:

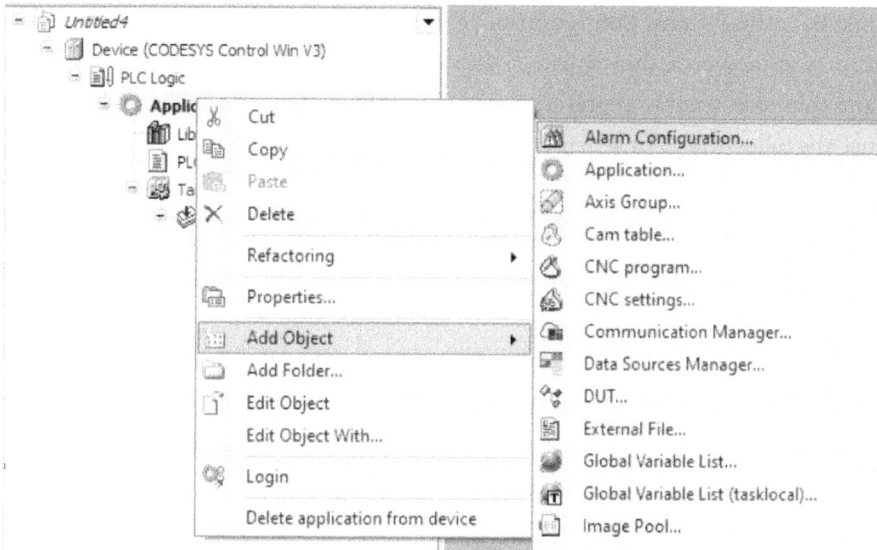

Figure 15.1: Path to add an alarm

This will bring you to a wizard like the one in the following screenshot. In the wizard, give the alarm configuration a name and click **Add**. For our alarm configuration, we will simply keep the default name. Once you click **Add**, you will see the **Error**, **Info**, and **Warning** attributes under **Alarm Configuration** in the project tree.

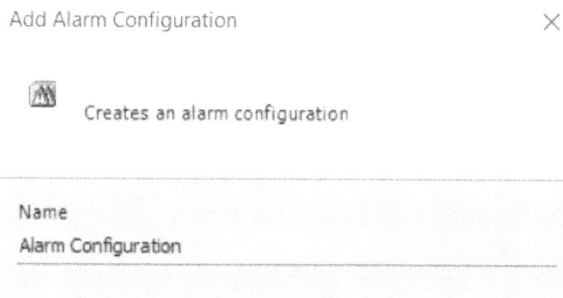

Figure 15.2: Alarm Configuration wizard

The **Info**, **Error**, and **Warning** attributes that are generated can be seen in the following screenshot:

Figure 15.3: Alarm Configuration tree

For alarms that utilize **Alarm Configuration**, each **Error**, **Info**, or **Warning** alarm will be identical in setup. In other words, they will have the same fonts, colors, and so on. You can technically choose any color you want to represent each of the states. However, in keeping with tradition, we will use the following color scheme:

- **Error**: red
- **Warning**: yellow
- **Info**: green

To set up the configuration attributes, you will need to click on the **Error**, **Info**, and **Warning** tabs individually.

The first step is setting the configuration for each of the alarm types. You will first double-click on them. Once you double-click any of the alarm types, you should see a configuration screen similar to the following:

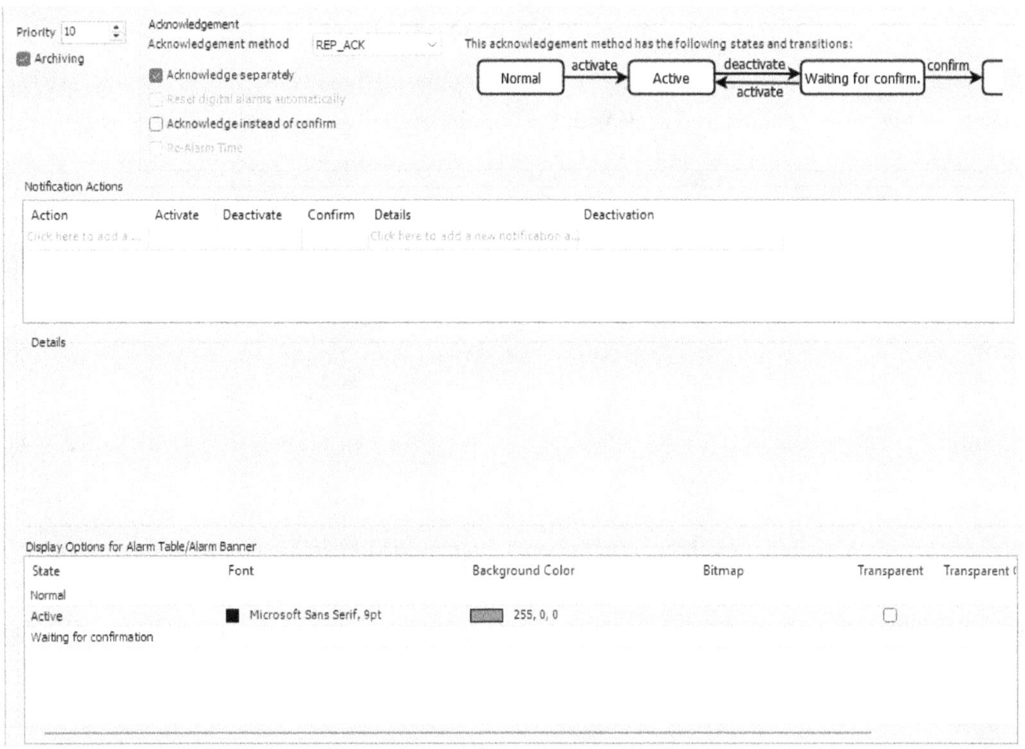

Figure 15.4: Error configuration menu

For this example, click the **Archiving** and **Acknowledge separately** checkboxes. Next, navigate to the bottom and select the font and background color. *Figure 15.4* shows the configuration that will be used for errors. You will do the same for the **Info** and **Warning** alarms, but those backgrounds will be green and yellow, respectively. After this is complete, you will need to set up your alarm groups.

Alarm groups

Alarm groups consume alarm configurations like the one that was just set up. To generate an alarm group, right-click on the **Alarm Configuration** button, select **Add Object**, then **Alarm Group...**, similar to what can be seen in the following screenshot:

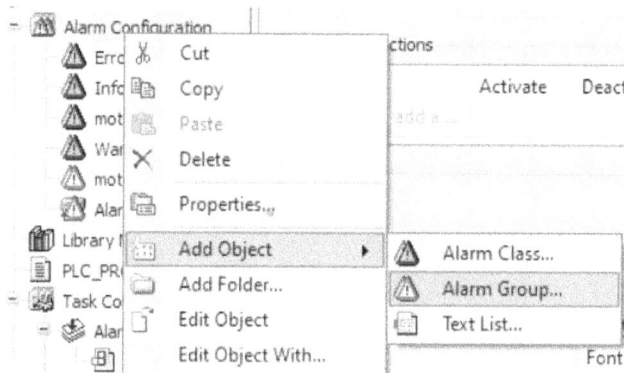

Figure 15.5: Alarm group generation

After you click **Alarm Group...**, you will be met with a wizard similar to the one in the following screenshot:

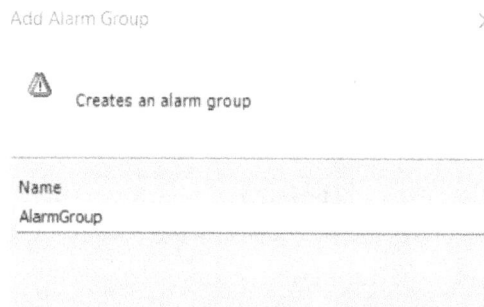

Figure 15.6: Add Alarm Group wizard

For our example, we are going to give the alarm group the name motor and click the **Add** button.
Once you do that, you should see the group appear in the tree, as in the following screenshot:

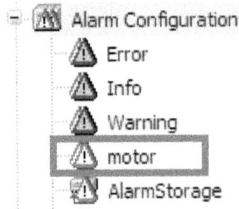

Figure 15.7: Alarm Configuration group

Before you can fully set up the **motor** group, you will need to implement the following variables
in the PLC_PRG POU file:

```
PROGRAM PLC_PRG
VAR
    info  : BOOL := TRUE;
    warn  : BOOL := FALSE;
    error : BOOL := FALSE;
END_VAR
```

In this case, we have three variables that will be responsible for showing the alarm message. Now
that those variables are set, double-click the **motor** alarm group. Once you click the group icon,
configure the screen similar to the following by selecting each field and adding the information
found in *Figure 15.8*:

ID	Observation Type	Details	Deactiva...	Class	Message
0	Digital	(PLC_PRG.error) = (TRUE)		Error	There is a motor error
1	Digital	(PLC_PRG.info) = (TRUE)		Info	All Good!
2	Digital	(PLC_PRG.warn) = (TRUE)		Warning	All Not so Good!

Figure 15.8: The motor group configuration

For our example, **Observation Type** will be set to **Digital**. Next is the **Details** column. This is the
logic that will fire the alarm class, and by extension, the message attached to it. This is where
the variables that we created come into play. As can be seen in *Figure 15.8*, the variables that we
created are used here. The next column we need to set is **Class**. This is essentially the alarm type
that will fire when the logic in the **Details** column is satisfied.

With all that, the alarm's configuration should be set up. However, as it stands, this is just the core logic and configuration. For this to be useful, we need to attach it to an HMI component so we can display our alarm on a screen.

Alarm HMI components

After we set up the alarm's configuration, we can drop in an HMI component. In terms of COD-ESYS, there are two types of HMI controls. One is the **alarm banner**, and the other is the **alarm table**. Consider the following:

- **Banner**: A banner shows one message at a time. It will prioritize alarms and only show the **Error**, **Warning**, or **Info** alarm, in that order, unless configured otherwise. In other words, a banner will display the most important alarm first. No matter the type of alarm, the alarm can be toggled by toggling the variable in the alarm group.

- **Table**: An alarm table will show active alarms for an alarm group. New alarms will show at the top of the table and can be toggled by toggling the variable they are tied to. Where banners are meant to be on every HMI screen, tables are usually set on a diagnostic screen. Compared to banners, a table can give more information as it will show more alarms, but it is usually tied to an exclusive screen.

A banner is arguably the most important part of an alarm, as it will allow you to see the most important information, while a table is similar to a log. Essentially, outside of the attention-getting gimmicks such as blinking lights and loud sounds, this will be the first point of contact to determine what's triggering the abnormal behavior. Therefore, these messages need to be very clear and concise. Depending on what system you're working with, you can prioritize your alarms to show the most vital ones first.

Where a banner will show only one alarm message, the table, on the other hand, can show the other alarms. In other words, the table is a more comprehensive component that shows a more detailed picture of what's going on in the machine. For example, a banner might read that the system is overheating. This is indeed an important clue, but it is very unlikely that the operator will be able to determine the root cause of the problem with that message alone. However, by reading the table, the operator can skim the other alarms and be warned of other situations, and get a better idea of the root cause of the problem.

> **Note**
>
> When possible, set alarms for things that could result in injury or death as the highest priority!

Depending on the design, HMI table inclusion will vary. However, at the very minimum, you're going to add a banner. The banner should be at the very top of each HMI screen, and it should be in a place where the operator can easily see it. Usually, you'll want to place the banner in the center at the top, as follows:

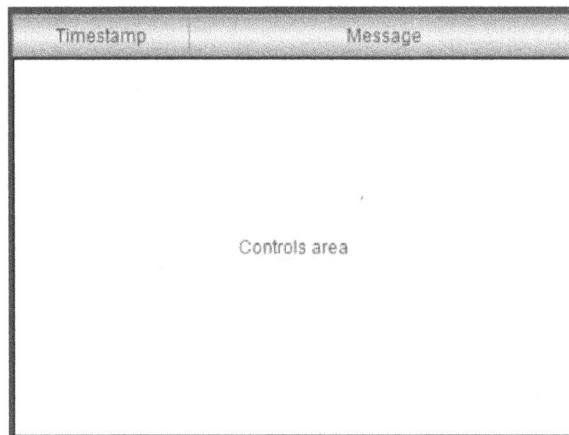

Figure 15.9: Mock layout

The preceding layout shows the alert bar at the top of the screen. Under the bar is the control area.

Setting up an alarm banner

To set up the bar, firstly, we will need to set up the alarm configuration. For this example, we will use the same value that we set up in **Alarm Configuration**. After that step is complete, you will want to set up the following variables in the PLC_PRG POU file:

```
PROGRAM PLC_PRG
VAR
    info  : BOOL := TRUE;
    warn  : BOOL := FALSE;
    error : BOOL := FALSE;
    ack   : BOOL;
END_VAR
```

This example will use four variables. The info, warn, and error variables will display the associated alarm when they are set to TRUE. The last variable, ack, will serve as the acknowledgment variable. The mechanics of acknowledging alarms are something we'll explore later; for now, just keep it as a placeholder.

Once the variables are set, the next thing to do is add an HMI screen to the project and add a banner to it. Keep the banner's parameters to the default settings and run the code.

> **Note**
>
> By default, banners will read from all the alarm groups. If you have more groups, the banner will pick up those alerts too. You can customize this setting by changing the **Alarms Group** property under **Alarm Configuration**.

This code configuration should result in *Figure 15.10*:

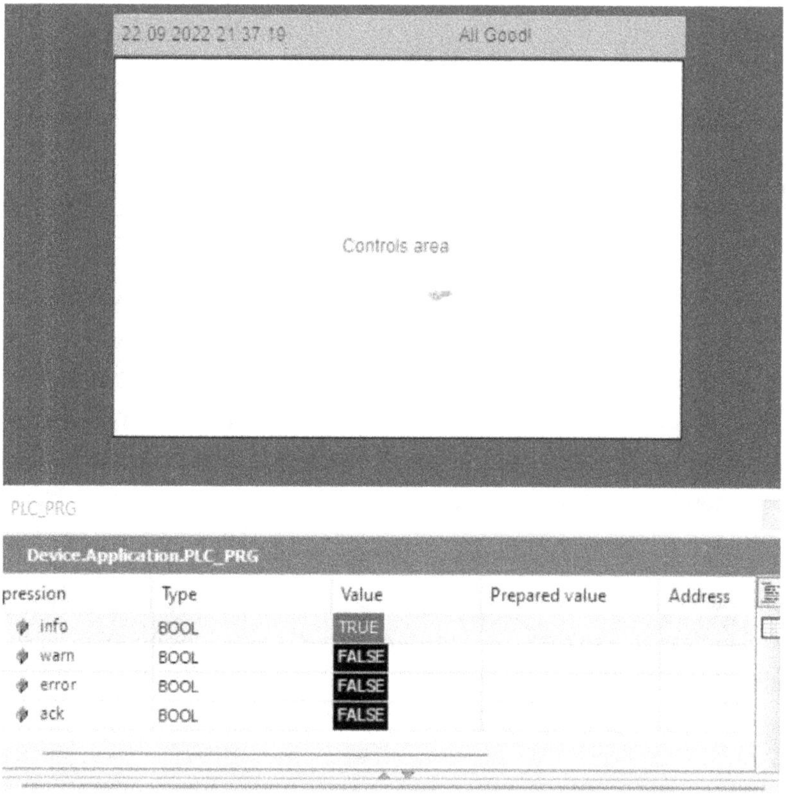

Figure 15.10: Info alert

To explore the other alarms, set the `info` variable to `FALSE` then set one of the other variables to `TRUE` and observe the banner. If all goes well, you should see the banner color along with the message change.

With a working banner example under our belt, we can now turn our attention to setting up alarm tables.

Setting up an alarm table

As we explored, an alarm table is a layout that will show multiple alarms. An example of an alarm table is shown in the following screenshot:

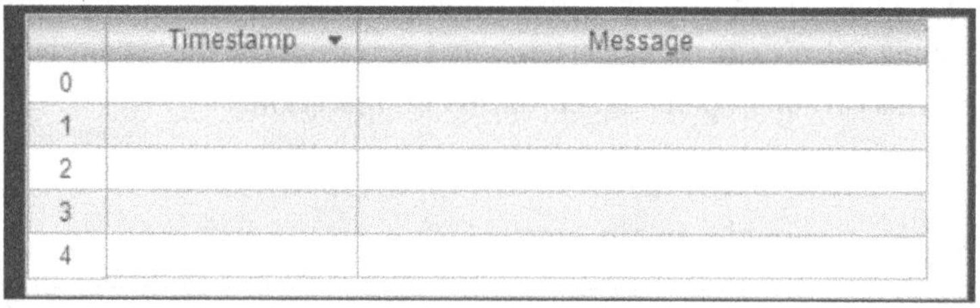

Figure 15.11: Alarm table

As can be seen, the alarm table is split into rows and columns. The table will autogenerate a **Timestamp** column and a **Message** column; however, you will have the option of adding extra columns as you see fit. This ability is available for the banner as well. For this example, we're going to keep it simple and only use the **Timestamp** and **Message** columns.

Where and how you use the alarm table is ultimately up to you; however, I prefer to create a specialized HMI screen for alarms and diagnostic purposes. In other words, I like to put alarm tables on maintenance screens that are used to display diagnostic information. Due to the nature of the table, I feel that these types of controls are best used on diagnostic-oriented screens to make it easier to locate issues that may arise in the system. However, that is a personal preference, and you may find it more suitable to go another route.

In terms of placement, since an alarm table is larger than a banner, I usually like to place these toward the middle of the HMI screen or off to one of the sides. Again, this placement is a personal preference. Regardless, due to the size of the table, it can easily look out of place depending on where you place it.

Setting up an alarm table is very similar to setting up an alarm banner. To use an alarm table, you'll have to create an alarm configuration and add it to the alarm table setup. To properly set up the alarm table, your settings should match the following:

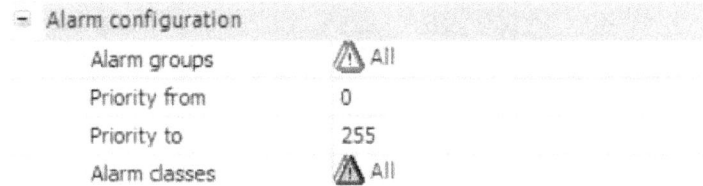

Figure 15.12: Alarm table configuration settings

For the alarm configuration and code, we can simply reuse everything from the banner example. For this example, we will remove the banner and simply place an alarm table on the screen. In short, all that is required for this example is to remove the banner and replace it with an alarm table. Once the program is running, set all the variables to TRUE, as in the following screenshot:

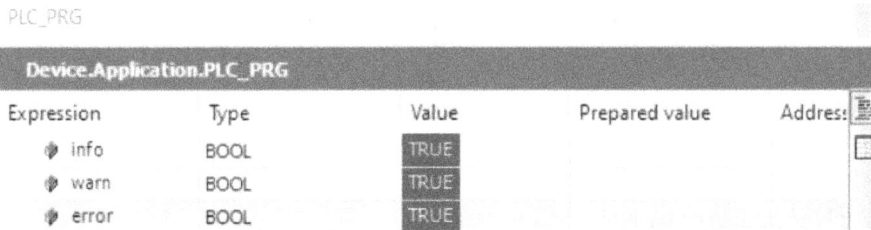

Figure 15.13: PLC variables

As seen in *Figure 15.14*, all the alarms are being shown at once, which is the expected behavior.

Figure 15.14: Alarm table output

Alarms can be removed from the table by simply turning them off. In other words, setting the TRUE variable to FALSE will remove the alarm from the table.

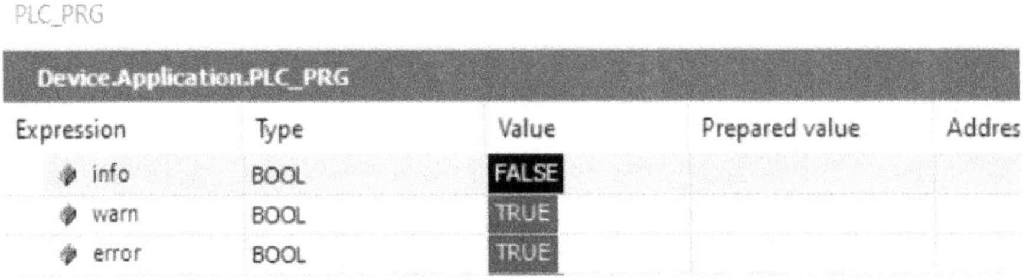

PLC_PRG

Device.Application.PLC_PRG				
Expression	Type	Value	Prepared value	Addres
info	BOOL	FALSE		
warn	BOOL	TRUE		
error	BOOL	TRUE		

Figure 15.15: The info variable set to FALSE

Once you've done this, consider the following screenshot. As you can see, the info alarm is now gone.

	Timestamp ▾	Message
0	24.09.2022 23:06:23	All Not so Good!
1	24.09.2022 23:05:36	There is a motor error

Figure 15.16: The info alarm removed

The preceding figure shows what happens when the variable that is tied to the alarm is turned off. Put simply, the alarm is removed from the table. To reverse this, you can turn the alarm back on by simply toggling the variable back to TRUE.

Alarm tables such as the one we're exploring in CODESYS can be a bit tricky to use. As we saw, setting a variable to FALSE essentially auto-acknowledged the alarm. As we will explore in the *Alarm acknowledgement* section, this can be bad, as the alarm may be cleared before an operator has a chance to read it. If there is a borderline part in the system, this could lead to a dangerous situation. As such, I recommend implementing PLC logic that will prevent the variables from automatically resetting themselves to FALSE without operator involvement.

So far, this chapter has been dedicated to setting up controls to display alarms on the HMI. However, we have only touched on the PLC logic. As we saw, triggering an alarm is usually as simple as setting a Boolean variable to TRUE or FALSE. In the next section, we're going to take a closer look at the PLC side.

PLC alarm logic

There is nothing fancy or complex about triggering an alarm. As we have seen, all we have to do is set a variable to TRUE or FALSE. However, understanding when to set the variable to the correct state can be tricky. For most things in automation, we use bounds or operating ranges to determine whether the part is in a healthy state or not. In other words, many things, such as heaters, motors, and so on, have an optimal operating range that they should always be in. Straying from the optimal range can easily affect the performance of the machine. Typically, alarms will be set when certain aspects of the machine begin operating outside of their optimal ranges, especially when those ranges can lead to injury to personnel or damage to the machine.

Though it is common to hardcode range values, it is usually better to set alarm bounds on something akin to an HMI calibration screen. An HMI screen will allow technicians to alter alarm parameters without needing to modify code. Hence, our examples are going to have an accompanying HMI.

For this example, we are going to simulate a series of pumps. We are going to set up an operating range that looks like the following:

- **Normal:** 0–50 PSI
- **Approaching limit:** 50–75 PSI
- **Over limit:** >75 PSI

The banner colors for this example will be as follows:

- **Normal range:** Green
- **Approaching limit:** Yellow
- **Over limit:** Red

The key to doing this will be the logic in the PLC code. To implement this program, we are going to use the following variables in the PLC_PRG file:

```
PROGRAM PLC_PRG
VAR
    info    : BOOL := FALSE;
    warn    : BOOL := FALSE;
    error   : BOOL := FALSE;
```

```
    info_pump  : BOOL := FALSE;
    warn_pump  : BOOL := FALSE;
    error_pump : BOOL := FALSE;

    good_range  : INT;
    warn_range  : INT;
    error_range : INT;

    psi : INT;
    ack : BOOL;
END_VAR
```

In this example, we are turning the alarms, including the info alarms, off by default.

Below the alarm variables are the range variables. These are the variables that will be tied to an HMI control. As the name suggests, these are the variables that will be used to set the good, warn, and error range, which will be tested against the psi variable.

A minimalist PLC program for this setup is as follows:

```
info_pump  := (psi < warn_range);
warn_pump  := (psi >= warn_range AND psi <= error_range);
error_pump := (psi > error_range);
```

For this example, we are going to add a new alarm configuration to give the appropriate message. In this case, we are going to add a new alarm group called pumps and configure it to match the following:

ID	Observation Type	Details	Deactiva...	Class	Message
0	⁰¹ Digital	(PLC_PRG.info_pump) = (TRUE)		⚠ Info	PSI Optimal
1	⁰¹ Digital	(PLC_PRG.warn_pump) = (TRUE)		⚠ Warning	PSI Approching Upperlimit
2	⁰¹ Digital	(PLC_PRG.error_pump) = (TRUE)		⚠ Error	PSI Over Range

Figure 15.17: Pump alarm group

After adding the new alarm group, the **Alarm Configuration** tree should look like the following figure:

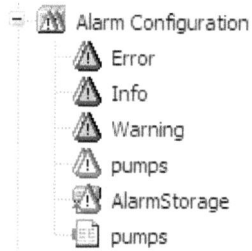

Figure 15.18: Alarm Configuration tree

We will use the following layout for the HMI:

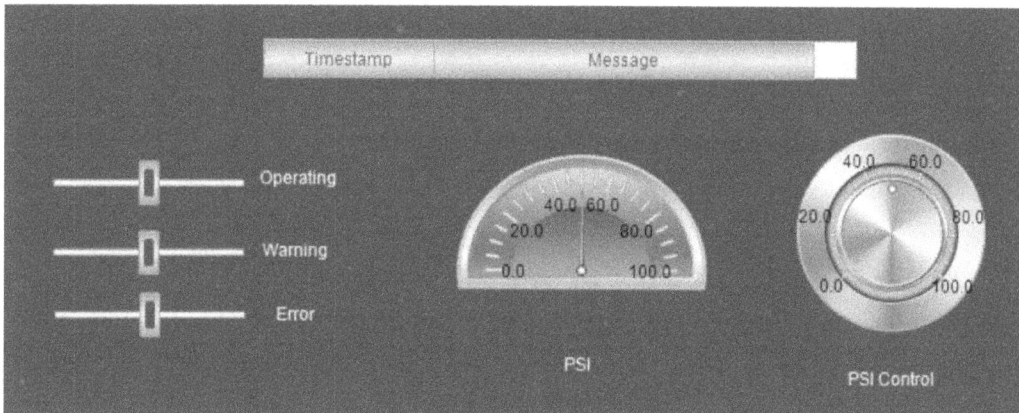

Figure 15.19: HMI layout

The HMI will allow us to set our limits via the sliders on the left of the screen. The variables will be assigned with the following pattern:

- **Operating**: good_range
- **Warning**: warn_range
- **Error**: error_range

The **PSI Control** pot will be attached to the PSI variable, as will the gauge. The gauge will be used to view the simulated PSI reading that is set with the pot. Finally, we are going to use the top alarm banner to display the alarms.

The next thing we need to do is set the scales on the sliders. We are going to set the **Operating** slider to what is shown in the following screenshot:

⊟ Scale	
Show scale	
Scale start	0
Scale end	50

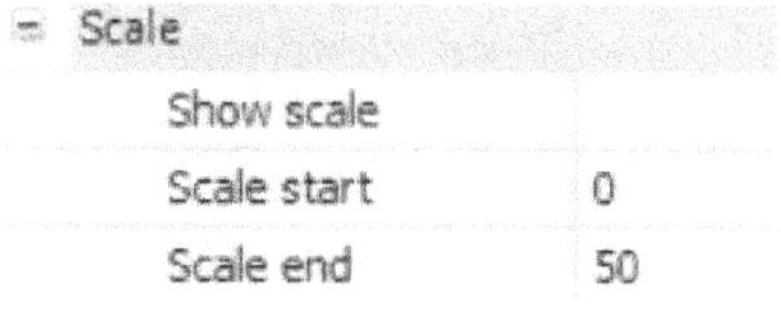

Figure 15.20: Operating slider scale

The **Warning** slider's scale values are depicted in *Figure 15.21*:

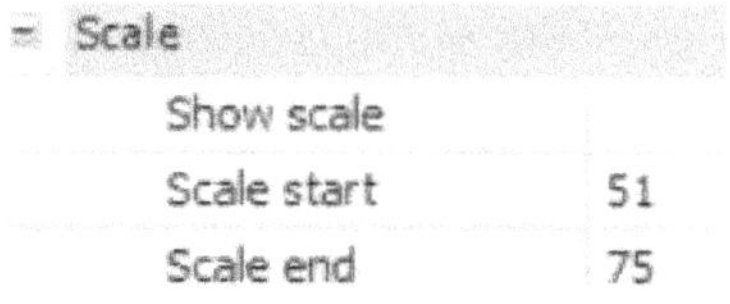

⊟ Scale	
Show scale	
Scale start	51
Scale end	75

Figure 15.21: Warning slider scale

Lastly, the scale for the **Error** slider can be seen in *Figure 15.22*:

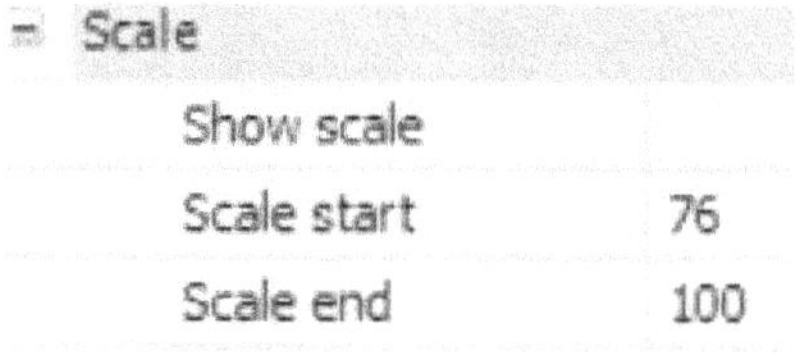

⊟ Scale	
Show scale	
Scale start	76
Scale end	100

Figure 15.22: Error slider scale

Next, we're going to set the pot. Since our scales are going to max out at 100 PSI, we're going to set the scale end to 120. We're going to do this just to give it some extra space for the error alarm. You will need to match the scale on the pot to the values shown in the following screenshot:

⊟ Scale	
Subscale position	Inside
Scale type	Lines
Scale start	0
Scale end	120

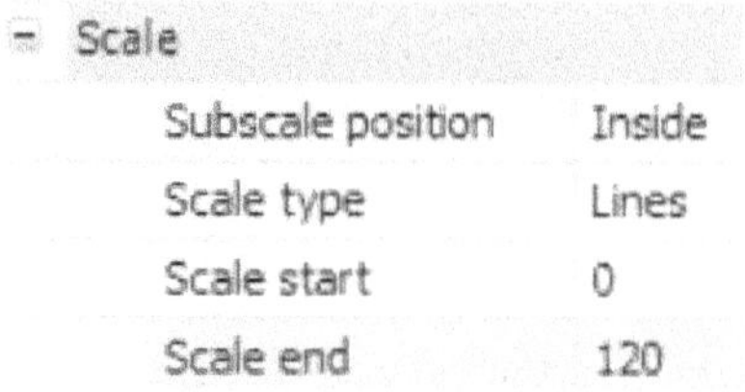

Figure 15.23: Pot scale

We're going to set the scale on the gauge to match the pot; that is, **Scale end** will be set to 120. We'll set the scale on the gauge the same way we have set the scale on every other HMI component.

Lastly, we need to set up the alarm banner. If you have not set up the alarm configuration for this example, you will need to do that now. You can use the same color schemes and fonts we used throughout this chapter.

After all these components are configured, you can run the program. Once you run the program, you'll want to set all the slides to the right of the screen, as shown in the following screenshot. Once you do that, you will see the banner turn green.

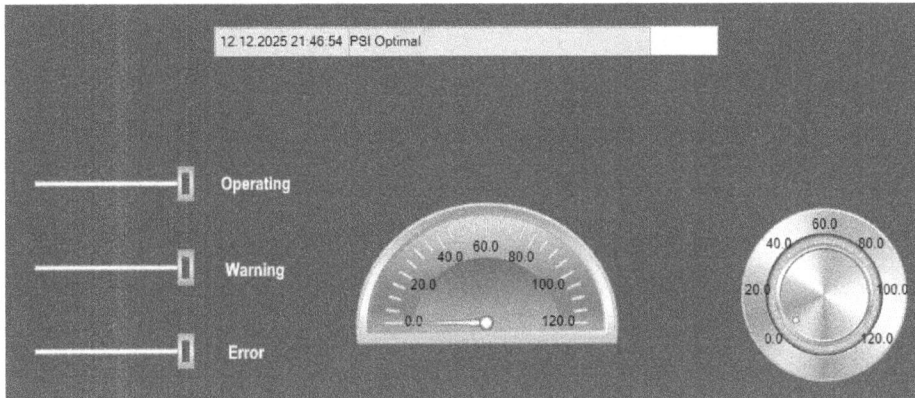

Figure 15.24: Green banner

Once the program is running, slowly rotate the pot to the right. Notice that once you get past 80 PSI, the banner will turn yellow:

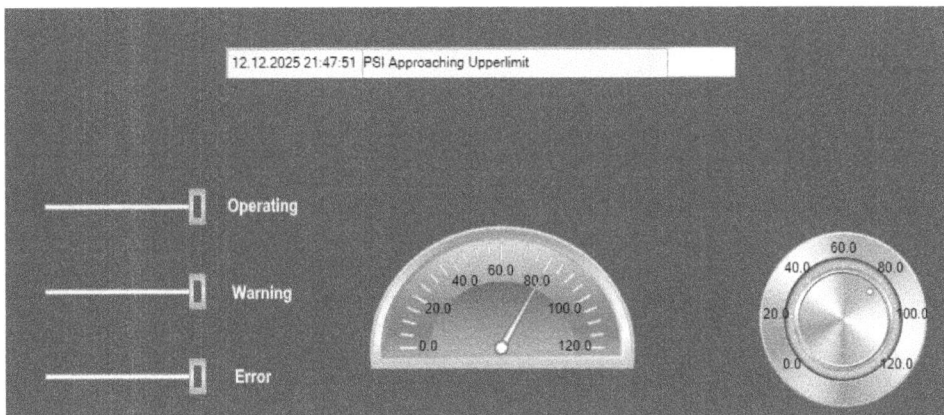

Figure 15.25: Yellow banner

Finally, if you max out the pot, you will see the banner turn red.

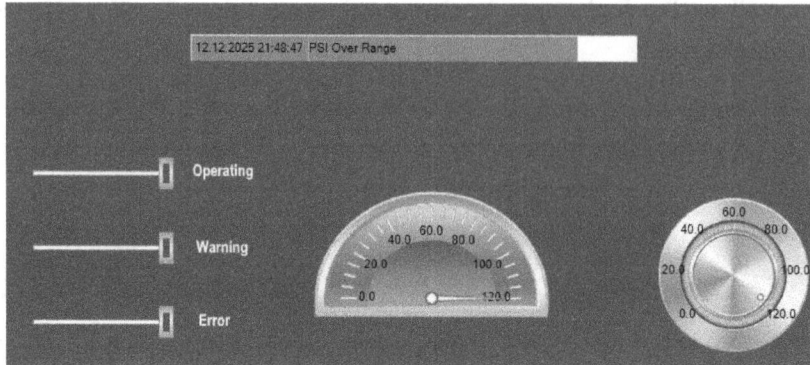

Figure 15.26: Red banner

In a real-world application, the data that dictates which banner to show will most likely be fed in by a sensor of some type. Generally, alarms need to be dynamic and read data. Nonetheless, you will most likely always use some type of input, such as sliders, to set the limits. Depending on what you're working on, it may be easier to trigger the alarms via the configuration in the **Alarm Configuration** menu as opposed to custom PLC logic. For our example, we triggered the alarm programmatically, which is acceptable in many cases; however, it is important to explore using the GUI as well. Consider the following screenshot:

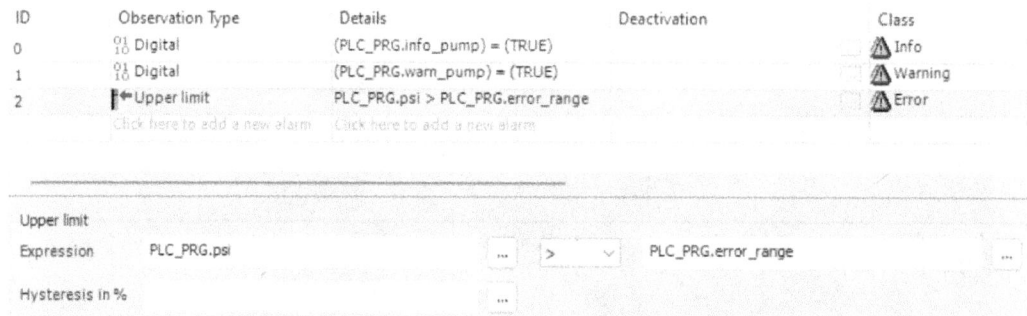

Figure 15.27: Setting error limit via alarm configuration

Essentially, this will set the **Upper limit** logic statement for the alarm to trigger; however, using this methodology can be somewhat restrictive if you need to perform machine operations, such as a machine shutdown. For simply triggering an alarm, the GUI configuration works fine, but if more complex operations are needed, it's best to use PLC code. With all that said, the next thing we need to look at is alarm acknowledgment.

Alarm acknowledgment

Conceptually, you can think of alarm acknowledgment as a signature. Similar to the way racecar drivers are typically required to sign a legal waiver before a race, acknowledging that they could get hurt, an operator acknowledging an alarm is them confirming to the machine that they understand there is some type of abnormal situation. Usually, there is a button or other control that is manually engaged by the operator to acknowledge the situation and confirm to the machine that they understand what's happening.

It is typically considered a best practice to require the operator to acknowledge each alarm manually and individually for safety-critical messages. Many development systems, including CODESYS, do provide a means to acknowledge all alarms with a single press or by using code. Though auto- and bulk-acknowledging alarms have their time and place, I recommend forcing the operator to acknowledge alarms individually, regardless of what they are for. By auto- or bulk-acknowledging alarms, the operator could miss an abnormal situation before it morphs into a dangerous one. In this section, we're going to explore how to acknowledge alarms.

Acknowledging alarms logic

The key to acknowledging alarms is the acknowledgment field. When the variable is TRUE, the text will clear out. For this example, we are going to add a **Push Switch** to the HMI, as in *Figure 15.28*:

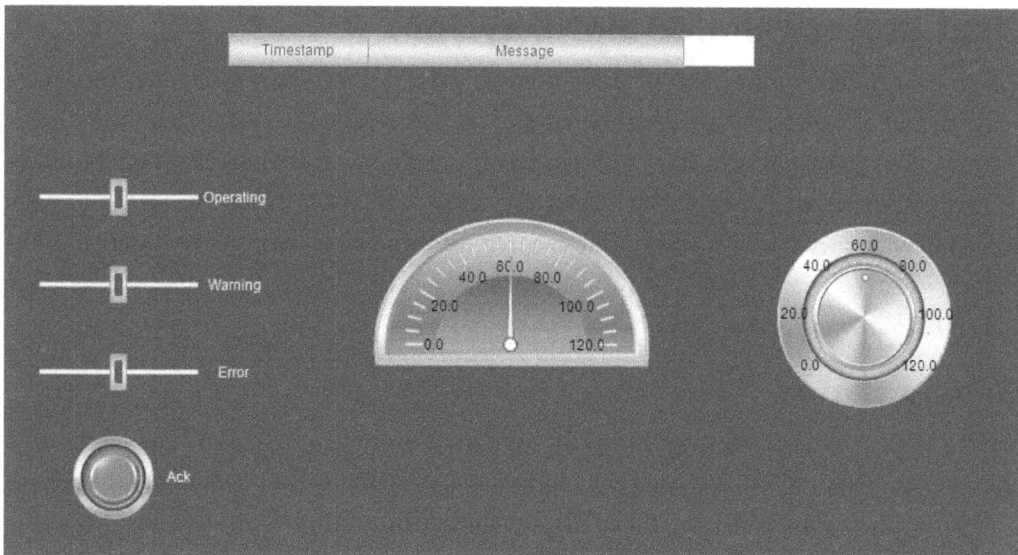

Figure 15.28: HMI with an Ack switch

Once you add the switch, add a variable called ack of the BOOL type to the PLC_PRG POU file. We will need to assign the button to the banner's **Acknowledge** variable field, as in the following screenshot:

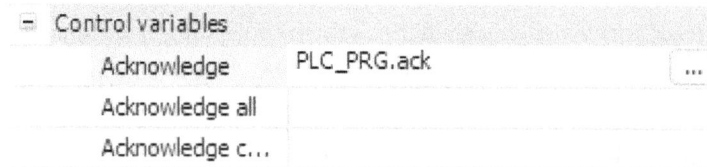

Figure 15.29: Banner's Acknowledge variable field

After configuring the banner, you will need to set up the **Push Switch** to toggle the variable. To configure the switch, simply assign the ack variable to the switch's Variable field. Once everything is set up, run the program and throw the error alarm. Once the alarm appears, you will want to dial back on the pot, preferably to zero, and click the **Ack** button. You should notice that the text is clear and that the appropriate alarm is displayed.

At this point, it is recommended that you swap out the banner for a table and experiment. Notice that with the table, you will have to select the alarm that you want to acknowledge. With all that being said, we now know enough to move on to our final project: building an alarm system for our motors.

Final project: motor alarm system

For the final project, we are going to create a motor alarm system. In the real world, motors are a pivotal part of automation. However, if a motor starts drawing too much or too little current, there could be a problem. Also, if the operating temperature is over or under range for the motor, there could be another type of problem. Therefore, we need alarms to indicate when these events occur and what they are. To round out the chapter, we are going to create an HMI similar to the one in the last section; however, we are going to add more alarms. So, the first thing we are going to do is lay out some requirements.

Getting started

Motors have an optimal operating range for temperature, drawn voltage, and communication between the driver and PLC. We need to monitor these, and if there is any abnormal behavior, we need to trigger an alarm. Also, since there can be multiple issues all at once, we need to log all the problems so a technician can search through them. Our software needs to trigger the following:

- A warning if the voltage is less than 10 volts or greater than 20 volts
- An error alarm if the voltage is less than 4 volts or greater than 25 volts

- A warning alarm if the temperature is less than 65°F or greater than 100°F

- An error alarm if the temperature is less than 60°F or greater than 110°F

- An error alarm if there is no communication from the drive

To do this, we are going to need to build an HMI that can simulate temperature, communication, and voltage.

Design/implementation of the HMI

The HMI for the final project can be viewed in *Figure 15.30*:

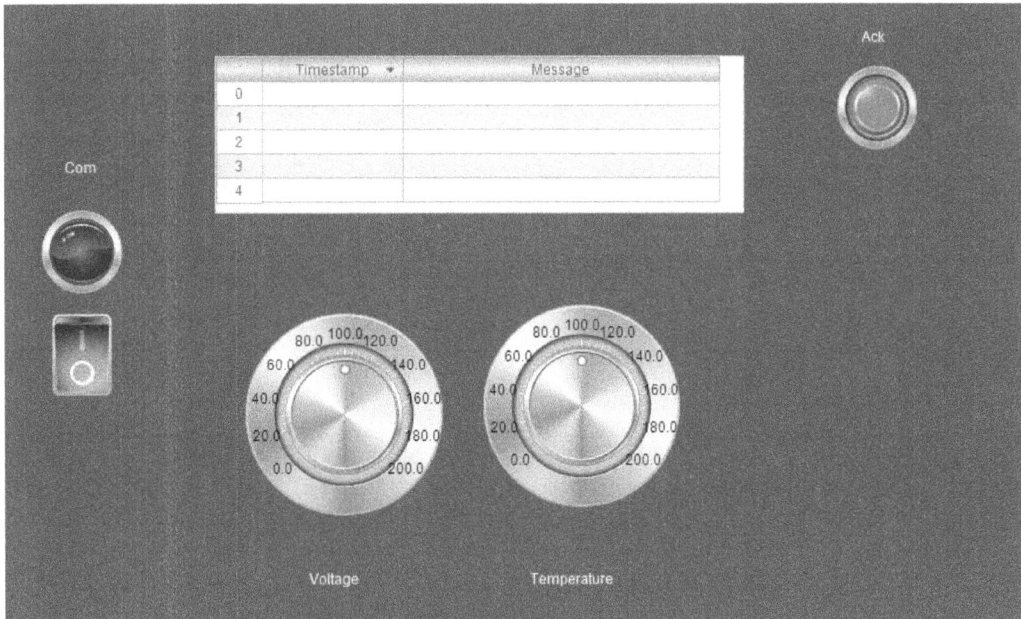

Figure 15.30: Motor HMI

For this HMI, scale the pots to a large number, such as 200 for the upper value.

The variables we are going to implement in the PLC_PRG POU file are as follows:

```
PROGRAM PLC_PRG
VAR
    overTempErr  : INT := 110;
    overTempWar  : INT := 100;
    underTempWar : INT := 65;
    underTempErr : INT := 60;
```

```
    overVoltWar  : INT := 20;
    overVoltErr  : INT := 25;
    underVoltWar : INT := 10;
    underVoltErr : INT := 4;

    pot_temp     : INT;  // temp pot
    pot_voltage  : INT;  // voltage pot

    com : BOOL;  // switch and LED
    ack : BOOL;  // ack switch
END_VAR
```

For motor alarms, you can sometimes get away with hardcoding many of the values, as long as you know in advance that the motor type won't change. For this example, we are going to assume that the motor type will be static and hardcode the values. After you implement the variables, assign them to their corresponding controls that are denoted by the comment. If the variable is not commented, it will be used in an alarm.

Next, we're going to create a new alarm group called `Final_Example`. Since we are hardcoding, we can set the alarm thresholds with the GUI. The alarm group configuration should look like the following screenshot:

ID	Observation...	Details	C	Class	Message	On
0	Upper limit	PLC_PRG.overTempWar >= PLC_PRG.overTempWar		Warning	Motor Nearing Over Temp	
1	Lower limit	PLC_PRG.overTempWar < PLC_PRG.overTempWar		Warning	Motor Nearing Under Temp	
2	Lower limit	PLC_PRG.pot_temp < PLC_PRG.underTempErr		Error	Motor Under Temp	
3	Digital	(PLC_PRG.com) = (FALSE)		Error	No Com With Motor	
4	Upper limit	PLC_PRG.pot_voltage > PLC_PRG.overVoltWar		Warning	MotorApproaching VoltLimits	
5	Lower limit	PLC_PRG.pot_voltage < PLC_PRG.underVoltWar		Warning	Motor Approaching Under ...	
6	Upper limit	PLC_PRG.pot_voltage > PLC_PRG.overVoltErr		Error	Motor Over Voltage	
7	Lower limit	PLC_PRG.pot_voltage < PLC_PRG.underVoltErr		Error	MotorUnderVoltage	
8	Upper limit	PLC_PRG.pot_temp > PLC_PRG.overVoltErr		Error	Motor Over Temp	
	Click here to a...	Click here to add a new alarm				

Figure 15.31: Alarm thresholds

Next, set the ack variable for the alarm table the same way that it is set in the following screenshot:

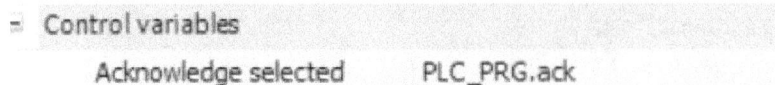

- Control variables

        Acknowledge selected   PLC_PRG.ack

Figure 15.32: Acknowledgment

You will also need to set the `Final_Example` alarm group and tie the `ack` variable to the switch, as we did in the PSI example. Once you do this, you are ready to run and play with the HMI. Launch the program and turn the pots. Observe which alarms appear on the table. When an alarm goes white, select it and click the **Ack** switch on the HMI, and watch how the text is cleared. Try triggering multiple different alarms with the HMI to really get a feel for how the system works.

Summary

In summary, this chapter has been a crash course on HMI alarms. We have covered the HMI and PLC side, as well as the setup for alarms. We have also explored how to acknowledge alarms. At this point, you should know the basics of alarm systems. Overall, you will need to understand this chapter to be an automation programmer, so please ensure that you understand the material.

The future is not set. The automation industry is experiencing its fourth revolution. Put bluntly, the days of the old manufacturing world are quickly evaporating. New technologies such as AI, IIoT, the cloud, and so on are drastically changing the automation landscape. In the next chapter, we're going to explore some of these new technologies and see how the industrial controls world can change!

Questions

1. What is an alarm?
2. What does a red alarm usually mean?
3. What does a green alarm mean?
4. What does a yellow alarm mean?
5. What is an alarm group?
6. What is an alarm acknowledgment?

Further reading

- *Alarm Management*: `https://content.helpme-codesys.com/en/CODESYS%20 Visualization/_visu_struct_alarm_management.html`
- *Visualization of the Alarm System*: `https://content.helpme-codesys.com/en/CODESYS%20 Visualization/_cds_f_visualize_alarms.html`

Get This Book's PDF Version and Exclusive Extras

Scan the QR code (or go to packtpub.com/unlock). Search for this book by name, confirm the edition, and then follow the steps on the page.

Note: Keep your invoice handy. Purchases made directly from Packt don't require an invoice.

Part 4

Putting Knowledge Into Action

In this section, you'll explore the broader landscape of industrial automation technologies and look ahead to the future of control systems. You'll gain an understanding of distributed control systems (DCSs), PLC networking fundamentals, and emerging trends driving Industry 4.0. This part concludes with a comprehensive capstone project in which you'll apply everything you've learned, design principles, programming, debugging, architecture, and more, to repair and enhance a realistic, broken codebase. By completing this project, you'll reinforce your mastery of the concepts and techniques covered throughout the book.

This part of the book includes the following chapters:

- *Chapter 16, DCSs, PLCs, and the Future*
- *Chapter 17, Putting It All Together: The Final Project*

16

DCSs, PLCs, and the Future

We are currently in the midst of the Fourth Industrial Revolution. This new revolution is blurring the lines between traditional computing technologies and industrial controls. The days of simply getting by with PLCs are coming to an end much the way the days of programming in pure Ladder Logic are.

The future is going to be much more integrated. Therefore, to understand how the Fourth Industrial Revolution is going to play out, we're going to explore the following concepts:

- Industry 4.0
- IoT
- Distributed computing and parallel computing
- Common network protocols
- DCS controls
- SCADA
- Cybersecurity
- Emerging technologies

This chapter is going to be theoretical on how integration works at a high level.

What is Industry 4.0?

Industry 4.0 is the integration of new and advanced technologies into automation projects. For the most part, the controls world has been relatively static for the past few decades. The only real paradigm-shifting changes over the past few decades are PLCs replacing relay logic and the introduction of OOP in 2013; however, this stagnation is rapidly coming to an end. As has been explored, technologies such as AI and IoT are revolutionizing how things are done.

The overall goal of Industry 4.0 is to create smart factories. What defines a smart factory will vary; however, they are typically described as factories and processes that can adapt and change to factors such as customer demand and market conditions. To accomplish these feats, technologies that have been seemingly relegated to the traditional IT world are now finding their way into the manufacturing landscape. Some of these technologies include the following:

- IoT
- AI
- Cloud computing
- Data analytics
- Containerization
- As well as many other advanced technologies

A smart factory will, as the name suggests, be smart. It will be able to give close to real-time feedback, allow for enhanced decision making, be allotted a degree of autonomy, and much more. Whole books could be written on the subject, and that would only begin to scratch the surface.

In a nutshell, it is fair to think of Industry 4.0 as smart factories. However, for the factories to be smart, they need devices to communicate, which is where IoT comes into play.

What is IoT?

IoT, as has been explored, are devices that allow for communication. More specifically, these are "smart" devices. What a smart device is can vary greatly depending on who you talk to and the context you're working in. However, a general consensus is that a smart device is a device that can send/receive data over a network and has basic computing capabilities. Common smart devices are as follows:

- Modern phones such as iPhones and Android devices
- Tablets

- Smart watches
- Sensors
- Thermostats

Essentially, anything that can communicate via a network connection is typically considered smart to some degree. In terms of automation, common IoT devices are sensors that can communicate via some type of network.

There are many reasons why things such as smart sensors are becoming more integrated in the modern manufacturing world, with the most prevalent being the collection and transmission of data. With IoT, a pair of wires can allow many devices to freely share data. When other technologies, such as wireless communications are factored in, even components as simple as wires can be eliminated. In either case, the true benefit will be data that can be transmitted across a factory or even across the planet with ease.

Many smart devices can stream data very efficiently across a network. This means that information such as manufacturing data can be streamlined to the cloud for storage, analyzed quickly, and even used to improve AI models. IoT can be a major strength when it comes to manufacturing, but it also comes with many risks.

Cybersecurity and IoT

Most IoT devices are not made with security in mind, if at all. It is not uncommon for components such as sensors to come with hardcoded passwords, not enforce strong authentication methods, leave traffic unencrypted, and so on. This means that nefarious actors, such as hackers, can use IoT devices to infiltrate a network.

When it comes to security, anything that has weak authentication can be easily hijacked with the right tools. Since these devices are connected to your network, once something such as a sensor is commandeered, it can be used as an entry point to pivot to other devices on the network. This means, when possible, you want to ensure that data coming to and from the smart device is encrypted, strong passwords are implemented, there are no hardcoded passwords on any device, devices have the most up-to-date firmware when possible, and so on. Another defense that can be used is to segment your network. That is, to isolate IoT devices from the rest of the network and use strong firewalls. This will ensure that if someone did manage to compromise a sensor, the infection couldn't spread very far.

> **Note**
>
> It is generally safe to assume that anything with an IP address can be exploited by a hacker for nefarious purposes.

IoT is just one technology that is making its way into Industry 4.0. However, if Industry 4.0 is a grand machine, IoT is but a cog. The next thing we need to explore is distributed and parallel computing!

Exploring distributed and parallel computing

As can be deduced, many of the techniques we explored in this chapter are well beyond the scope of a traditional PLC. This is because Industry 4.0 is adopting technologies that are not only PLC-based but will incorporate traditional computing to a much higher degree than is currently used. Essentially, this means that the future of automation is going to be a hybrid of advanced control systems, as well as traditional computing systems such as PCs and servers. In the next section we're going to take a look at one of these techniques, distributed computing.

Understanding distributed computing

Distributed computing is not necessarily the same thing as a **distributed control system** (DCS). In traditional computing, a distributed system is a series of networked computers that work towards solving a common goal. A common application of distributed computing is to speed up and provide redundancy for a system.

Another common use case for distributed computing in Industry 4.0 is IoT devices. Remember that many IoT devices do have some processing capabilities. This means that many of these devices can help enable real-time processing. For example, in certain cases, if a smart sensor detects that 10 items have passed in front of it, the device can automatically react and send a signal to shut off the assembly line. At the other end of the plant, another smart device may receive the signal from the first smart device and count 10 more items and repeat the cycle.

AI and data analytics are also making a splash in the industrial realm. These systems require large amounts of data and copious amounts of computing power. Systems such as Apache Spark are used for such applications. When large amounts of manufacturing data must be analyzed, systems such as Apache Spark or Hadoop are commonly used to assist in making real-time decisions, among other things.

Typically, these systems will require multiple computer nodes to analyze data from smart sensors, PLCs, and other data sources. At the end of the process, meaningful data will be produced that can be used for a variety of tasks.

Understanding parallel computing

Parallel computing is where a task is split across multiple cores in a CPU. Where distributed computing is usually done by splitting a task into smaller chunks and having them executed on multiple computing devices, parallel computing is done all on the same device. This is usually done by splitting up a task, sending it to different cores on the CPU, and having the processor solve the chunks at the same time.

For typical systems, **Graphics Processing Units (GPUs)** are used to achieve parallelism. A GPU will typically have more cores than a CPU, but the cores are less powerful. This means that the cores on them are typically not good enough to do everyday computing tasks, but they are good enough to help solve chunks of a large problem simultaneously. This may seem out of place in the context of a PLC programming book; however, it will make more sense when we explore emerging technologies.

Concurrency

Parallel computing may sound a lot like concurrency; however, they are two different concepts. Concurrency is when a CPU switches between multiple tasks. For example, a CPU may work on Task A for a bit, then stop and work on Task B for a while, and repeat. Concurrency is supported by some advanced PLCs. However, it is not all that common in lower-end controllers. It is also important to remember that special language extensions and other plugins may be needed.

Understanding processing is great; however, for it to be useful, we need data. In modern applications, we need data to come from multiple sources. Therefore, we're going to switch gears and explore networking!

Exploring networking

The key to smart factories is **device communication**. That is, the core of a smart factory is all the control devices, equipment, IoT devices, sensors, and more being able to relay data to each other. As you can deduce, these systems do not speak English. This means that shared communication protocols are needed for the devices to share data. In the next section, we're going to explore a couple of very common communication protocols that are the basis for many more proprietary forms of communication.

TCP/IP

One of the most common forms of computer communication is called the **Transmission Control Protocol (TCP)**. TCP is one of the main transport protocols of the **Internet Protocol (IP)** suite and is often referred to as TCP/IP. TCP and IP are two individual protocols; however, they are often used together. Compared to many other protocols, TCP is very reliable but slow in comparison.

For TCP to work, it requires a three-way handshake between two devices. In a typical client-server model, the device that initiates communication and sends the initial packet is called the client, and the other device is called the server. Essentially, when the two devices connect, the client will send a synchronization (**SYN**) request to the server; in turn, the server will send a synchronization/acknowledgment (**SYN/ACK**) signal back to the client, and finally, the client will send a final acknowledgment (**ACK**) to the server. The process can be seen in *Figure 16.1*:

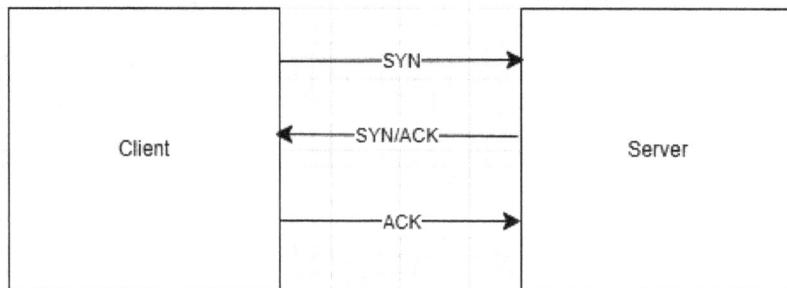

Figure 16.1: Three-way handshake

Compared to other communication protocols, such as the **User Datagram Protocol (UDP)**, which will be explored next, TCP is slower.

To understand why TCP is so much slower than other communication protocols, you must first understand that TCP is much more reliable than many other communication protocols. In short, outside of the three-way handshake, when data is transmitted via TCP, it will sequence the data packets, perform acknowledgments, perform error detection, and, lastly, perform corrections. In all, this means that TCP will (more or less) ensure that the data is transmitted successfully and in the correct order.

For many applications, TCP will be either too slow or simply unnecessary. Another alternative that can be used when TCP is either too slow or unnecessary is UDP.

UDP

Compared to TCP, UDP is much, much faster but much less reliable. Much like TCP, UDP allows a client and a server to communicate with each other; however, unlike TCP, there is no handshake. One device will just send data across the line as soon as it is told to do so. Unlike TCP, UDP will not perform any error checking, acknowledgments, sequencing, or so on. It will, however, conduct a checksum to ensure the integrity of the data, and if a data packet is damaged, the packet will be dropped. With UDP, you simply send and receive data; there is no guarantee that the data will arrive in the correct order, whether the data packet is corrupted or not, or whether the data packet will even arrive at all.

The process of sending and receiving data for a UDP system can be viewed in *Figure 16.2*:

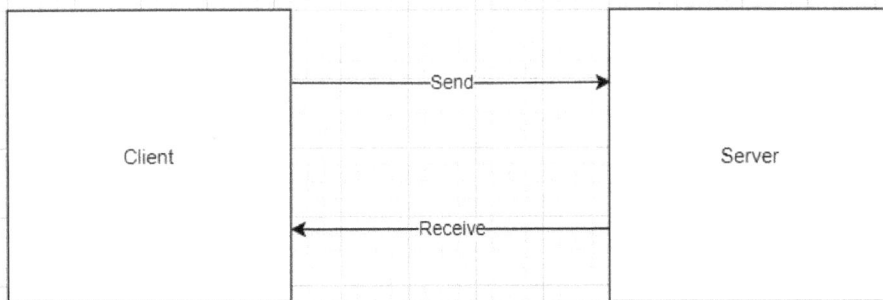

Figure 16.2: UDP send/receive process

As can be seen, the UDP process is nothing more than sending and receiving data between the two devices. There are no intermittent steps; all the system is doing is sending and/or receiving data. The speed that UDP offers stems from the very simple transmission sequence and the fact that nothing is guaranteed.

When I was first starting out in the IT field, and I learned about UDP and how unreliable it was, I couldn't fathom what it could be used for. For the life of me, I couldn't understand why anyone would want to use something as unreliable as UDP. However, I soon came to understand that there are many uses for UDP. UDP is used for applications that do not depend on each data packet. This may seem a bit odd at first, as it may be hard to think of applications that do not depend on each data packet, but a couple are digital streaming and digital communications.

To conceptualize this, consider streaming a movie. If a data packet is lost, the worst that will happen is you will experience a blip in the movie. In the case of streaming, it is more important to try to keep a smooth streaming experience. On the other hand, consider a video call. If you were to use TCP, the lag would make the call almost unintelligible. Again, with UDP, you may lose a few packets of data, which, at worst, would cause a blip or two, but you would still have a relatively smooth call.

As odd as it may sound, UDP is also used quite a bit in automation programming. Many devices use UDP as a communication method. I have seen UDP used for many different things. One area in which I have seen UDP used frequently is in PLC-to-device communication. By this, I mean the PLC talking to devices such as external power supplies and other devices that the PLC may need to control.

TCP and UDP are used in many different IT applications and should be thought of as general IT transport protocols. Though these protocols are general, there are many other proprietary communication protocols that are designed specifically for automation.

PLC/automation device communication

UDP and TCP are general communication protocols. By this, I mean that they are used for many different types of IT applications, such as internet applications, common computer networks, and so on. However, many of the PLC manufacturers produce their own communication systems to be used with their PLCs and various types of industrial components. Some of these systems are very similar and use the same physical connectors as standard computers do – for example, Ethernet cables. However, some use exotic connectors and will be unique for certain devices. The first communication protocol we are going to discuss is one of the most popular, which is called Modbus.

Modbus

Modbus is an industrial communication protocol. Modbus is a little different than the other protocols that we have discussed thus far. Where TCP and UDP are more agnostic in terms of IT applications, Modbus was developed in the late 70s for use in PLC communications by the company Modicon, which is now Schneider Electric. Modbus is what is known as an open protocol. This means that even though it was developed by Modicon, the specs on how the protocol works are openly published and can be used in accordance with a license or for free. For the most part, Modbus is the standard for industrial communications.

Modbus works off what is known as a master/slave configuration. Master/slave systems are very common for industrial communication. For these systems, the master will either query a slave or node device for information such as a sensor reading. The master can also request that the node device do something such as toggle a valve, turn on a motor, or the like. With Modbus, only the master can initiate communication with the node devices; however, the node devices cannot initiate communication with the master device.

Modbus can be used for many different things. One thing that Modbus is used for is HMI communication. For example, there are third-party C# and Java libraries that can be used to orchestrate Modbus communication between devices. For a device such as the Velocio PLC, Modbus can be used for communication between the PLC and a C# HMI.

It is important to know that there are many different types of Modbus implementations. For example, there is Modbus ASCII and Modbus RTU. Both RTU and ASCII are serial connections. Though both will ultimately do the same job, they do differ in how they work. In terms of Modbus RTU, there is a 3.5-character space between messages. In other words, the 3.5-character space is used as a delimiter. On the other hand, ASCII uses two ASCII characters to distinguish messages. RTU uses a binary form to transmit data, whereas ASCII transmits data in ASCII form. This means that although ASCII Modbus is more readable, it is less efficient than RTU.

When setting up a Modbus network, each node will have a unique ID. Relevant IDs can be viewed in *Table 16.1*.

Role	Modbus Address
Slave	1-247
Reserved by standard (no device allowed to be assigned)	248-255
Master	None
Broadcast (function performed by master)	0

Table 16.1: Modbus IDs

A slave may have any ID between 1 and 247; however, as standard, no device can have an address in the range of 248-255, as these are reserved addresses. Also, the broadcast "device" isn't an actual device. The broadcast is a function done by the master. A broadcast is the master sending out one message to all the slaves. This could be to initialize the slaves at startup, reset them, or perform any type of group action.

It is important to understand that there are different flavors of the Modbus protocol. You need to ensure you choose the proper hardware and are developing the correct software for compatibility with your chosen Modbus flavor. Another common implementation of Modbus is Modbus TCP/IP, which is Modbus wrapped in Ethernet IP, a.k.a., a TCP frame payload. In the case of Modbus TCP/IP, you can use standard switches and Ethernet cables for communication. Generally, Modbus TCP/IP is becoming more popular in newer systems as it is a newer technology.

Though Modbus is a very common protocol that is often touted as the industry's de facto protocol, as stated before, it is not the only one. There are many other protocols, and the next one we are going to explore is called Profibus.

Profibus

Another very common communication protocol for automation controllers is **Profibus**. Profibus was developed and promoted by Siemens to network things such as sensors to a controller. Profibus works off a master/slave network configuration. Usually, the master device will be some type of controller, such as a PLC. On the other hand, the slave nodes will be devices such as sensors, drives, and so on. Profibus networks can experience speeds of up to 12 Mbps; however, most systems are set to a significantly lower speed, usually around the 1.5 Mbps range.

Unlike many other communication systems, Profibus requires the use of a specialized cable. Usually, the cable is a shielded purple single-pair RS-485 cable with a DB-9 connector at the end instead of something such as a standard Ethernet cable. At first glance, the connector on the cable can seem odd, as it has an on/off switch on it. This switch connects to a terminating resistor that, when placed in the on position, denotes the end of the device chain. This can cause issues because if a switch is in the incorrect state, the chain can be prematurely cut short. If you do opt to use Profibus and you do encounter device communication issues, one of the first places you should look at is the terminating switches.

Another major difference between Profibus and Ethernet networks is that Profibus will usually support larger networks. However, great care must be taken when selecting the length of a Profibus cable. On the short end, it is recommended that there be a minimum cable length of about 3 feet (or 1 meter) between each of the nodes. A cable length of anything shorter can result in communication issues. It is common, even if the nodes are next to each other in the same cabinet, to have 3 feet of cable between each node. On the other hand, the length of the cable will dictate how fast you can transfer data. In terms of Profibus, the maximum length you want to use is about 1,200 meters, which will allow up to about 9.6 Kbps. At the other end of the spectrum, you can get up to 12,000 Kbps with a length of 100 meters. The shorter the cable is, the faster the data transfer rate can be. This is a very important concept to remember when developing a Profibus network, as you will have to weigh the pros and cons of having longer cables but slower transmission speeds, and vice versa.

There is also a limit to the number of devices that can be on a Profibus network. In short, each device on a Profibus network must have a unique device address. The drawback is that devices on a Profibus network can range from 1 to 127. At most, you can have 126 devices on the network. The address will either be set with a physical dip switch on the device or via the configuration software.

Profibus is a very common communication system, and it is still widely used. However, there is another type of Profi network called Profinet, which utilizes new Ethernet-based technologies. The next section will be dedicated to exploring Profinet.

Profinet

Siemens also offers another major protocol, called **Profinet**. Compared to Profibus, Profinet is based on newer, Ethernet-based technology. Profinet shares many similarities with Ethernet, even down to the cabling. Most who employ the communication system will use an industrial version of an Ethernet cable. Normally, you will be able to spot a Profinet cable due to its green color. However, it is common for some to use a standard Ethernet cable when in a pinch or for troubleshooting purposes.

Outside of being able to use off-the-shelf cables, Profinet is also faster than Profibus. The extra speed characteristic stems from its Ethernet roots. Similar to Profibus, Profinet also has length limitations. A Profinet cable can be up to 100 meters in length. However, Profinet is, on average, faster than Profibus. Usually, the standard operating speed for a Profinet network is 100 Mbps. Generally, Profinet is favored in newer applications that require faster communication speeds and response times.

Note

Though you can sometimes get away with using a standard Ethernet cable, you should use the recommended cabling when possible. Using cheap cables can introduce noise into the system, and they are sometimes more prone to breaking.

EtherCAT

Similar to Modbus, Profinet, and Profibus, **EtherCAT** is another proprietary communication protocol developed by Beckhoff. EtherCAT stands for **Ethernet for Control Automation Technology**. Similar to Profinet, it is an Ethernet-based communication protocol. EtherCAT is a communication system that is used for a wide range of applications, including industrial machinery, medical equipment, mobile machines, and a variety of other applications. Similar to Profinet, the physical connection is standard Ethernet cabling. This means that, much like Profinet, off-the-shelf Ethernet cables can be used when troubleshooting or in a pinch. Though the cabling is the same, the underlying communication system is different.

The way the EtherCAT system works is unique compared to the other communication protocols that we have explored thus far. Essentially, the EtherCAT master will send a data packet known as a frame to all the nodes on the network. The nodes will read the frame and will perform the instructions that were meant for it and ignore the instructions that were meant for the other devices on the network. The devices will also add their information to the frame. EtherCAT devices typically have two Ethernet ports. One of the ports is for sending data, and the other is used for receiving.

Typically, the network is configured in a ring-like topology similar to what can be seen in *Figure 16.3*.

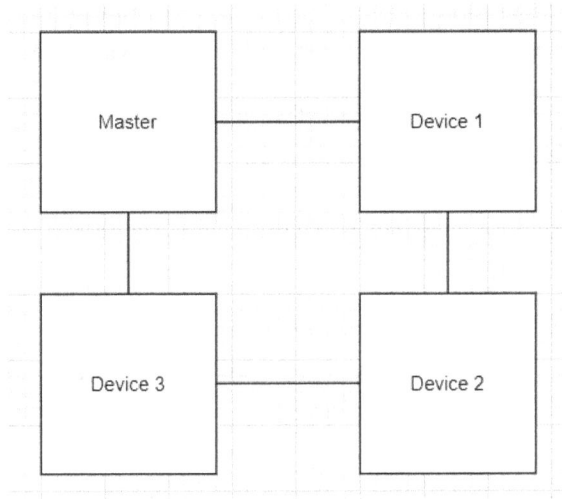

Figure 16.3: Typical ring EtherCAT configuration

With this configuration, as long as the communication hardware is intact and working, the frame will circulate throughout the network. Overall, EtherCAT in this configuration will provide you with the following:

- The ability to use off-the-shelf Ethernet cables
- Allows for processing on the fly
- No need for hardware such as switches and so on, as with Profinet
- A downed node will not necessarily kill the communication chain

In all, EtherCAT is a very powerful and robust communication protocol that, due to using the lowest two layers of the Ethernet protocol, is significantly faster than Modbus or Profinet, which makes it very suitable for real-time applications. Another common industrial system is EtherNet/IP.

EtherNet/IP

EtherNet/IP is an industrial fieldbus protocol that runs on standard Ethernet hardware. This protocol uses TCP/IP and UDP/IP for the transport layers and CIP as its application layer. For this protocol, the IP does not stand for Internet Protocol as it did with TCP/IP. For EtherNet/IP, the **IP** stands for **Industrial Protocol**. The CIP protocol is the same protocol family used by other common communication systems, such as DeviceNet, ControlNet, and CompoNet.

This protocol doesn't force a single topology, such as a star or ring topology. Since it uses standard Ethernet, you can use the same network layouts that are used in traditional IT. Though you are not limited to a specific network layout, the most common topology for this protocol is the star network. In this configuration, there is a central node in the middle of the network that directs traffic to other devices.

Unlike with Modbus and the Profi-networks, EtherNet/IP does not have a fixed device count, at least in theory. You can technically add as many devices as you want to the network. In reality, there are practical limits such as the PLCs having CIP connection limits, limited network switch bandwidth, and so on.

Communication is only half the battle; the next concept that we're going to explore in terms of smart factories is **Distributed Control Systems (DCSs)**.

Exploring DCSs

DCSs are very popular when there are multiple processes that need to be coordinated. Essentially, a DCS is a network of controllers that use some form of communication to coordinate processes.

> Note
>
> Though not always used, it is very common for a DCS system to have a master computing cluster that acts as a conductor to help coordinate the other controllers.

Generally, many of the DCSs that I have worked on in the past can be conceptualized as in *Figure 16.4*:

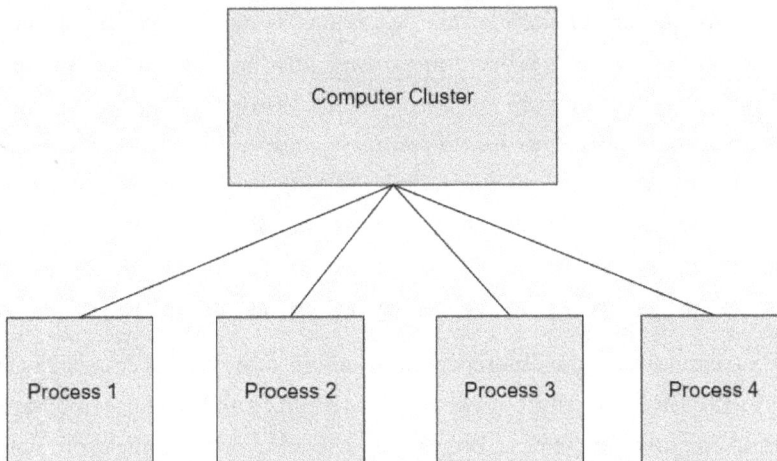

Figure 16.4: DCS layout

In *Figure 16.4*, the central cluster coordinates four processes. At its heart, a DCS can simply be thought of as a central controller that supervises multiple processes. The cluster can be composed of multiple computing systems, such as operator terminals and so on. In all, a DCS is a supervisory system for a whole process or facility.

DCS applications

A logical question is where are DCSs used? The answer is pretty much anywhere. Some common areas where DCSs are used are as follows:

- Smart factories
- Traffic control systems
- Chemical plants
- Manufacturing facilities
- Nuclear power plants
- Agricultural environments
- Some space launch systems

DCSs can be used anywhere where whole processes need to be monitored or coordinated.

Understanding the difference between PLCs and DCSs

The line between PLCs and DCSs is beginning to blur. In short, you can think of DCSs and PLCs as two separate types of controllers. In all fairness, what constitutes a PLC, DCS, and the various types of controllers on the market is beginning to blur. A DCS in this context is a system such as the Emerson DeltaV, which allows I/O to be spread out. However, it is not unheard of to have a DCS be a cluster of computing devices that control/orchestrate PLCs. For example, if you have a number of welding robots that are controlled by a PLC, you can have a DCS coordinate the robots by sending control signals to the PLC. In cases like this, a DCS wouldn't do a good job of controlling the robot directly, but it could be used to orchestrate the robots. Put simply, a PLC is designed to control a thing, while a DCS is designed to orchestrate a process.

The best way to demonstrate the differences between the two types of controllers is to remember what they are used for. In short, if you need scalability and the ability to control multiple processes, a DCS is probably the best. You can use the following definitions to help select the appropriate system:

PLC: A PLC is used when there is a need for a fast response time and the process that it controls is singular. You may also use a PLC when the application is not geographically dispersed. Put simply, a PLC is designed to control a single system, such as a single machine.

DCS: A DCS is used when you need to supervise a whole process. Think of a DCS as a conductor of an orchestra. The same way a conductor does not play an instrument in the band, a DCS is usually not responsible for a single process.

Since the lines between a DCS and PLC are blurring, especially when computers that can support parallel and distributed computing are in the mix, a PLC can sometimes coordinate multiple machines, which, to a degree, is a form of DCS. This is typically done with advanced PLCs that can support techniques such as concurrency.

Exploring SCADA

A DCS is somewhat conceptually similar to a SCADA system, and to make matters worse, some newer engineers will often consider an HMI to be a SCADA. However, a SCADA system is more of an overarching monitoring system. SCADA systems are large systems that are designed to control and monitor whole processes. A SCADA system will usually control a whole plant and allow remote access for people who are not on site. A SCADA system is composed of many different modules, such as PLCs, HMIs, RTUs, sensors, and so on. In other words, SCADA systems are high-level supervisory systems that tell other modules what to do.

They will also perform actions such as logging data into databases. Whereas an HMI is just a UI, SCADA can best be thought of as a system that includes HMIs, PLCs, sensors, and so on. In short, a SCADA system is a remote monitoring system that is composed of many different hardware and software components, while an HMI is a software component that is local to a machine or set of machines.

These new integrated environments are juicy targets for attackers. With smart factories becoming the norm, the interconnected nature of factories is becoming a prime target for bad actors. This makes sense if you think about it. If you're a bad actor and want to cause havoc, you could, in theory, cripple a smart factory and cost the company millions of dollars in lost revenue. Therefore, in the next section, we're going to explore some basics to help secure your networks.

Exploring cybersecurity

Cybersecurity is something that, up until recently, wasn't considered so much in the day-to-day life of a typical automation engineer. For smaller businesses, simply programming a PLC or having basic password management will suffice. However, with the integration of new technologies and the sweeping nature of Industry 4.0, cybersecurity is going to be a must. To understand how to defend your system, you need to first understand how an attacker thinks.

Understanding reconnaissance

Before an attacker can attack your system, they need to understand your system. This can be done in a couple of ways, one of which is actually very legal. When an attacker puts your system in their crosshairs, they will try to learn as much as they can about your organization, the system, your coworkers, and even you. This is called **passive recon**. In this phase, the attacker will use resources such as social media and other publicly available resources to find as much info as they can. A common place that attackers can scope for information is LinkedIn or job posting sites, as it is common for people to list what technologies they work with or what the company uses. As a result, the attacker can get a feel for the systems being used, and they can research vulnerabilities in them. Keep in mind, this is perfectly legal because the attacker is never engaging with the system. On the other hand, **active recon** requires the attacker to directly interact with the target systems. This could be someone scanning for open ports, trying to get information about the target's OS, and more. If no permission is given, this action is typically illegal.

Note

Avoid posting too much about yourself or your position online. This information could be used to hack either your home or workplace. One thing that can really help prevent attacks is to avoid posting what PLCs, SCADA, HMIs, and other technologies you're working with.

Avoiding the use of default passwords

In automation, many systems come with **default passwords**; that is, passwords that are preloaded onto a device. These passwords are not meant to be secure. They simply serve the purpose of allowing the integrator to log in and configure the device. Under no circumstances should you ever deploy a system with a default password. Though this may seem like common sense, it happens way too often. All devices should have their passwords changed as soon as possible. The password should be changed to a secure password that is at least 12 characters and contains an assortment of letters, numbers, and special characters. This is especially true for IoT devices.

Note

Do not use passwords that are related to you, such as the name of a pet, significant other, child, or so on. It is best to create a fake persona and use those names/dates for your passwords.

Configuring firewalls

A **firewall** is a piece of software or hardware that can block traffic. This is very useful because it can block unwanted users, such as hackers, from accessing your networks. Typically, a firewall works off a set of rules to allow or block traffic. A data packet can only pass through the firewall if the rules allow it.

A common type of attack a firewall can block is a **Denial of Service (DOS)** or its more advanced counterpart, a **Distributed Denial of Service (DDoS)**. These attacks attempt to flood a network with junk traffic that can render it inoperable. However, if you put a firewall up, it will usually block the junk packets and prevent the flood.

Whitelisting and blacklisting

A common way to configure your firewalls is to use **whitelisting** or **blacklisting**. Whitelisting is a technique where you block all traffic that is not on a list. Whitelisting makes your system very safe because you can, in essence, prevent any traffic that you do not explicitly state. However, for large enterprise systems, this can be cumbersome and inefficient. You'll often use whitelisting in what's called a Zero Trust environment. In this type of environment, you essentially treat all traffic as suspicious.

A more common way of vetting traffic is using what's called blacklisting. Blacklisting is the opposite of whitelisting in a way. Where whitelisting only allows traffic from a vetted list, blacklisting will block traffic that is on the list and allows all other traffic.

Though SCADA security has evolved a lot, SCADA systems are often a juicy target in terms of security. Firewalls are often placed in front of SCADA systems to help vet traffic. Doing so adds a layer of defense against would-be hijackers.

Implementing encryption

Encryption is key! Encryption is essentially a way to scramble your data so that if a hacker does get hold of it, they won't understand it. Data can exist in three states:

- **Data at rest**: Data is sitting on a storage device. It is not moving or being used.
- **Data in transit**: Data is moving through a network. It is transported from point A to point B.
- **Data in use**: Data is being used by some process.

To protect your data, you want to encrypt it as much as possible.

IoT devices sometimes do not encrypt data, or at least not very well. This means that attackers who intercept the data will be able to understand it very easily. This may not seem like a big deal, but it is. If a hacker can read the data, they can gain meaningful information from that data. Worse yet, they can alter the integrity of the data. This means they could cause havoc, such as throwing off the number of parts being made, altering values such as temperature parameters, which can damage parts, and many other nefarious tasks.

To prevent this, it is highly recommended to encrypt IoT data. To do this, you will typically have to employ advanced programming and network technologies. There will be extra effort involved, but it will be worth it in the long run.

Turning off unused ports

Ports are like doors into your system. If someone is trying to attack your network, the first thing the attacker will look for is open ports. This is usually what they do during active recon. If they find an open port, they can usually use that as an access point. In a network, there are some basic ports that are always used. These ports are called well-known ports, and their numbers range from 0 to 1023. These ports support services such as SSH, FTP, HTTP, HTTPS, and so on. Ports 1024-49151 are known as registered ports. These ports are registered for certain services, such as certain databases, and so on. Finally, there are what are called dynamic ports. These ports range from 49152-65535 and are usually used by programmers for their apps.

> Note
>
> Dynamic ports are sometimes called ephemeral or private ports.

Unfortunately, if you use these services, you need to have the port on. The general rule of thumb is to turn off any service you are not using and to disable any ports you are not using for your system. It is generally recommended that you scan your network with a tool such as **Nmap**, which can be used to return all open ports, running services, OS fingerprints, and other types of information that can be used to attack your system.

Exploring segmentation

It is usually a good idea to segment your networks. This will usually require the assistance of a network engineer. **Segmentation** is essentially isolating devices into their own networks. This can be done in many ways, and by doing so, you are ensuring that there are a limited number of ways in and out of that part of the network, which means attackers will have fewer avenues to attack you.

At the very least, you want to isolate your industrial network from your users' network. This means your factory or whatever environment you're in should not be connected to the main office network. If you do have the two networks integrated, you can open yourself up to attacks. If an office network device, office user, or anything else becomes compromised, your factory could be attacked. This is actually a very common avenue that attackers will use.

In some cases, some office users will need to be able to access the network. For example, managers, engineers, and so on. In these cases, certain techniques can be used, such as **Virtual Private Networks (VPNs)**, and so on; however, this should be limited and allowed on an as-needed basis.

These are just some high-level techniques and concepts that are related to cybersecurity. In reality, there is a lot to the field. A lot goes into securing a network, and this section just touches the surface. Cybersecurity for industrial automation is a fairly new concept, but a very important one, especially in the context of Industry 4.0, where everything is connected via a network. It is highly recommended that you design your system with security in mind. This may include hiring a specialized cybersecurity engineer or consultant. In summary, cybersecurity is a journey, not a destination!

Though it can be argued that cybersecurity is an emerging field in automation, it is not the only one. There are many other emerging technologies that are being integrated into automation. In the next section, we're going to explore some of the concepts and how they will play out in the automation realm!

Emerging technologies

The automation landscape of tomorrow is more closely going to resemble the IT landscape of today. In fact, outside of PLC programming, the two fields may be indistinguishable from one another. The advanced computing that will be required to power the industrial landscape is going to require many technologies and techniques that are just now emerging in the IT landscape. Throughout this book, we have explored some concepts, such as AI, cloud computing, DevOps, and so on, that are not usually associated with industrial technologies. In this section, we're going to delve a little deeper into emerging technologies to see how Industry 4.0 may play out, starting with microservices.

Exploring microservices

Microservices are common in larger software systems. Many organizations use them, and they are the norm. To be fair, industrial systems use this architecture as well; however, the architecture is not as common in automation as it is in the traditional IT landscape. However, for advanced applications, such as the ones explored in this book, they will become the norm.

In short, microservices are an architecture where responsibilities are broken down into services. For example, a service can be thought of as a part of a system that does a task. In this case, a service might be a program that monitors the temperature of something, such as an oven. In this architecture, a service is responsible for only one thing. By doing this, if a service becomes unavailable, it won't necessarily kill the whole system.

Prompt engineering

As we have explored, LLMs such as ChatGPT are here to stay. **Prompt engineering** is going to become a key skill for developers as these systems can and usually are used to help generate code, troubleshoot code, and more. Many systems offer an **Application Programming Interface (API)**. This means, by using a traditional programming language such as C#, Java, Python, Node.js, or the like, you can integrate these LLMs into your system.

ChatGPT cannot run directly on your PLC or control board; however, you can still use it. To do this, you can network a PLC to a PC. The PLC can send data to the PC that is running some type of software that can send data to the AI using an API call. From there, the operator could write a prompt such as *Why is Line 3 running slow?* and, assuming that the necessary data is sent to the AI, it could give you valuable feedback. This means that AI systems could help with the following:

- Optimization
- Troubleshooting
- Production advice

Moreover, this software, which would be used to read data from the PLC and pass it to the AI, would be a prime example of when to use a microservice.

It should be noted that AI such as ChatGPT is not fully fleshed out, and it's not fully understood what they can and cannot do. However, depending on the model, data, and so on, you can greatly increase the intelligence of your system by introducing AI and prompts into it.

Understanding digital twins

With the widespread adoption of parallel and distributed computing, simulations are going to be integral in the future. A key type of simulation is a **digital twin**. A digital twin can be thought of as a computerized model of a machine or series of machines that simulates a process. The physical machine(s) will typically be outfitted with IoT sensors that feed data into a computer, and the physical device can be simulated in a digital landscape. These digital twins can then be used as a testbed for different parameters. For example, the results of altering the speed of several machines can be modeled and analyzed with digital twins.

Digital twins open up very interesting avenues to explore with advanced AI systems such as ChatGPT and the like. The data collected from the digital twins can be fed into an AI system, and prompts can be employed to help analyze the data. For example, you could ask, *What happens if we decrease the coolant on machine 4?* and, using data from the digital twin, the AI could provide insight into what could happen.

Digital twins can require copious amounts of computing resources. You will need GPUs and to employ parallel processing for all the complex math and data processing that will be required. This can be very costly and difficult to maintain on-site. A workaround is to use the cloud.

The cloud in industrial settings

We have touched on what the cloud is. Essentially, cloud service providers sell users resources such as compute resources, storage, databases, and more. Most cloud providers also offer services to operate your IoT infrastructure. To effectively utilize these technologies in the future, the cloud is going to become pivotal. In short, a cloud service provider such as AWS will give you a place to deploy your microservices to, provide the advanced computing resources needed to power digital twins, house the copious amounts of data that are usually generated by modern systems, and more. Though you will have to pay for these services, and to have the full shebang of all the cool emerging technologies will cost a pretty penny, it will be worth it. Time is money, and in automation, having access to this real-time data and the necessary computing power to make sense of it is vital. The cloud and automation will become more integrated in the future. If you want a competitive edge, exploring a few cloud providers, such as AWS, Azure, Google Cloud, Oracle, or any other, is worth it!

Summary

This chapter explored emerging technologies, SCADA, DCSs, cybersecurity, and more. In short, this chapter was an overview of Industry 4.0 and what it could bring. How Industry 4.0 will play out is yet to be fully known. Where the old landscape had automation on one end and the traditional IT world on the other, the new industrial landscape will be a hybrid of the two. Much like how the days of only needing to know Ladder Logic are coming to an end, so too are the days of simply knowing how to program a PLC. The factories that are emerging are going to more closely resemble traditional IT systems than their PLC-only based counterparts of the past. If I could give any advice to a would-be engineer, it would be this: research the emerging technologies and current computing techniques of today and learn to apply those to whatever project you're working on. With that, we're going to move on to our final chapter!

Questions

1. What's a DCS for?
2. What is a digital twin?
3. Can ChatGPT be run on a PLC?
4. What is the difference between whitelisting and blacklisting?
5. What is a firewall?
6. What is a DDoS?
7. Are IoT devices built with security in mind?
8. What is encryption?
9. What is a microservice?
10. What is parallel computing?
11. What is distributed computing?
12. What is the difference between UDP and TCP?
13. What is the recommended network architecture for EtherCat?

Join our community on Discord

Join our community's Discord space for discussions with the authors and other readers: `https://packt.link/embeddedsystems`

17

Putting It All Together: The Final Project

Important!

The goal of this chapter is for you to use what we have explored throughout the book and apply it. Therefore, there will be no prebuilt code. This project will be all hands-on, which means you will have to type it out and troubleshoot it as you would in the real world. To simulate a real project, you will be met with broken code, red herrings, and faulty logic that you will need to troubleshoot using the concepts we have explored. Nonetheless, there will be hints to help you along!

Congratulations on making it this far in the book. Hopefully, by this point, you have a good grasp on the more modern and advanced concepts of PLC programming and software engineering in general. By this point, you should have become not only a better PLC programmer but also a better software developer in general. Thus far, we have explored OOP, advanced Structured Text, alarms, HMIs, the SDLC, and much, much more. In all, at this point, if you understand most of the material we covered, you're probably light years ahead of the average automation programmer.

As far as programming and HMI development are concerned, we have reached a point where we can combine all these concepts into a fully working project. This chapter will be unlike the other chapters as we will not be exploring new concepts. Instead, we are going to apply the concepts we have learned throughout the book to make a simulated industrial oven. In a nutshell, we are going to cover the following:

- Project overview
- Requirement gathering
- HMI design
- HMI implementation
- PLC code design
- Implementing the PLC code
- Testing the application

The goal of this project is to integrate many of the concepts that we have learned in the previous chapters to form an industrial oven. Ovens are very common PLC-driven devices as they are used in many different manufacturing processes.

However, unlike most of the other projects we have built thus far, the code will be written in such a manner that it should be improved upon. In other words, the code in this chapter will be the first draft of a program and won't work as intended. Key details to look for are as follows:

- Bad names
- Incomplete methods, function blocks, and so on.
- Incorrectly declared variables.
- Potential safety issues.

There will be other issues to look for as well. This twist stems from the fact that most software will usually need to be debugged, cleaned, and refactored before release. Also, as with normal automation programming, there will be some red herrings along the way that are very similar to ones you will encounter in the real world! Though we will apply skills we learned throughout the book, you as the reader should be constantly on the lookout for a way to improve and, when necessary, debug the software as you would do for a real-world application.

This chapter will attempt to follow the full SDLC in a Waterfall-like method. Since this is a learning example, the exact process that one would use for a real-world project will likely differ. Nonetheless, we're going to keep the workflow as real-world as possible.

Technical requirements

To complete this project, you will need a working copy of CODESYS. However, no code for this project will be provided. This project is designed to be erroneous and new versions of the programming environment can introduce other bugs that stem from the newer system as opposed to the engineered defects.

This chapter will require a comprehensive knowledge of all the topics covered in the previous chapters. In other words, you can think of this chapter as the final boss chapter. If you have been skipping around the book, it is best to go back and read the chapters that you skipped. If you feel comfortable with the material already covered, you are free to proceed.

Project overview

For this project, we're going to create an industrial oven. Industrial ovens are common PLC applications; they also need to have a number of safety controls to help avoid injury or death. In the real world, industrial ovens can be used for applications like the following:

- Curing paint
- Preheating parts
- Drying parts
- As well as many other applications.

For this project, our simulated customer is requesting an oven system for drying metal fixtures after a washing cycle. The way the manufacturing process works is that once a part is washed, it is placed in the oven for a variable amount of time depending on the fixture so that all excess moisture can be burned off. We have to be careful because there are rubber O-rings in the fixtures that will melt if the O-rings experience temperatures above their rated limit. The customer will want to be able to dry different parts that will require different dry times, and each fixture will have an O-ring with a different temperature limit. With all that in mind, we can now move on to gathering our requirements.

Gathering the requirements

Based on the run down above, we can establish the project features needed with the following user stories:

- As an operator, I want to be able to manually set the optimal temperature of the oven so that I can use the oven for different fixtures.

- As an operator, I want to know when the oven is too hot to enter so that I know not to enter the heated area.

- As an operator, I want the door to automatically lock and unlock so I don't accidentally enter the oven.

- As an operator, I want to know when the oven's temperature is at room temperature so that I can safely enter the oven to remove the dry fixtures.

- As an operator, I want to view an alarm when the temperature is over the O-rings' rated temperature so that I know when the O-ring has been compromised.

- As an operator, I want the PLC to automatically shut down when the oven's temperature reaches 10°F over the O-rings' rated temperature so that I can retrieve the parts as quickly as possible.

These are high-level requirements, as we start developing the project we will probably run into more questions as we start implementing the user stories. However, these requirements are adequate for us to start hammering out the PLC and HMI side of the system.

> **Note**
>
> User stories that are similar to what we have are typically high-level requirements. As you start drilling down and completing work you will often need to either refine these requirements or generate more granular feature requirements.

Chances are, if you were developing this application for an organization, the workload would probably be split between PLC programmer(s) and HMI developer(s); however, in cases such as this one, you will be responsible for both the HMI and PLC side of the project.

Depending on your thought process, you may want to start working on either the HMI or PLC side of the application first. For me, it has always been easier to work from the HMI side to the PLC code. This is mainly because once you have a decent outline of what the HMI is responsible for, it is easier to hammer out the PLC code. However, this is a personal preference, and you may find it easier to do the work in the opposite manner. In real life, you can plan out your workflow any way you want. With that, the first thing we are going to do is lay out the design of our HMI.

HMI design

The first thing we should do is lay out our HMI. Based on the requirements, at the minimum we're going to need,

- An alarm table
- A series of input fields to allow the user to enter the temperature of the oven
- A gauge to show the current temperature of the oven
- A power switch and LED for the oven
- An alarm acknowledgment button

With the requirements, we can lay out our HMI to look like the following screenshot:

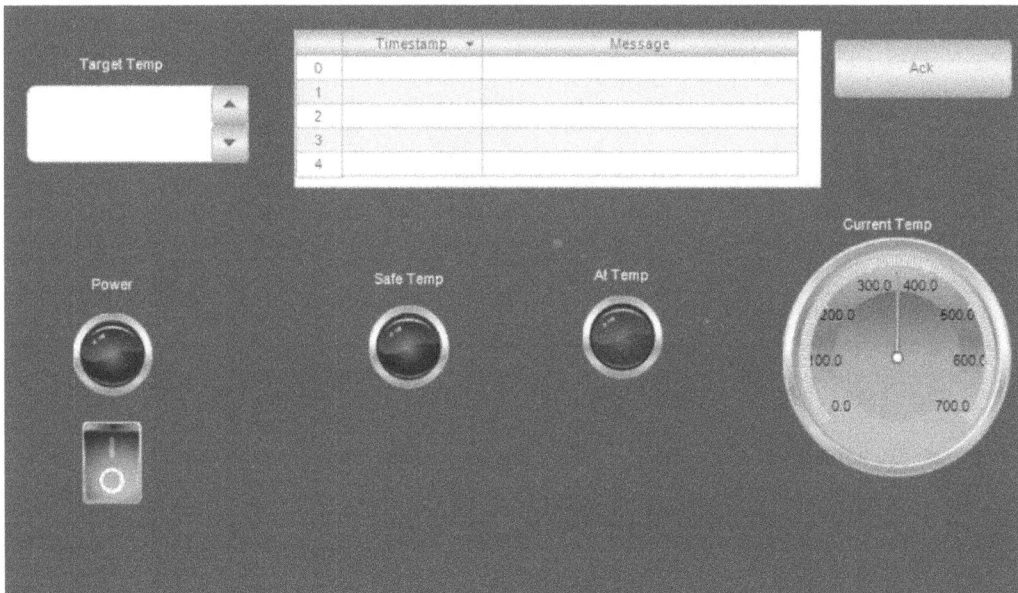

Figure 17.1: Oven HMI

This is a simple HMI layout for our project. We have a simple **Power** button in the lower left-hand corner and a target temperature spinner above it. In the center of the screen, we have an alarm table for our alarm readout as well as three LEDs to indicate that the oven is on, the temperature of the oven is safe to enter, and a final LED that indicates the temperature of the oven is at the target temperature. We also have a temperature gauge to read the exact temperature of the oven and an acknowledgment button to acknowledge the alarms. Now that we have a rough layout for the HMI, we can continue and start implementing the logic for it.

HMI implementation

The first thing we need to do is start declaring variables. For this example, we are going to put all the variables that control the HMI in a **global variable list** (**GVL**) called vars for ease of use. The first set of variables we are going to implement are the LED variables.

> Improvement!
>
> Consider the name of the GVL. Is vars a good name for it? Should you refactor the GVL with a better name?

LED variables

We have three LEDs that are used as temperature indicators and one LED that is used as a power indicator. We are going to create four Boolean variables, as follows:

```
PROGRAM PLC_PRG
VAR
    //LEDs
    power       : BOOL;
    safe_temp   : BOOL;
    target_temp : BOOL;
END_VAR
```

> Flaw
>
> Are these LED variables declared in the correct file?

The following indicates which variables should be mapped to which LEDs:

- The power variable will be assigned to the switch and the power LED
- The safe_temp variable will be assigned to the safe_temp LED
- The target_temp variable will be assigned to the target_temp LED

Once you are complete with hooking up those variables, you can move on to the declaration and assignments of the acknowledgment variable.

Acknowledgment variable

The next variable that we need to set up is the acknowledgment variable. As in the past chapters, we will create a Boolean variable called ack, as follows:

```
ack : BOOL;
```

This variable will need to be assigned to the following:

- The **Ack** button
- The **Acknowledgement** field in the alarm table

The button configuration should look like this:

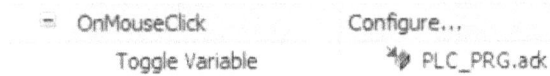

Figure 17.2: Button setup

As the preceding screenshot depicts, you will want to toggle the ack variable when the button is clicked. As for the alarm table, you will need to set up a field similar to what is shown in the following screenshot:

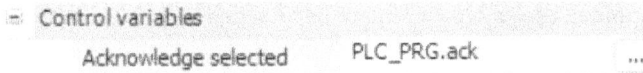

Figure 17.3: Alarm table acknowledgment configuration

If you followed the steps correctly, you should now have the button and part of the alarm table set up and ready to go. This means you can move on to implementing the spinner.

> Design improvement
>
> Can we use a simpler control for the acknowledgement button?

Spinner variables/setup

The spinner variable is going to be an integer. The variable will be responsible for providing a target temperature for the oven. The variable will look like this:

```
target_temp_value  : INT;
```

The `target_temp_value` variable will be assigned to the target temperature spinner.

> Where to implement?
>
> Notice it was not specified where to implement the `target_temp_value`. Using what we've learned and explored, where should you implement it?

We also need to set the range on the spinners as well. For the sake of simplicity, we're going to set the range on the target temperature spinner to `500`, as in the following screenshot:

Value range		
Minimum value	100	...
Maximum value	500	

Figure 17.4: Target temperature value range

For this example, we're going to set a minimum value of 100°F and a maximum temperature of 500°F. Once you complete these operations, you can move on to creating the variable for the gauge.

Gauge variable/setup

Much as with the spinner, the gauge is going to be attached to an integer as well. We're going to use the following variable for the gauge:

```
oven_temp : INT;
```

In a real-world application, all the values would be floating points such as a `REAL` data type. However, for this project, we are going to use `INT` data types to avoid using decimals for the sake of simplicity. Much as with the spinner, we will also need to set the range on the gauge as well. We're going to set the range to 700°F to indicate overheating. The extra 200° is an arbitrary number; however, when you're working with things such as gauges, you will usually want to set the range over the maximum value just in case the part experiences values over the expected maximum. You will want to set the values as shown in the following screenshot:

Scale	
Subscale position	Outside
Scale type	Lines
Scale start	0
Scale end	700
Main scale	100

Figure 17.5: Gauge configuration

In this case, we set the maximum value on the gauge to 700; however, we also adjusted the main scale to 100. This is so the gauge lines are not bunched up, and the gauge is not cluttered. After you complete these operations, your gauge should look like the one shown in *Figure 17.6*:

Figure 17.6: Configured gauge

The final component that we need to set up is the alarm table. Once you're sure you are done with the gauge, you can move on to the alarm table.

Alarm table variables/configuration

To configure the alarm table, the first thing we're going to do is create an `Alarm_configuration` object and add an alarm group called **Temperature**. When you're done, your alarm configuration tree should look like this:

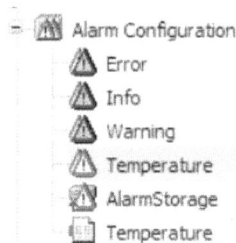

Figure 17.7: Alarm configuration tree

In the case of this example, we're going to trigger the alarm with a series of Boolean variables that will be set in the PLC code. This means we're going to need to declare three more variables, as follows:

```
oven_overTemp : BOOL;
oven_atTemp   : BOOL;
oven_safeTemp : INT;
```

Bug

Do you see a wrong data type in the code?

From the variables, we can see that there will be an info alarm that will tell the operator that the oven is safe to enter, a warning variable that will tell the operator when the oven is at the set temperature, and finally an error alarm that will tell the operator that the oven is overheating.

In the variables above we have oven_safeTemp and in the LED section we have safe_temp. In theory both variables should be set to the same state when the oven is at a safe temperature to enter. Now, there is a tradeoff to doing this. For starters, if the code is not properly implemented the variables could lose sync with each other, meaning that one may be on and the other off; however, we will have more granular control over the variables. If we condensed everything to use one variable, for example, safe_temp, we wouldn't have to worry about losing sync between the two variables, but we would lose granular control. In my opinion, it is important to keep the granular control in case we have future expansions that need it.

After you declare these variables, you will need to set up the alarm configuration. As such, you will want to double-click **Temperature** and match the setup to the following:

ID	Observation Type	Details	D...	Class	Message
0	Digital	(vars.oven_overTemp) = (TRUE)		Error	Oven is overheating
1	Digital	(vars.oven_atTemp) = (TRUE)		Warning	Oven is at temp
2	Digital	(vars.oven_safeTemp) = (TRUE)		Info	Oven is safe to enter

Figure 17.8: Alarm configuration setting

Once that is done, we will need to configure our error, warning, and info classes.

Error class setup

The error class will consist of the following configuration:

State	Font	Background Color
Normal		
Active	■ Microsoft Sans Serif, 9.75pt, style=Bold	Red
Waiting for confirmation		

Figure 17.9: Error class configuration

Once you have completed the error setup, double-click on the warning class.

Warning class configuration

The warning class will consist of the following configuration:

State	Font	Background Color
Normal		
Active	■ Microsoft Sans Serif, 9.75pt, style=Bold ☐	255, 255, 0

Figure 17.10: Warning class configuration

Once you finish the configuration for this class, you will need to set up the final class, which is the info class.

Info class configuration

The final class that we will need to set up is the info class. This class will consist of the following settings:

State	Font	Background Color
Normal		
Active	■ Microsoft Sans Serif, 9.75pt, style=Bold ▓	0, 255, 0

Figure 17.11: Info class configuration

After you complete the configuration for this class, you can move on to assigning the alarm group to the alarm table.

Alarm table configuration

The steps to assign the alarm group to the table will be the same as the ones outlined in *Chapter 15*. Your table configuration should match the following screenshot:

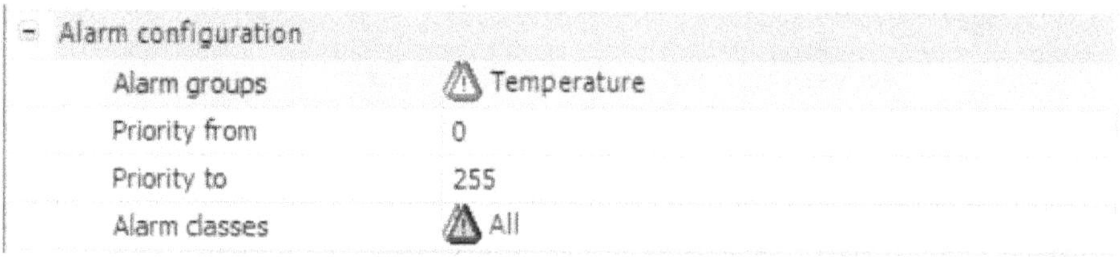

− Alarm configuration	
Alarm groups	⚠ Temperature
Priority from	0
Priority to	255
Alarm classes	⚠ All

Figure 17.12: Alarm table configuration

At this point, the control HMI should be complete. All of the controls should be hooked up and configured. Therefore, the next phase in the development cycle is to implement the PLC code.

PLC code design

Since we are going to begin writing the PLC code, we need to start fleshing out a design. To keep the design simple, let's break the project down into the following function blocks:

1. Oven: This function block will handle turning the oven on and off, as well as ramping the oven up to temperature.

2. Alarms: This function block will trigger error, warning, and info alarms.

3. Door: This function block will be responsible for locking and unlocking the oven door.

You can see an illustration of the function blocks in the following diagram:

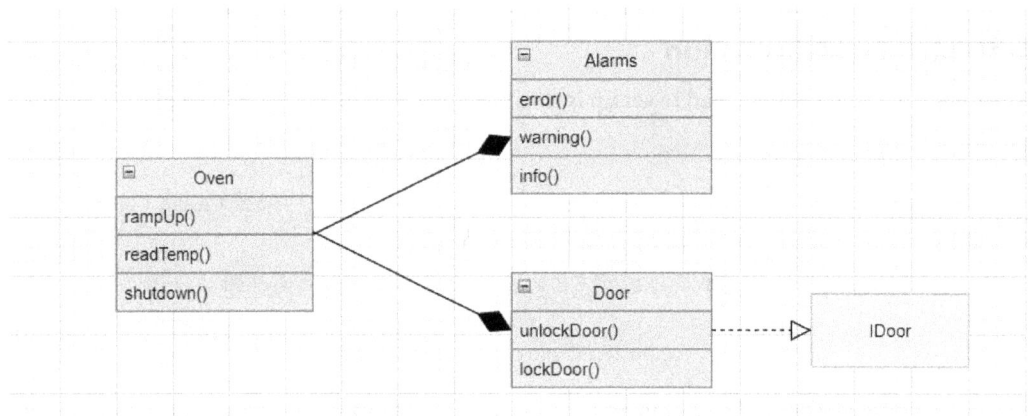

Figure 17.13: PLC code UML

As can be seen in the preceding diagram, the PLC side will consist of an Oven, Alarms, and Door function block as well as a Door interface. The Oven function block will be the workhorse of the PLC program. Essentially, the PLC side will be built around the composition principle. In short, we can justify this with the following statements:

- The oven has a series of alarms that need to be triggered
- The oven has a door that needs to be locked and unlocked

Finally, the purpose of the Door interface is for us to create a model of a door. There are many types of doors that we can use but they will all automatically lock and unlock. Therefore, to properly model the door, we will use an interface and simply implement the methods for the specific door that we use.

This design is very simple and requires minimal code. Also, since we are using composition and all the function blocks/methods are following the **single-responsibility principle (SRP)**, this design will allow for future expansion and easy maintenance.

Though simple, the PLC design will be quality enough to implement. As with the theme of this book, since we have a decent design, we should be able to easily implement the code. With all that being said, we can now implement the PLC code.

Implementing the PLC code

The code implementation should be relatively minimal. The first thing we are going to do is declare our function blocks. For this, we are going to create a folder named FunctionBlocks and use it to house the Oven, Alarms, and Door function blocks. When all the function blocks and methods are implemented, your tree should look like this:

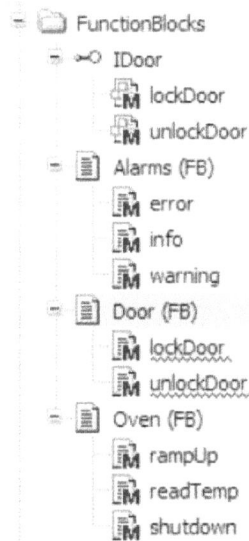

Figure 17.14: Function blocks

Implementation

Notice that parameters or return types for the methods were not specified. Before you move on, modify the methods with all the parameters and return types you think they need.

To start implementing code, we're going to implement the PLC_PRG POU file.

PLC_PRG file

The first place we're going to start implementing code is in the PLC_PRG file. Since it is our entry point, we're going to put our starting logic here. In the spinner section, a number of variable implementation locations were left up to you. I recommend placing the variables in the vars GVL or in a separate GVL. For my implementation I'm going to put them in the vars GVL for simplicity. If you opted to use another GVL or location modify the book code accordingly.

> **Do you agree?**
>
> Did you put your variables in the vars GVL? If not, why? Try to come up with some pros and cons for putting them in the vars GVL verses somewhere else.

The reference variable for the Oven, Alarms, and Door function blocks will be defined in the PLC_PRG location, like in the following:

```
PROGRAM PLC_PRG
VAR
    oven   : Oven;
    alarms : Alarms;
    door   : Door;
END_VAR
```

Once you add the oven variable, you should only need to add the following code to the file:

```
vars.safe_temp := TRUE;
alarms.info();
IF vars.power = TRUE THEN
    oven.readTemp();
    door.lockDoor();
END_IF
//shutdown
IF vars.power = FALSE THEN
    oven.shutdown();
END_IF
```

For this program, we are going to make an assumption for the sake of the simulated project. When the program starts, we are going to assume it is safe to enter, hence setting the `safe_temp` variable to TRUE. If the power is on, a warning message will be displayed on the alarm table. The oven at temp LED will be displayed, and the *door locked* message will trigger as well. If the power is off, the oven.shutdown method will be called, and if the temperature is below 85°F, the door will unlock and the safe LED will turn on. An alarm will also be displayed saying the oven is safe to enter. With this complete, we can now move on to implementing the `Alarms` function block.

Flaws

Examine the code. Are there fundamental flaws? For example, does the code factor in the oven temperature before setting certain variables to TRUE? If it does figure out a way to fix the flaw. Also, are all the lines of code necessary? Is there any dead code in the snippet? Finally look at the LED naming conventions, are those all correct?

Alarms function block

The code for the `Method` blocks should be relatively simple. Essentially, whichever method is called will set the appropriate alarm. The `error` method's code should look like the following:

```
vars.oven_overTemp := TRUE;
vars.oven_safeTemp := FALSE;
vars.oven_atTemp   := FALSE;
```

With that, we can move on to implementing the `info` method with the following code:

```
vars.oven_overTemp := FALSE;
vars.oven_safeTemp := TRUE;
vars.oven_atTemp   := FALSE;
```

Safety Flaw

There is a safety flaw with the `safeTemp` variable. We shouldn't always set this to TRUE. What should you do to ensure the oven is actually safe to enter? *Hint:* you can put a control statement for a check.

As can be deduced from the error and info methods, all we are doing is setting the correct variable to TRUE. Finally, the warning method should look like the following code snippet:

```
vars.oven_overTemp := FALSE;
vars.oven_safeTemp := FALSE;
vars.oven_atTemp   := TRUE;
```

Safety Flaw

Is there a safety flaw here as well? Why or why not? What can you do to improve this to make it safer and smarter?

The implementation of the methods in the Alarms function block is probably not the most effective. The code can be simplified to one method. The methods were designed like that on purpose so that you, the reader, can modify and improve upon the code. After you finish implementing the project, come back to this section and try to condense and improve upon the code.

After you fully implement this code, we are going to move on to implementing the other function blocks. Moving down the tree, we are going to implement the method in the Door function block.

Door function block

For things such as doors, it is common to have a large light on the outside of the door as a safety feature. It is also common to put an LED on the HMI; however, since this is a simulation, for now, we are only going to have a variable to indicate the door is locked. We are going to add the following variable to the vars GVL file:

```
door_status : WSTRING;
```

After you add that variable to the GVL file, you can start to implement the methods in the Door function block. For the current iteration of the project, we are only going to display the status of the door, so the code for these two methods will also be relatively simple. With that being said, the unlockDoor method should look like the following code snippet:

```
vars.door_status := "Door is unlocked";
```

As can be deduced by looking at the unlockDoor method implementation, the doorLock method will be equally simple, with the following implementation:

```
vars.door_status := "Door locked";
```

This should, for the most part, do it for the Door function block. However, much as with the alarm class, see if you can modify this. See if you can address the following:

- Can you condense these two methods into a single method using control statements? Does doing this make more or less sense?
- How can you modify the HMI and PLC code to support an LED on the HMI screen?
- Should you create a separate HMI visualization for the door?
- Are the door messages consistent? Should these be cleaned up?

Before you address these questions, let's move on and implement the Oven function block logic.

Oven function block

The final function block that we have to implement is the Oven function block. This function block will be more complex than the other function blocks as this will be the workhorse of the program. Ensure that you are carefully following along and keep an eye out for bugs!

The first method that we are going to implement is the rampUp method. In a real-world application, you assume that when the oven is on, it is dangerous. Similarly, you will want to turn on the red LED and turn off the green one. This will signal to the operator that the oven is potentially hot and not to handle any dangerous areas. To accommodate this, we are going to implement the following two lines of code to simulate this:

```
vars.ovenOn := TRUE; //Sends hypothetical signal to oven heater
vars.safe_temp := FALSE; //at a safe to enter temperature
vars.target_temp := TRUE; //At target temperature
```

Bugs

Did you notice any bugs?

Once that logic is in place, we need to move on to our readTemp method. This method is essentially going to be the workhorse of the program. This method will be responsible for firing alarms to give the temperature status to the operator as well as triggering the rampUp phase when the oven is not already at temperature. The readTemp method will simply consist of a series of control statements, as follows:

```
METHOD PUBLIC readTemp : BOOL
VAR
```

```
    alarms : Alarms;
    door   : Door;
END_VAR
VAR_INPUT
END_VAR
```

Once you create the `alarms` variable, you can move on to implementing the logic for the rest of the method with the following code:

```
IF vars.oven_temp < vars.target_temp_value THEN
    rampUp();
    alarms.warning();
ELSIF vars.oven_temp = vars.target_temp_value THEN
    alarms.warning();
ELSIF vars.oven_temp > vars.target_temp_value THEN
    alarms.error();
END_IF
```

Flaws

There are a few flaws in the blocks. Take a look at the code and return type, what can be improved?

The final function will simply be responsible for putting the oven back into a shutdown mode. Depending on the type of oven and the shutdown sequence, this method will vary. However, for this project, we are going to keep it simple; we will reset the red LED to *off* and the green one to *on* when the temperature of the oven is less than 85°F.

To accomplish this, the variables should look like the ones in the following snippet:

```
METHOD PUBLIC shutdown : BOOL
VAR
    door : Door;
END_VAR
VAR_INPUT
END_VAR
```

This method will also unlock the door. In real life, there are things such as motion detectors and so on in the oven to prevent the oven from heating up in case someone or something is inside. For this project, we are going to keep it simple and ignore that; however, it is recommended that you go back and add a similar feature to enhance the project. The code to do this will look like the following:

```
vars.ovenOn := FALSE;
IF vars.oven_temp < 85 THEN
    vars.safe_temp := TRUE;
    vars.target_temp := FALSE;
    door.unlockDoor();
ELSE
    vars.safe_temp := FALSE;
    door.lockDoor();
END_IF
```

> **Best practice flaw**
>
> Notice we have a hard coded value, what can we do to improve this?

Once you have the code in what you feel is a stable state, you can move on to testing the functionality of the program.

Testing the application

Now that we have the code implemented, we can run a few test cases to see if the code works as expected. If you look at the code, we have an oven_temp variable that in real life would be tied to some type of thermal sensor. For our purposes, we are going to control it manually to simulate the conditions inside the oven. In real-world automation programming, this is a common technique. We don't always want to heat the oven to the target temperature until we know for sure the software is working. To control the temperature, we can add a spinner or slider to the HMI. We could also control the simulated temperature by writing values to the variable.

Testing the door lock

When testing an industrial device, it is typically a good idea to start with testing the safety features first. This is because at the very least we want to ensure that the machine is safe to work on.

We are going to start with the most basic and safety-critical part: testing the door. To do this we're going to write out a test case. There are many templates and programs you can download from the internet. However, for this chapter we're simply going to use an Excel spreadsheet that is formatted like *Figure 17.15*.

We explored the concept of unit testing in prior chapters. What we have not explored is how to write a unit test on paper. There are many different formats you can use to write unit test cases; however, all that's important is to capture the following:

- The functionality the unit test is supposed to capture.
- The values we're going to affect like inputs, values, and the like.
- The expected behavior.
- The actual behavior.
- The date (optional but recommended)
- If the test passed or not.

The goal for this test is to ensure the door is locking and unlocking properly. For this, we are going to use the following test case:

Functionality	Input	Expected Value	Actual Value	Date	Pass(y/n)
door lock	power on	door locked		10/21/2022	

Figure 17.15: Test case

The test case in the preceding screenshot is relatively simple as we are testing a Boolean state. In other words, the door is either locked or unlocked.

When we run the program and switch the power on, you should get the following if the code works:

door_status	WSTRING	"Doorlocked"

Figure 17.16: Actual door output

If your code passes the test, the full test case should look like the following:

Functionality	Input	Expected Value	Actual Value	Date	Pass(y/n)
door lock	power on	door locked	door locked	10/21/2022	y

Figure 17.17: Completed test case

The next capability to test is if the door unlocks properly. Testing if the door unlocks will be a bit more in-depth as the temperature will be a factor. We will need to create a few test cases to ensure the door unlocks properly. To test the functionality, we can use the test cases in the following screenshot:

Functionality	Input / temp	Expected Value	Actual Value	Date	Pass(y/n)
door unlock	power off / 100	door locked		10/21/2022	
door unlock	power off / 82	door locked		10/21/2022	
door unlock	power on / 80	door locked			

Figure 17.18: Unlock test cases

To test this functionality, we're going to turn the power variable on, then set the oven_temp variable to 100, and then finally write the power variable back to FALSE. If your code works, you should see the door in a locked state, similar to what can be seen in the following screenshot:

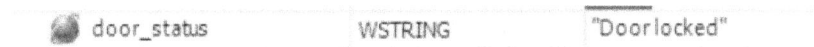

door_status	WSTRING	"Door locked"

Figure 17.19: Door state for the first test

You can repeat the process with the other temperatures:

Functionality	Input / temp	Expected Value	Actual Value	Date	Pass(y/n)
door unlock	power off / 100	door locked	door locked	10/21/2022	y
door unlock	power off / 82	door locked	door locked	10/21/2022	y
door unlock	power on / 80	door locked	door locked	10/21/2022	y

Figure 17.20: Door test cases

Once your code passes the door tests, move on to testing the gauge.

Testing the gauge

Another vital safety component of the oven is the gauge. The gauge is of vital importance as it will keep the operator informed of the internal temperature of the oven. In theory, the gauge should show the temperature of the oven. In other words, the gauge should match what the oven_temp variable is set to. We can come up with a few test cases to verify the functionality of the gauge. Essentially, what we want to verify is that the value we set the oven_temp variable to is the same value that is displayed on the gauge.

With the test criteria established, we can use the test cases in the following screenshot:

Functionality	temp	Expected Value	Actual Value	Date	Pass(y/n)
gauge	200	200		10/21/2022	
gauge	500	500		10/21/2022	
gauge	100	100		10/21/2022	

Figure 17.21: Gauge test cases

To execute the test, we will set the oven_temp variable and observe the gauge. When you set the variable to 100, your gauge should match the following reading:

Figure 17.22: Gauge reading for 100°F test case

Repeat the process with the other values to ensure the gauge reflects the proper values.

Functionality	temp	Expected Value	Actual Value	Date	Pass(y/n)
gauge	200	200	200	10/21/2022	y
gauge	500	500	500	10/21/2022	y
gauge	100	100	100	10/21/2022	y

Figure 17.23: Completed gauge test cases

Next, we're going to test the alarm system. This is another safety-critical functionality as it will alert the operator to issues.

We should get the following messages in these situations:

- An info message when the oven is safe to enter
- A warning message when the oven is at the target temperature
- An error message when the oven is 10°F over the target temperature

With this, we are going to create three basic test cases to test this functionality. Now, in the real world, you would want at least a few cases for each alarm. This will be up to you to apply what you've learned thus far and apply it to create more cases for the alarm. For this example, we are going to create three test cases, as follows:

Method	target temp	Oven temp	Expected Value	Actual Value	Date	Pass(y/n)
error	90	95	No change		10/22/2022	
info	70	80	Oven is safe to enter		10/22/2022	
warn	90	90	Oven is at temp		10/22/2022	

Figure 17.24: Alarm test cases

For the sake of practice, only the first test case will be run. You will be responsible for running the remainder of the test cases.

According to our requirements, the error alarm should only be on if the oven is 10° over the set value. When we run the values, we get the following output:

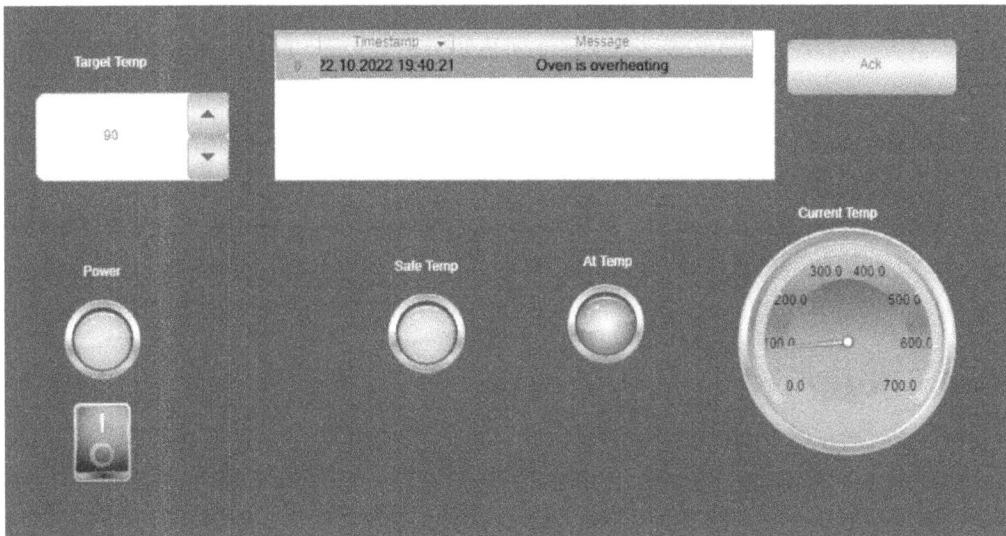

Figure 17.25: HMI status for error test case

As we can see with the default code, these values cause a failure for the test case. The error alarm should only trigger when the oven_temp variable is at least 10° over the set value, not 5. Therefore, we have at least one bug in the program. Now that we have found one bug, perform the rest of the test cases to see if there are bugs in the software. Moving forward, we have not tested the LED status. Observe *Figure 17.25*: do the LEDs seem to work as one would expect? If not, do you think there is a bug there?

Upgrades

Nothing in automation stays static. This project was a rough draft of a simulated PLC production code. There are many ways we can improve the project, for example, we could,

- Add the door's lock status to the HMI.
- Digitally track the oven's temperature with a textbox.
- Replace the spinner with a keypad input.
- Add a blinking LED to indicate the oven is over temperature.
- Add logic to track if the door is open or closed.
- These are just a few ideas. When it comes to projects like these your imagination is the only limit!

Summary

Congratulations! You have now completed the book! In this chapter, we have explored creating a simple oven. We have built this project using a Waterfall-like methodology, and we have gone through most of the SDLC sections.

In this chapter, we have built the code and HMI for a simulated real-world project. Now, we have found bugs in the code, and you will be responsible for finding more and retesting fixes for them. There are no right or wrong ways to solve these bugs and test cases; you are free to use your intuition and what we have covered to fix them. If you are completely stuck, I would recommend looking at the questions for a punch list of things to fix and a few more test cases to create.

Final thoughts

This book was an introduction to advanced PLC programming. Industry 4.0 is going to drastically change the automation landscape by introducing new technologies. The key to mastering these technologies is to understand how to properly architect code and systems. The key takeaway is that Industry 4.0 is going to force PLC programmers to adjust their hardware first attitudes and treat software as a first-class citizen. If I could offer any advice to a new PLC programmer, I would recommend learning technologies and techniques that are used in the traditional IT space. OOP is only one modern technique you should learn as the future of automation will evolve faster and be faster paced.

Fix it up!

Still unable to fix the buggy code?

Join our Discord space at `https://packt.link/embeddedsystems` to have a chat with the author and resolve your doubts.

Get This Book's PDF Version and Exclusive Extras

UNLOCK NOW

Scan the QR code (or go to `packtpub.com/unlock`). Search for this book by name, confirm the edition, and then follow the steps on the page.

Note: Keep your invoice handy. Purchases made directly from Packt don't require an invoice.

18

Unlock Your Exclusive Benefits

Your copy of this book includes the following exclusive benefits:

- ☁ Next-gen Packt Reader
- 📄 DRM-free PDF/ePub downloads

Follow the guide below to unlock them. The process takes only a few minutes and needs to be completed once.

Unlock this Book's Free Benefits in 3 Easy Steps

Step 1

Keep your purchase invoice ready for *Step 3*. If you have a physical copy, scan it using your phone and save it as a PDF, JPG, or PNG.

For more help on finding your invoice, visit https://www.packtpub.com/unlock-benefits/help.

> **Note:** If you bought this book directly from Packt, no invoice is required. After *Step 2*, you can access your exclusive content right away.

Step 2

Scan the QR code or go to `packtpub.com/unlock`.

On the page that opens (similar to *Figure 25.1* on desktop), search for this book by name and select the correct edition.

<packt>   Q Search...   Subscription 🛒 👤

Explore Products   Best Sellers   New Releases   Books   Videos   Audiobooks   Learning Hub   Newsletter Hub   Free Learning

Discover and unlock your book's exclusive benefits

Bought a Packt book? Your purchase may come with free bonus benefits designed to maximise your learning. Discover and unlock them here

Discover Benefits   Sign Up/In   Upload Invoice

Need Help?

✦ 1. Discover your book's exclusive benefits   ∧

  Q Search by title or ISBN

  CONTINUE TO STEP 2

👤 2. Login or sign up for free   ∨

☁ 3. Upload your invoice and unlock   ∨

Figure 25.1: Packt unlock landing page on desktop

Step 3

After selecting your book, sign in to your Packt account or create one for free. Then upload your invoice (PDF, PNG, or JPG, up to 10 MB). Follow the on-screen instructions to finish the process.

Need help?

If you get stuck and need help, visit `https://www.packtpub.com/unlock-benefits/help` for a detailed FAQ on how to find your invoices and more. This QR code will take you to the help page.

Note: If you are still facing issues, reach out to `customercare@packt.com`.

Answer Sheet

Chapter 1 - Advanced Structured Text: Programming a PLC in Easy-to-Read English

1. Try, catch and finally

2. A basic AI that uses a series of IF-THEN statements

3. A fact is something that machine believes to be true a rule or currently knows. A Rule is IF-THEN logic that uses facts to infer new facts or actions

4. When a state machines transition from one state to another

5. Finite State Machine

6. To catch errors that can crash a program

7. Diagnostics, inference, etc.

Chapter 2 - Complex Variable Declaration: Using Variables to Their Fullest

1. A Global Variable List that will allow any POU to access the variables declared in them.

2. GVL

3. 3

4. A variable that cannot be changed during the program's execution

5. The constant is immutable

6. A DUT that models something

7. A GVL can be accessed by any POU a struct cannot

8. Struct

9. GVLs can introduce bugs because they can be accessed by any POU and they can make programs hard to troubleshoot due to this reason.

Chapter 3 - Functions: Making Code Modular and Maintainable

1. A callable block of code

2. A preset parameter

3. Default initialization

4. A parameter value that is mapped by name.

5. Without default parameters arguments are mapped in a one-to-one manner.

6. The amount of code necessary to complete a single task that does not include the word and.

7. The data type a function returns.

8. Yes

9. A function that can hide and streamline complexity.

10. Modular code helps keep code more organized and increases it maintainability.

11. Yes

Chapter 4 - Object-Oriented Programming: Reducing, Reusing, and Recycling Code

1. Class

2. A method that can call itself

3. A function block pointer that points to its own function block

4. Get and Set

5. Get will retrieve an internal value and Set will set it.

6. A function that lives in a function block

Chapter 5 - OOP: The Power of Objects

1. Abstraction, encapsulation, inheritance, and polymorphism

2. No

3. One

4. PUBLIC attributes can be used by a POU that has a reference to the function block. PRIVATE attributes can only be used by attributes internal to the function block

5. PROTECTED can only be used by internal attributes or by derived function block attributes

6. No

7. When two function blocks use an "*is-a*" relationship

8. When two blocks use an "*has-a*" relationship

Chapter 6 - Best Practices for Writing Incredible Code

1. Keep it simple, stupid. Essentially, keep your projects as simple as possible

2. Code that does not contribute to the success of a program

3. Code that cannot be ran in the program

4. `robotElbowJoint`

5. A short summary of a key piece of information

6. An unneeded summary of what something does

7. Any time you need to manipulate or compare a value

8. All letters of first word are lower case, and the first letter of each subsequent word is upper case. Example, `camelCaseNaming`

9. The first letter of each word is upper case. Example, `PascalCaseNaming`

10. Each word is separated by an underscore. Example, `snake_case`.

Chapter 7 - Libraries: Write Once, Use Anywhere

1. Prebuilt code that can augment your project.

2. It is the necessary literature/instructions for how to use the library.

3. MIT, BSD, Apache

4. Façade; however, other good patterns that were not explored are Factories and Singletons

Chapter 8 - Getting Started with Git

1. The main branch checked out into an isolated area that allows you to modify it without corrupting the main branch

2. A version control system

3. Git is a CLI tool that is used in conjunction with a code repository management like GitLab

4. `git clone <url>`

5. Local repos live on your computer remote repos live in the repository management system

6.   Pull code from the remote repository, think of this like an update

7.   git branch -a will show all the repos you have access to while git branch -r will only show the remote branches you have access to

8.   `git checkout -b <branch>` or `git switch -c <branch>`

9.   Stages all the files in the directory for committal

Chapter 9 - SDLC: Navigating the SDLC to Create Great Code

1.   An iterative approach to completing a software project

2.   The main ceremonies are as follows:

    - Sprint planning

    - Daily standup

    - Sprint review

    - Retrospective

    - Backlog refinement

3.   A set of steps that need to be followed to implement a software project

4.   Typically, 5-6, but this can vary

5.   A rigid and sequential methodology to implement the SDLC

6.   Requirements

7.   A basic SDLC outline will include:

    a.   Requirements

    b.   Design

    c.   Implementation/code

    d.   Testing

    e.   Deployment

    f.   Maintenance

8.   The phase where the program's structure is defined, diagrams are produced, and more

Chapter 10 - Architecting Code with UML

1. Unified Modeling Language

2. Any of these

    -   Sequence diagrams

    -   Object diagrams

    -   Component diagrams

    -   Activity diagrams

    -   Timing diagrams

    -   Communication diagrams

    -   Package diagrams

    -   Profile diagrams

    -   Use case diagrams

    -   State machine diagrams

3. Name -> variables -> methods

4. An arrow

5. A diamond

6. It helps flesh out a design, convey information, and find errors/oversights.

7. To model class/function blocks and their relationships.

8. -

9. +

10. #

Chapter 11 - Testing and Troubleshooting

1. Using message variables to trace the flow of program

2. Using a debugger

3. A series of steps that can be used to help find and eliminate bugs

4. You can use prompts to formulate a query

Chapter 12 - Advanced Coding: Using SOLID to Make Solid Code

1. Functions, methods, function blocks, classes, structs, interfaces, microservices.

2. Break what comes out after the and into a module of its own. For example, break out a method with the word and in a sentence into two methods.

3. Use slimmer, more specific interfaces over larger ones.

4. The SOLID principles are as follows:

   a. **S**: Single-responsibility principle

   b. **O**: Open-closed principle

   c. **L**: Liskov substitution principle

   d. **I**: Interface segregation principle

   e. **D**: Dependency inversion principle

Chapter 13 - Industrial Controls: User Inputs and Outputs

1. A control that can be pressed to perform an action.

2. Arrays

3. Yes

4. We can but it is not recommended.

Chapter 14 - Layouts: Making HMIs User-Friendly

1. Green = Good or Go, Yellow = Warning, Red = Error or Stop

2. Gray

3. One

4. Home screen

5. Set the control's invisible field to TRUE

Chapter 15 - Alarms: Avoiding Catastrophic Issues with Alarms

1. Anything that is designed to get the operator's attention

2. Error

3. Info/All good

4. Warning

5. Alarm configuration for logically related alarms

6. An operator confirming they see the alarm

Chapter 16 - DCSs, PLCs, and the Future

1. Distributed Control System: Device that can oversee a distributed process

2. Digital simulation of something

3. No

4. Whitelisting only allows notated traffic through a network. Blacklisting blocks notated traffic from flowing though the network

5. A program/device that can block traffic

6. Distributed Denial of Service: An attack that is meant to render your network inoperable

7. No

8. Scrambling your data so it cannot be read by hackers

9. A small, self-contained portion of a larger system

10. Solving tasks in parallel normally on a GPU

11. Using multiple computers to do a task

12. UDP will essentially spray data, and nothing is guaranteed UDP is faster than TCP. TCP is slower but ensures data arrives correctly. TCP uses a three-way handshake UDP does not

13. Ring

‹packt›

packtpub.com

Subscribe to our online digital library for full access to over 7,000 books and videos, as well as industry leading tools to help you plan your personal development and advance your career. For more information, please visit our website.

Why subscribe?

- Spend less time learning and more time coding with practical eBooks and Videos from over 4,000 industry professionals
- Improve your learning with Skill Plans built especially for you
- Get a free eBook or video every month
- Fully searchable for easy access to vital information
- Copy and paste, print, and bookmark content

At www.packtpub.com, you can also read a collection of free technical articles, sign up for a range of free newsletters, and receive exclusive discounts and offers on Packt books and eBooks.

Other Books You May Enjoy

If you enjoyed this book, you may be interested in these other books by Packt:

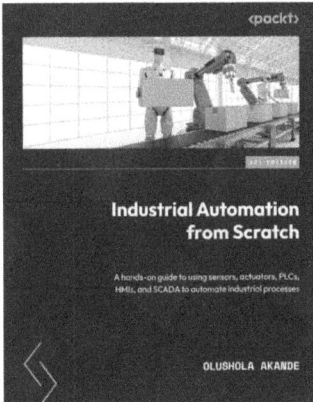

Industrial Automation from Scratch

Olushola Akande

ISBN: 978-1-80056-938-6

- Get to grips with the essentials of industrial automation and control
- Find out how to use industry-based sensors and actuators
- Know about the AC, DC, servo, and stepper motors
- Get a solid understanding of VFDs, PLCs, HMIs, and SCADA and their applications
- Explore hands-on process control systems including analog signal processing with PLCs
- Get familiarized with industrial network and communication protocols, wired and wireless networks, and 5G
- Explore current trends in manufacturing such as smart factory, IoT, AI, and robotics

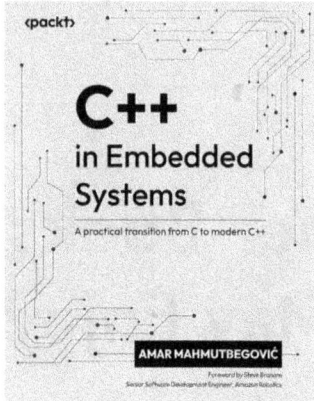

C++ in Embedded Systems

Amar Mahmutbegović

ISBN: 978-1-83588-114-9

- Debunk myths and misconceptions about using C++ in embedded systems
- Set up build automation tailored for C++ in constrained environments
- Leverage strong typing to improve type safety
- Apply modern C++ techniques, such as Resource Acquisition Is Initialization (RAII)
- Use Domain Specific Language (DSL) with a practical example using Boost SML
- Implement software development best practices, including the SOLID principle, in embedded development

Packt is searching for authors like you

If you're interested in becoming an author for Packt, please visit authors.packt.com and apply today. We have worked with thousands of developers and tech professionals, just like you, to help them share their insight with the global tech community. You can make a general application, apply for a specific hot topic that we are recruiting an author for, or submit your own idea.

Share your thoughts

Now you've finished *Mastering PLC Programming, Second Edition*, we'd love to hear your thoughts! Scan the QR code below to go straight to the Amazon review page for this book and share your feedback or leave a review on the site that you purchased it from.

https://packt.link/r/1836642555

Your review is important to us and the tech community and will help us make sure we're delivering excellent quality content.

Index